T0093818

THE INTERNET OF ENERGY

*A Pragmatic Approach Towards
Sustainable Development*

THE INTERNET OF ENERGY

A Pragmatic Approach Towards Sustainable Development

Edited by
Sheila Mahapatra, PhD
Mohan Krishna S., PhD
B. Chandra Sekhar, PhD
Saurav Raj, PhD

First edition published 2024

Apple Academic Press Inc.
1265 Goldenrod Circle, NE,
Palm Bay, FL 32905 USA

760 Laurentian Drive, Unit 19,
Burlington, ON L7N 0A4, CANADA

CRC Press
2385 NW Executive Center Drive,
Suite 320, Boca Raton FL 33431

4 Park Square, Milton Park,
Abingdon, Oxon, OX14 4RN UK

Library and Archives Canada Cataloguing in Publication

Title: The internet of energy : a pragmatic approach towards sustainable development / edited by Sheila Mahapatra, PhD, Mohan Krishna S., PhD, B. Chandra Sekhar, PhD, Saurav Raj, PhD.
Names: Mahapatra, Sheila, editor. | Krishna, Mohan, 1987- editor. | Sekhar, B. Chandra, editor. | Raj, Saurav, editor.
Description: First edition. | Includes bibliographical references and index.
Identifiers: Canadiana (print) 20230539572 | Canadiana (ebook) 20230539610 | ISBN 9781774914182 (hardcover) | ISBN 9781774914199 (softcover) | ISBN 9781003399827 (ebook)
Subjects: LCSH: Electric power systems—Automatic control. | LCSH: Electric power distribution—Data processing. | LCSH: Energy conservation—Data processing. | LCSH: Internet of things.
Classification: LCC TK1007 .I58 2024 | DDC 621.310285/4678—dc23

Library of Congress Cataloging-in-Publication Data

CIP data on file with US Library of Congress

ISBN: 978-1-77491-418-2 (hbk)
ISBN: 978-1-77491-419-9 (pbk)
ISBN: 978-1-00339-982-7 (ebk)

About the Editors

Sheila Mahapatra, PhD

Professor and Head, Electrical and Electronics Engineering Department, Assistant Dean (Research), Alliance College of Engineering and Design, Alliance University, Bangalore, Karnataka, India

Sheila Mahapatra, PhD, is currently working as a Professor and Head, Electrical and Electronics Engineering Department, and Assistant Dean (Research), Alliance College of Engineering and Design, Alliance University, Bangalore, Karnataka, India. She has a total experience of 18 years in the field of engineering, academics, and research. Her areas of interest include power systems optimization, FACTS devices, renewable energy, energy economics, and sustainability. To her credit, she has many research publications in SCI/ SCI-E, Scopus-indexed journals, book chapters, and conference proceedings. CIE, and Scopus-indexed journals as well as book chapters and conference proceedings to her credit. She serves as a reviewer for reputed journals including *IEEE Transactions on Power Delivery, IET Generation, Transmission and Distribution, Journal of Power Technologies, International Journal of Modelling and Simulation, Journal of Engineering Science and Technology, Journal of Concurrency and Computation, Practice and Experience,* and several others. She is a life member of reputed professional societies, including IEEE, the Institution of Engineering and Technology, and the International Society for Technology in Education. She was awarded a PhD in Electrical Engineering (FACTS controllers' implementation in power transmission system) from The North Cap University, Gurugram, India. She received her BTech degree in Electrical Engineering from Utkal University, Orissa, and MTech in Power Systems and Automation from Andhra University, A.P., India, in 2002 and 2008, respectively. She was the batch topper while pursuing her master's degree from Andhra University, A.P., India.

Mohan Krishna S., PhD

Associate Professor, Department of Electrical and Electronics Engineering, Alliance College of Engineering and Design, Alliance University, Bangalore, India

Mohan Krishna S., PhD, is an Associate Professor in the Department of Electrical and Electronics Engineering at Alliance College of Engineering and Design, Alliance University, Bangalore, India. His research interests include, electric vehicles, smart homes and IoT-based building energy management systems, and state observers for induction motors energy economics and sustainability. He has published his research in SCI/SCIE and Scopus-indexed journals as well as book chapters and conference proceedings. He is an advisory board member to the energy section of *Heliyon* (Elsevier). He also serves as Associate Editor of *the International Journal of Smart Vehicles and Smart Transportation*. Additionally, he is a member of the editorial review board of the *International Journal of Energy Optimization and Engineering.* He is also editing several books in the domain of electric vehicles, smart grids, and energy sustainability. He was awarded a PhD in Electrical Engineering (sensorless control of induction motor drives for EV applications) from Vellore Institute of Technology (VIT), India, and his BTech and MTech degrees from Amrita Vishwa Vidyapeetham, Coimbatore, India. He also acquired a domain-specific MBA (Power Management) from the University of Petroleum and Energy Studies (UPES), Dehradun, India.

B. Chandra Sekhar, PhD

Specialist–EV in Transportation Business Unit at L&T Technology Services Limited (LTTS), Bangalore, India

B. Chandra Sekhar, PhD, is currently working as a Specialist–EV in Transportation Business Unit at L&T Technology Services Limited (LTTS), Bangalore. Prior to joining LTTS, he worked as a technical lead in Automotive Business Division at TATA Consultancy Services Limited (TCS), Bangalore. Subsequently, he worked as a R&D Engineer in Solar Division at Power One Micro-Systems Pvt. Ltd, Bangalore. He received his PhD degree in Electrical Engineering from the Central Power Research Institute (CPRI) Research Center, Bangalore, Karnataka. He is a Professional Member of IEEE (94812129). He is a Board of Studies (BoS) committee member for New Horizon College of Engineering (NHCE), Bengaluru; SJB Institute of Technology, Bengaluru; NBKR Institute of Technology, Nellore, Andhra Pradesh; Global Academy of Technology, Bengaluru; Bapatla Engineering College, Bapatla, Andhra Pradesh; and Basaveshwar Engineering College (BEC), Bagalkot, Karnataka. He has published over 21 papers in peer-reviewed journals and conferences (international journals [4], international conferences [8], national journal [7], and national conferences [2]). His research areas are hybrid electric vehicles, EV chargers, battery management systems (BMS), power electronics converters, DC microgrids, LED drivers, and hybrid renewable energy systems.

Saurav Raj, PhD

Assistant Professor, Electrical Engineering Department, Institute of Chemical Technology, Mumbai, India

Saurav Raj, PhD, is an Assistant Professor in the Electrical Engineering Department at the Institute of Chemical Technology Mumbai, Marathwada Campus, Jalna, India. He has also worked as Assistant Professor at Alliance University, Bangalore, India. He has been an Associate Editor of the European Journal of Electrical Engineering and Editorial Board Member of the SCIREA Journal of Electrical Engineering. He has also been serving as a reviewer for reputed journals and conferences, including IET-GTD, Elsevier, Springer, Taylor & Francis, IGI Global USA, and several other national/international conferences of repute. He has done extensive research work in power system optimization, reactive power planning, swarm, and evolutionary optimization techniques, FACTS devices, computational intelligence, and renewable energy. He graduated from R. P. Sharma Institute of Technology, Patna, India, in Electrical and Electronics Engineering. He received his PhD from IIT (ISM) Dhanbad, India.

Contents

Contributors .... *xi*

Abbreviations .... *xv*

Preface .... *xxi*

1. **Design and Control Using an ISOGI-Q Algorithm for Grid Integration of PV-Array** 1
    Alka Singh and Praveen Bansal

2. **Thermal Efficiency Investigation of a Solar Parabolic Trough Collector System Using Two Dissimilar Reflectors** 21
    S. N. Vijayan, S. Sendhil Kumar, and S. Karthik

3. **Smart Buildings Using IoT and Green Engineering** 37
    Amit Kumar Singh, Anshul Gaur, and Pradeep Kumar

4. **A Novel Electrical Load Forecasting Model Using a Deep Learning Approach** 67
    Neelapala Anil Kumar, Ravuri Daniel, and Prudhvi Kiran Pasam

5. **Battery Plant Model Development for BMS Application** 91
    G. N. Dhanya and K. V. Abhinand

6. **Modeling of Constant Voltage Control in Synchronous Buck and Boost Converters Using MATLAB/Simulink for Point-of-Load Application** 103
    Sumukh Surya and Vineeth Patil

7. **Determination of Open Loop Responses of Switched DC–DC Converters Using Various Modeling Techniques** 125
    Sumukh Surya

8. **Evolution of Hybrid Ultracapacitors in Solar Microgrids** 155
    J. Pradeep Kumar Rao and H. N. Nagamani

9. **Speed Sensorless Model Predictive Current Control of Induction Motor-Driven Electric Vehicle** 179
    Karuna Kiran

10. **Performance Analysis of Hybrid Off-Grid Two-Axis Photovoltaic Tracking System/Fuel Cell Energy System Incorporating High-Efficiency Solar Cell** 195
    Shubhashish Bhakta, Mesfin Megra, Pikaso Pal, and Ashebir Berhanu

11. **Efficient Integration of Distributed Generation in Radial Distribution Network for Voltage Profile Improvement and Power Loss Minimization via Particle Swarm Optimization** 207
    Pragya Guru, Nitin Malik, and Sheila Mahapatra

12. **Classical and Predictive Control of Interior Permanent Magnet Synchronous Motor for Railway Application** 227
    Mannan Hassan, Muhammad Suhail Shaikh, Muhammad Shahid Mastoi, Rao Atif, Muhammad Farhan, Muhammad Amjad, Muhammad Bilal Shahid, and Abdul Latif Shah

13. **Electric Mobility** 251
    Ashwini Kumar Sharma, Srimanti Roychoudhury, and Sunam Saha

14. **Power Loss Reduction and Voltage Improvement Through Capacitors and Their Optimization for the Distribution System** 267
    Shubash Kumar, Chandar Kumar, Muhammad Suhail Shaikh, Anwar Ali Sahito, and Zahid Ali Arain

15. **Solar Photovoltaic Powered Automatic Irrigation System for the Agriculture Sector** 295
    Himanshu Sharma and Pankaj Kumar

16. **Optimal Parameter Estimation of 3-Phase Transmission Line Using a Grey Wolf Optimization Algorithm** 313
    Muhammad Suhail Shaikh, Abdul Latif Shah, Shafiq Ur Rehman Massan, Rabia Ali Khan, Munsif Ali Jatoi, Shubash Kumar, and Mannan Hassan

17. **Network Reconfiguration-Based Outage Management for Reliability Enhancement of Microgrid: A Hardware in Loop Approach** 337
    Shruti Prajapati, Sonal, Sourav Kumar Sahu, and Debomita Ghosh

18. **A Net Energy Meter-Based Approach for Islanding Detection in Modern Distribution Systems** 359
    Soham Dutta, Akash Kumar Pandey, Sourav Kumar Sahu, and Pradip Kumar Sadhu

Index..*385*

Contributors

K. V. Abhinand
Software Engineer, e-Powertrain KPIT, Bangalore, Karnataka, India

Muhammad Amjad
Department of Electronic Engineering, Faculty of Engineering, The Islamia University of Bahawalpur, Pakistan

Zahid Ali Arain
Department of Sciences and Technology, Indus University, Karachi, Sindh, Pakistan

Rao Atif
Ministry of Education Key Laboratory of Magnetic Suspension Technology and Maglev Vehicle, School of Electrical Engineering, Southwest Jiaotong University, Chengdu, China

Praveen Bansal
Department of Electrical Engineering, Delhi Technological University, Delhi, India

Ashebir Berhanu
Department of Electrical Power and Control Engineering, Adama Science and Technology University, Adama City, Ethiopia

Shubhashish Bhakta
Department of Electrical Power and Control Engineering, Adama Science and Technology University, Adama City, Ethiopia

Ravuri Daniel
Department of Computer Science and Engineering, Bapatla Engineering College, Bapatla, Andhra Pradesh, India

G. N. Dhanya
Software Engineer, e-Powertrain KPIT, Bangalore, Karnataka, India

Soham Dutta
Department of Electrical and Electronics Engineering, Manipal Institute of Technology, Manipal Academy of Higher Education, Manipal, Karnataka, India

Muhammad Farhan
Department of Electrical Engineering and Technology, Government College University Faisalabad, Pakistan

Anshul Gaur
Electronics and Communication Department, Uttarakhand Technical University, Uttarakhand, India

Debomita Ghosh
Department of Electrical and Electronics Engineering, BIT Mesra, Ranchi, Jharkhand, India

Pragya Guru
School of Engineering and Technology, The NorthCap University, Gurgaon, Haryana, India

Mannan Hassan
Ministry of Education Key Laboratory of Magnetic Suspension Technology and Maglev Vehicle, School of Electrical Engineering, Southwest Jiaotong University, Chengdu, Republic of China

Munsif Ali Jatoi
Salim Habib University, Karachi, Pakistan

S. Karthik
Assistant Professor, Department of Mechanical Engineering, Sri Krishna College of Engineering and Technology, Coimbatore, Tamil Nadu, India

Rabia Ali Khan
Newports Institute of Communications and Economics, Karachi, Pakistan

Karuna Kiran
Indian Institute of Technology (Indian School of Mines), Dhanbad, Jharkhand, India

Chandar Kumar
Department of Sciences and Technology, Indus University, Karachi, Sindh, Pakistan

Neelapala Anil Kumar
Department of Electronics and Communication Engineering (ACED), Alliance University, Bangalore, Karnataka, India

Pankaj Kumar
Department of EEE, SRMIST, Delhi NCR Campus, Ghaziabad, Uttar Pradesh, India

Pradeep Kumar
Electrical Engineering Department, Ambalika Institute of Technology and Management, Lucknow, Uttar Pradesh, India

S. Sendhil Kumar
Professor, Department of Aeronautical Engineering, Annasaheb Dange College of Engineering and Technology, Ashta, Sangli, Maharashtra, India

Shubash Kumar
School of Electrical Engineering, Yanshan University, Qinhuangdao, P. R. China; Department of Sciences and Technology, Indus University, Karachi, Sindh, Pakistan

Sheila Mahapatra
School of Engineering and Technology, Alliance College of Engineering and Design, Bangalore, Karnataka, India

Nitin Malik
School of Engineering and Technology, The NorthCap University, Gurgaon, Haryana, India

Shafiq Ur Rehman Massan
Newports Institute of Communications and Economics, Karachi, Pakistan

Muhammad Shahid Mastoi
Ministry of Education Key Laboratory of Magnetic Suspension Technology and Maglev Vehicle, School of Electrical Engineering, Southwest Jiaotong University, Chengdu, China

Mesfin Megra
Department of Electrical Power and Control Engineering, Adama Science and Technology University, Adama City, Ethiopia

H. N. Nagamani
Addl. Director (Retd.), Central Power Research Institute, Bangalore, Karnataka, India

Pikaso Pal
Department of Electrical Engineering, Indian Institute of Technology (Indian School of Mines), Dhanbad, Jharkhand, India

Akash Kumar Pandey
Design Engineer, Larsen and Toubro, Chennai, India

Prudhvi Kiran Pasam
Department of Information Technology, SRKR Engineering College, Bhimavaram, Andhra Pradesh, India

Vineeth Patil
Department of Electrical and Electronics Engineering, Manipal Institute of Technology, Manipal Academy of Higher Education, Manipal, Karnataka, India

Shruti Prajapati
Department of Electrical and Electronics Engineering, BIT Mesra, Ranchi, Jharkhand, India

J. Pradeep Kumar Rao
Department of Engineering (R&D), Central Power Research Institute, Bangalore, Karnataka, India

Srimanti Roychoudhury
School of Engineering and Technology, Adamas University, Kolkata, West Bengal, India

Pradip Kumar Sadhu
Department of Electrical Engineering, IIT(ISM) Dhanbad, Jharkhand, India

Sunam Saha
School of Engineering and Technology, Adamas University, Kolkata, West Bengal, India

Anwar Ali Sahito
Department of Electrical Engineering, Mehran University of Engineering and Technology, Jamshoro, Sindh, Pakistan

Sourav Kumar Sahu
Department of Electrical and Electronics Engineering, BIT Mesra, Ranchi, Jharkhand, India

Abdul Latif Shah
Newports Institute of Communications and Economics, Karachi, Pakistan

Muhammad Bilal Shahid
Ministry of Education Key Laboratory of Magnetic Suspension Technology and Maglev Vehicle, School of Electrical Engineering, Southwest Jiaotong University, Chengdu, China; Department of Electronic Engineering, Faculty of Engineering, The Islamia University of Bahawalpur, Pakistan

Muhammad Suhail Shaikh
School of Physics and Electronic Engineering, Hanshan Normal University, Guangdong, China

Ashwini Kumar Sharma
Graphic Era, Deemed to be University, Dehradun, India

Himanshu Sharma
Department of EEE, SRMIST, Delhi NCR Campus, Ghaziabad, Uttar Pradesh, India

Alka Singh
Department of Electrical Engineering, Delhi Technological University, Delhi, India

Amit Kumar Singh
Biomedical Engineering Department, VSB Engineering College, Karur, Tamil Nadu, India

Sonal
Department of Electrical and Electronics Engineering, BIT Mesra, Ranchi, Jharkhand, India

Sumukh Surya
Senior Engineer, Bosch Global Software Technologies Private Limited, Bangalore, Karnataka, India

S. N. Vijayan
Assistant Professor, Department of Mechanical Engineering, Karpagam Institute of Technology,
Coimbatore, Tamil Nadu, India

Abbreviations

ABC	artificial bee colony
AC	alternating current
ACO	ant colony optimization
AEN	average energy not supplied
AHHO	ameliorated Harris Hawks optimization
AI	artificial intelligence
ALO	ant lion optimizer
ANN	artificial neural network
BA	bat algorithm
BA	bat-inspired
BBO	biogeography-based optimization
BEMS	building energy management system
BEVs	battery electric vehicles
BIBC	bus injection to branch current
BioCI	bioinformatics CI
BIPV	building-integrated photovoltaic
BIPV/L	building-integrated PV/light
BIPV/T	building-integrated PV/thermal
BLE	Bluetooth low energy
BMIS	building management and information system
BMS	battery management system
BPSO	binary particle swarm optimization
C	capacitor
CA	circuit averaging
CAIDI	consumer average interruption duration index
CAS	compare and swap algorithm
CC	constant current
CCCV	constant current constant voltage
CCM	continuous conduction mode
CdS	cadmium sulfide
CH_4	methane
CHB-MLI	cascaded H-bridge inverter
ChmCI	chemistry CI
CI	computational intelligence

CO	carbon mono-oxide
CO_2	carbon dioxide
COD	chemical oxygen demand
CPRI	Central Power Research Institute
CS	cuckoo search
CSA	cuckoo search algorithm
CV	constant voltage
DA	dragonfly algorithm
DC	direct current
DCM	discontinuous conduction mode
DCMLI	Diode clamped multilevel inverter
DE	differential evolution
DE	dolphin echolocation
DERs	distributed energy resources
DFIG	doubly-fed induction generator
DG	distributed generation
DLF	direct load flow
DLF	distribution load flow
DMF	dimethylformamide
DNI	direct normal irradiance
DoD	depth of discharge
DSTATCOM	distributed static compensator
ECM	equivalent circuit model
EDLC	electrical-double-layer-capacitor
EMSOGI	enhanced multilayer second-order generalized integrator
EP	evolution programming
ES	energy sources
ESR	equivalent series resistance
EV	electric vehicle
FAME	faster adoption and manufacturing of electric
FCMLI	flying capacitor inverter
FOA	fruit fly optimization algorithm
FPA	flower pollination algorithm
GA	genetic algorithm
GHI	global horizontal irradiance
G-IoT	green internet of things
GM	gain margin
GP	genetic programming
GSA	gravitational search algorithm

GWO	grey wolf optimizer
GWOSCACSA	grey wolf optimizer sine cosine and crow search algorithm
H.T.	high tension
HESCO	Hyderabad electric supply company
HEVs	hybrid electric vehicles
HHO	Harris Hawks optimization
HIL	hardware in loop
HOMER	hybrid optimization of multiple energy resources
HS	Hirschberg–Sinclair algorithm
HVAC	heating, ventilation, and air-conditioning
ICA	imperialistic competitive algorithm
IEEE	Institute of Electrical and Electronics Engineers
IEQ	indoor environment quality
IGBT	insulated gate bipolar transistor
IM	induction motor
IoE	internet of energy
IoT	Internet of Things
IPMSM	interior permanent-magnet synchronous machines
IR	infrared
ISOGI-Q	Improved Second-Order Generalized Integrator
KH	krill herd
kV	kilo volt
KVA	kilo volt-ampere
KVAR	kilo volt ampere reactive
KW	kilowatt
LaRC	Langley research center
LF	load flow
LoRa	long-range
LQR	linear–quadratic regulator
LSTM	long short term memory
LT Network	long term network
LTE	long-term evolution
LTE-M	long term evolution for machines
MAPE	mean absolute percentage
MathCI	mathematics CI
MCA	modified cultural algorithm
MCS	Monte-Carlo simulation
MFO	moth–flame optimization
MGs	microgrids

ML   machine learning
MLP   multilayer perceptron
MPC   model predictive control
MPCC   model-predictive-current control
MTLBO   modified teaching-learning-based optimization
MTPA   maximum torque per ampere
MTPV   maximum torque per voltage
NASA   National Aeronautics and Space Administration
NB-IoT   narrow band-internet of things
NDZ   non-detection zone
NEM   net energy meter
NTC   negative temperature coefficient
NTLs   non-technical losses
OCP   open circuit potential
OCV   open circuit voltage
OFDM   orthogonal frequency-division multiplexing
OGWO   oppositional grey wolf optimization
OHHO   oppositional Harris Hawk optimization
OHTL   overhead transmission line system
P   photovoltaic
P.F.   power factor
Pb-C HUC   lead carbon hybrid ultracapacitor
PBIL   probabilistic incremental learning
PbO_2 lead oxide
PCC   point of common coupling
PFCC   power factor correction capacitor
PHEVs   plugin hybrid electric vehicles
PhyCI   physicist CI
PI   proportional–integral
PIV   peak inverse voltage
PLL   phase lock loops
PM   permanent magnet
PM   phase margin
PMU   phasor measuring unit
POL   point of load
PQ   power quality
PRV   pressure regulating valve
PSO   particle swarm optimization
PSSSINCAL   power system simulator siemens calculation

PTC	parabolic trough collector
PV	photovoltaic
PV	photovoltaic array
PVDF	polyvinylidene fluoride
PWM	pulse width modulation
R	feeder resistance
R	resistor
RDN	radial distribution network
RESs	renewable energy resources
RF	radiofrequency
RHP	right half-plane
RMSE	root mean squared
RNN	recurrent neural networks
RTD	resistance temperature detectors
S	apparent power
S&H	sample and hold circuit
SA	simulated annealing
SAIDI	system average interruption duration index
SAIFI	system average interruption frequency index
SAW	surface acoustic wave
SCA	sine–cosine algorithm
SCADA	supervisory control and data acquisition
SLFA	shuffled frog leap algorithms
SMPS	switch mode power supplies
SoC	state of charge
SOC	state of charge
SOGI	second-order generalized integrator
SPMSM	surface permanent magnet synchronous machine
SPV	solar photovoltaic
SRFT	synchronous reference frame theory
SSA	state space averaging
SSO	shark smell optimization
STATCOM	static compensator
STLF	short-term load forecasting
STSs	solar thermal systems
SVM	space vector modulation
THD	total harmonic distortion
TLBO	teaching-learning-based optimization
TS	tabu search

TSC	thyristor-switched capacitors
UPS	uninterruptible power supplies
USB	universal serial bus
UWB	ultra-wideband
VCRPP	voltage-constrained reactive power planning
V_{nl}	voltage at no load
VR	Voltage Regulation
V_{rated}	rated voltage
VSC	voltage source converter
WCA	water cycle algorithm
WHT	Walsh Hadamard transform
Wi-Fi	wireless fidelity
WOA	whale optimization algorithm
WUSB	wireless USB
X_L	feeder reactance
Z	feeder impedance
ZCD	zero current detector circuit

Preface

As a result of deregulation in the electrical sector, energy is being treated as a commodity. Further, with the integration of the renewable sector, smart grid technology and electric vehicle (EV) technology are a force to reckon with as a futuristic goal. So, it has now become imperative to collect, organize and analyze the data from interconnected smart grid infrastructure elements spread across generation, loads, storage units, and self-regulating apparatus in distribution centers. Internet of Energy (IoE) is poised to become a necessity with the substantial amplification in data and time required to analyze information where the SCADA system has its limitations. It paves the way toward a smarter world that drives energy management to be more resilient, responsive, and efficient.

Internet of Energy is a technical adage primarily referring to ameliorating and automating the electrical infrastructure for energy producers and manufacturers. In simple words, IoE is an intelligent interface between people, processes, data, and things around us. IoE focuses on the realization of the technicality of the Internet of Things in distributed energy sectors for optimizing energy efficacy and ruination in electrical infrastructure.

In the decentralized electrical system, innovative networking technologies will become increasingly important. Some vital reasons why we need to transform our old grids into an Internet of Energy include: enhanced global demand for clean energy, a radical shift in the energy sector, which presents tremendous challenges that require innovative means in energy management, centralized grids are no longer adequate and smart grids ensure more reliability.

IoE will act as an enabler for coordinating between consumers and producers to adequately meet demand and supply in an automated environment facilitated by smart, intelligent forecasting systems in a pursuit to envisage future energy demand. This stage is crucial for the operational optimization of the grid and enhancing system management for adequate handling of contingencies, storage monitoring, and load shedding.

IoE would play a pivotal role in the cost-effective integration of renewables to the grid, precise planning for energy markets, and grid extension, paving the way toward higher profitability and trading, but the most important challenge would be enhancing grid cybersecurity standards. As the future of the smart grid is digitalization, and so the call of the hour is to prepare ourselves as IoE would possibly be the answer to the integration of all these and develop skills for the next generation workforce where AI and quantum computing is transforming the energy sector, which requires innovative networking technologies.

This book is an attempt by authors to showcase contemporary work in IoE. We endeavor to project the book as a valuable tool for an extensive set of audiences to provide immense opportunities in hardware and software integration as well as data communication for future electric grids. It attempts to provide innovative solutions for interfacing the internet with the power grid for smart cities and helping transport to be more efficient, clean, and safe. The Internet of Energy is more relevant now as decarbonization, decentralization, and digitization are the three new global trends are transforming the energy sector and revisiting the way we produce, distribute, and consume electric power.

CHAPTER 1

Design and Control Using an ISOGI-Q Algorithm for Grid Integration of PV-Array

ALKA SINGH and PRAVEEN BANSAL

Department of Electrical Engineering, Delhi Technological University, Delhi, India

ABSTRACT

This chapter proposes the grid integration of a single-stage photovoltaic array (PV) via a voltage source converter (VSC). The VSC is controlled for active power injection as well as a distributed static compensator (DSTATCOM). The control of DSTATCOM is designed using an improved second-order generalized integrator (ISOGI-Q) based algorithm. In the last two decades, extensive use of nonlinear loads has polluted the grid severely. Various power quality (PQ) issues like harmonics injection, overheating of conductors, power factor, voltage regulation, voltage sag, swell, etc., are quite common problems.

Mitigation of PQ issues can be easily achieved using DSTATCOM, which can be designed to inject a current to cancel the harmonics generated by the nonlinear loads. This chapter discusses a five-level cascaded H-bridge multilevel inverter (CHB-MLI) realized as DSTATCOM Such a configuration shows several advantages over conventional two-level inverters. The CHB-MLI is always preferred due to its modular structure, ease of a developing prototype, compact size, and no requirement of clamping diodes and flying capacitors as in conventional multilevel inverters.

The controller for the CHB-MLI is designed on a widely used generalized integrator (SOGI) algorithm whose performance is further enhanced to provide good performance in the case of DC offset and high harmonics.

The Internet of Energy: A Pragmatic Approach Towards Sustainable Development. Sheila Mahapatra, Mohan Krishna S., B. Chandra Sekhar, & Saurav Raj (Eds.)
© 2024 Apple Academic Press, Inc. Co-published with CRC Press (Taylor & Francis)

The chapter discusses the ISOGI-Q control scheme implemented to extract the fundamental load current and eradicate all harmonics generated by nonlinear loads. Further, to utilize the full potential of PV-DSTATCOM, the proposed system is simulated to work in two modes viz daytime (Mode-I) and nighttime (Mode-II). Mode-I refers to the daytime operation when the active power injection is a priority, while in mode-II, reactive power is fed to the load during the night for power quality enhancement.

The proposed system is modeled and simulated in MATLAB/Simulink software. Phase shift pulse width modulation scheme is used to generate the firing pulses for DSTATCOM. Simulation and experimental results are discussed in detail, which depicts the satisfactory performance of the system. Results are highlighted, which show the harmonic content in source current brought within IEEE stipulated limits.

1.1   INTRODUCTION

The use of renewable energy is always preferred by power engineers and researchers. Renewable energy sources (ES) are becoming increasingly important all around the world. The challenges connected with conventional energy resources, such as energy shortages, pollution from fossil fuel combustion, and resource scarcity in the near future, have prompted a massive hunt for innovative alternatives in many countries. Although there are other non-conventional ES available today, solar and wind energy now dominate the market [1]. A fine example of a viable and independent energy generation system is a grid-connected PV system [2]. Energy insufficiency and significant import dependence on oil and coal imply serious problems with energy security. Energy production and consumption should be balanced. To minimize environmental changes, local energy generation must be established in the distribution network through the use of rooftop photovoltaic (PV) arrays. As a result, rooftop PV array utilization in small-scale applications is on the upswing [3]. However, it is also noteworthy that a higher proportion of nonlinear loads in the distribution network is the primary cause of power quality (PQ) problems.

Nonlinear loads in the distribution network result in poor power quality. Extensive use of power electronics controlled variable speed drives, Uninterruptible Power Supplies (UPS), arc furnaces, and residential loads like computers and printers directly inject harmonics in the grid and pollute it [4]. In Ref. [5], authors have proposed a different design configuration of PV integrated voltage source inverters. Wu et al. in Ref. [6] consider losses of single-stage and double-stage grid-connected PV-tied systems. The single-stage grid connection shows various advantages over the double stage in terms of the component cost of capacitors, diodes, switches, etc., and also, the single stage shows the advantage of higher efficiency, lower losses, and enhanced utilization of solar PV array. The improvement of power quality is primarily based on the selection of an inverter.

Conventional 2-level inverters suffer from higher switching losses, especially in medium and high voltage systems, and also have high peak inverse voltage (PIV) ratings of switches; therefore, the power switches suffer from high dv/dt stress. Therefore, nowadays, MLI is widely used for medium and high-power distribution systems [7], although it has been introduced lately. It possesses the capability to handle more power with reduced PIV rating of switches and lower stress, reduced filter, and reduced THD in output voltage and current.

MLI is broadly classified into three categories viz. (i) diode clamped inverter (DCMLI); (ii) flying capacitor inverter (FCMLI); and (iii) cascaded H-bridge inverter (CHB-MLI). In this chapter, CHB-MLI is used as a DSTATCOM unit [8] because of its modular structure, easy assembly, availability of redundant states, and no requirement for any clamping diodes and flying capacitors.

The synchronization and control of the proposed system as a compensator are important. Synchronization requires the use of commonly used circuits, viz. Phase Lock Loops (PLL). The amplitude, phase angle, and frequency of the input signal are all computed by PLLs. The most popular synchronization technique is Synchronous Reference Frame Theory (SRFT), proposed by Golestan et al. [9]; and Panda, Pathak, & Srivastava [10]. The SRF-PLL works satisfactorily in normal grid conditions, but its performance deteriorates under abnormal grid conditions. Hence, there is a need to modify or redesign the synchronizing circuits to handle abnormalities in the grid. In this chapter [11], the authors have implemented the SRF-PLL for a grid-tied PV array. Many new and adaptive PLL schemes are now reported in the literature, and generalized integrators such as Second Order Generalized Integrator (SOGI-PLL) based PLL have been proposed by Rodriguez et al. [12]. Further authors in Ref. [13] have presented detailed comparisons of various PLL circuits like Enhanced PLL, delay, and SOGI-PLL-based harmonic compensation for single-phase PV integrated grid-connected systems. The two-level inverter faces several issues, as discussed above; hence an alternative solution is to use MLIs, especially CHB-MLI. In Ref. [14], authors have discussed a comprehensive review of 88 papers on PV array integration-based CHB-MLI. In Ref. [15], authors have experimentally investigated the integration of two PV arrays with 5-level MLI for a large-scale grid system. In Ref. [16], the authors have implemented an enhanced multilayer second-order generalized integrator (EMSOGI) double-stage single-phase system, which gives better performance than conventional SOGI in terms DClink settlement and harmonics mitigation.

The conventional SOGI-PLL suffers from poor harmonic elimination, and DC offset issues. To overcome these problems, this chapter realizes an ISOGI-Q adaptive PLL algorithm to mitigate the harmonics generated by the load current under normal grid conditions. The major contributions of the chapter are as follows:

- The extradition of the fundamental component of load current using the ISOGI-Q algorithm.
- Enhanced utilization of PV arrays due to two DC link voltages on the DSTATCOM side.
- Self-balancing of DC link voltages without the need for complex DC voltage balancing circuits.
- Under dynamic conditions such as solar irradiance, a feed-forward PV component is used in the control system to make proper balance in the system.
- Proposed system is operated in two modes, viz. (i) To provide active power to the grid and (ii) to provide the reactive compensation and achieve unity power factor operation.
- The designed ISOGI-Q algorithm ensures better filtering capabilities as compared to conventional SRFT and SOGI algorithms.
- The proposed system is tested via MATLAB/Simulink under dynamic load conditions.

1.2   DESIGN AND CONFIGURATION OF PROPOSED SYSTEM

The proposed system considers a single-phase, single-stage grid-connected system with two PV arrays of 1 kW; each capacity interfaced at the DC link side of 5-level CHB-MLI. The CHB-MLI is realized as a DSTATCOM unit. It is controlled to suppress the current harmonics generated by the nonlinear load modeled in the form of a diode rectifier feeding a resistive-inductive load. The complete system layout of the proposed system is shown in Figure 1.1.

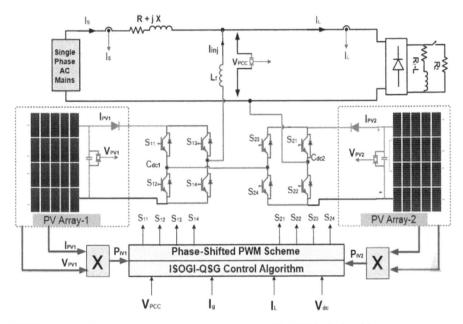

FIGURE 1.1 Complete block diagram of a system with ISOGI-Q algorithm.

1.2.1 DESIGN OF DC LINK VOLTAGE REFERENCE

A single phase, 110 V (AC rms) supply is connected at the PCC. The reference DC link voltage required is calculated by using the given relation [17].

$$E_{DC} = \sum_{j=1}^{N} E_{DC,j} = \frac{\sqrt{2} \times V_g}{m_i} \qquad (1)$$

where; m_i is the modulation index = 0.9, the grid voltage is V_g is considered to be 110 V (rms), then the calculated DC link reference voltage is 172.82 V, and in the simulation, it is approximated as 200 V, for both the H-bridges connected in a cascaded fashion.

1.2.2 DESIGN OF INTERFACING INDUCTOR

The value of the interfacing inductor is calculated [17] as follows:

$$L_{inf} = \frac{E_{o,rms}}{8 \times g \times f_r \times \Delta I_g} = \frac{155.54}{8*1.2*2500*25.70*0.1} = 2.52mH \qquad (2)$$

The estimated value of the inductor is 2.52 mH, and in the simulation, it is considered as 3 mH. Here, $E_{o,rms}$ is the rms output voltage of DSTATCOM, g is the overloading factor taken as 1.2, f_r is the switching frequency taken as 2.5 KHz, and the ripple current is considered to be 10% of the peak grid current. The grid current is calculated as follows:

$$I_g = \frac{P_{PV(max)}}{V_g} = \frac{2000}{110} = 18.18A \qquad (3)$$

The peak value of grid current is calculated as:

$$I_{peak} = \sqrt{2} \times 18.18 = 25.70A \qquad (4)$$

1.2.3   DESIGN OF DC LINK CAPACITORS

In the proposed system, there are two DC link capacitors are used with 100 V each; their value can be calculated as

$$C_{DC,j} = \frac{P_{DCj} / E_{DCj}}{2 \times \omega_r \times E_{DC-ripple}} = \frac{1000/100}{2 \times 314 \times 0.05 \times 100} = 3184.7133 \mu F \qquad (5)$$

The estimated value of the inductor is 3184.7133 µF, and in the simulation, it is considered as 3 µF each here; for the jth PV array, the DC power is $P_{DCj} = 1000$, and DC link voltage is $E_{DCj} = 100$ V each, ω_r is the angular frequency taken as 314 rad/sec and the DC voltage ripple is $E_{DC-ripple}$ and it is considered as 5% of each DC link voltage.

1.3   CONTROL ALGORITHM

The overall implementation of the ISOGI-Q control algorithm is depicted in Figure 1.2. As shown in Figure 1.2, the fundamental estimated component of load current (I_f) is extradited using the ISOGI-Q control algorithm. The overall control structure involves various calculations to achieve stable closed loop operations such as unit vector generation, DC link voltage control under varying load conditions, extraction of the real component of load current, determining reference current, and then finally, generation of PWM pulses for the firing of Insulated Gate bipolar junction transistors (IGBT). The proposed system also comprises a sample and hold circuit (S&H) and a zero current detector circuit (ZCD). The extracted fundamental load current is fed to the S&H circuit, and it is synchronized with the ZCD; when the unit synchronizing template crosses the zero, the ZCD generates the triggering signal, which is further fed to the S&H circuit. The S&H logic circuit captures the samples of the sensed load current once it receives the signal from the ZCD circuit. As a result, an accurate and fast estimation of the signal is achieved.

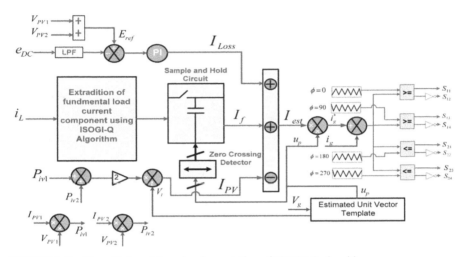

FIGURE 1.2   Overall closed-loop implementation of ISOGI-Q algorithm.

For a nonlinear load distributed network, the fundamental load component of current is extracted from the nonlinear load by using the ISOGI-Q algorithm. As shown in Figure 1.3, the in-phase and quadrature components are $i_{L\alpha}$ and $i_{L\beta}$ respectively. The transfer function can be represented as

$$\frac{i_{L\alpha}}{i_L} = \frac{k_{s\beta} \times \omega \times s^2}{s^3 + (k_{s\beta} + k_{DC})\omega s^2 + \omega^2 s + k_{DC}\omega^3} \tag{7}$$

$$\frac{i_{DC}}{i_L} = \frac{k_{DC} \times \omega \times (s^2 + \omega^2)}{s^3 + (k_{s\beta} + k_{DC})\omega s^2 + \omega^2 s + k_{DC}\omega^3} \tag{8}$$

Here from Eqns. (7) and (8), to obtain the system stability, the value of $k_{s\beta}$ and k_{DC} must be tuned using the Routh-Hurwitz stability criterion. All roots must have real parts and be equal.

1.3.1   GENERATION OF UNIT VECTOR TEMPLATE

The unit vector template or synchronizing template is generated from grid voltage V_g. The grid voltage is passed through a delay of 90°, as shown in

Figure 1.3. The in-phase component is considered as V_{gp}, and the quadrature component is V_{gq}, these voltage vectors are further used to generate V_t. Now the synchronizing template (u_p) and quadrature synchronizing template (u_q) are calculated as

$$u_p = \frac{V_{gp}}{V_t}; \quad u_q = \frac{V_{gq}}{V_t} \; ; \; V_t = \sqrt{V_{gp}^2 + V_{gq}^2} \tag{9}$$

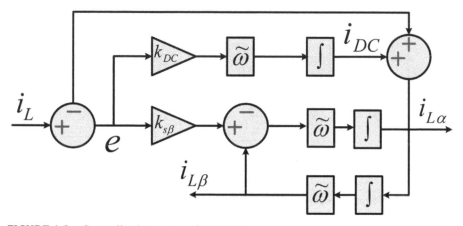

FIGURE 1.3   Generalized structure of ISOGI-Q.

1.3.2   DC LOSS CALCULATION AND GENERATION OF REFERENCE CURRENT

The proposed structure serves two purposes: (i) providing active power to the AC grid; (ii) compensating the harmonics generated by the nonlinear load and correcting the power factor of the supply side to unity. For the effective operation of the proposed system under both conditions, there is a necessity to control the fluctuations in DC link voltages obtained across PV arrays. Therefore, a proportional–integral (PI) control is used to control the DC link voltages. The DC link error can be estimated as

$$E_{DCe} = E_{DC-ref} - E_{DC} \tag{10}$$

The error signal is fed to the PI controller, and I_{loss} is calculated, as shown in Figure 1.2. Mathematically it can be represented as

$$I_{loss}(n+1) = I_{loss}(n) + k_p \{E_{DCe}(n+1) - E_{DCe}(n)\} + k_i E_{DCe}(n+1) \tag{11}$$

where; k_p and k_i denote the proportional and integral gains, respectively, and the system dynamic performance of current and voltage can be improved by the feed-forward term P_{PV} can be estimated as:

$$I_{PV} = \frac{2(P_{PV1} + P_{PV2})}{V_t} \tag{12}$$

The reference current is generated by multiplying the unit synchronizing template with the estimated load current, represented as:

$$i_{gr}^* = u_p I_{est} \tag{13}$$

$$I_{est} = I_{loss} + I_f - I_{PV} \tag{14}$$

The fundamental estimated load current is I_{est}, the fundamental load current is I_f, and PV feed-forward current is I_{PV}. The generated reference current is subtracted from the actual grid current, and further, the signal is compared with phase-shifted PWM techniques to generate firing pulses for 5-Level CHB-MLI.

1.4 SIMULATION RESULTS

The proposed system is simulated using MATLAB 2020R with two different modes under steady state and dynamic load and solar irradiation variations. The parameters used for simulation are given in Table 1.1.

TABLE 1.1   Parameter Used in Simulation

SL. No.	Parameters	Symbol	Value Used in Simulation
1.	Single-phase AC power supply	Vg(V)	110 V, RMS
2.	DC link capacitors	C_{DC1} and C_{DC2}	3,000 µF each
3.	Interfacing Inductor	$I_{'f}$	3 mH
4.	Reference DC link voltage	E_{ref}	200 V
5.	PI gains	K_p and K_i	1.22, 0.6
6.	Open circuit voltage	V_{OC1} and V_{OC2}	70.6 V each
7.	Short circuit current	I_{SC1} and I_{SC2}	18.4 A

1.4.1 STEADY STATE PERFORMANCE OF PV-DSTATCOM IN MODE-1

During Mode-1 operation, the system is working with PV-DSTATCOM and providing active power to the grid. Figure 1.4 depicts the steady state results of grid voltage V_g(V), grid current I_g(A), load current I_L(A), and total DC link voltage V_{DC}(V) with fixed solar irradiance of 1,000 W/m².

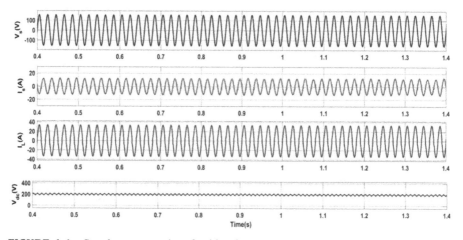

FIGURE 1.4   Steady-state results of grid voltage V_g(V), grid current I_g(A), load current I_L(A), and total DC link voltage V_{DC}(V).

Figure 1.6 shows the waveforms of PV array parameters such as currents (I_{PV1} & I_{PV2}), voltages (V_{PV1} & V_{PV2}), and power (P_{PV1} & P_{PV2}). The active and reactive power of grid P_g(W), reactive power Q_g(VAR), and the load active and reactive power, i.e., P_L(W) and Q_L(W), are represented in Figure 1.7. It is observed the grid current is out of phase with grid voltage; it shows the PV-DSTATCOM supplying active power to the grid. Therefore, the grid power is negative, as shown in Figure 1.6. The THD obtained in the grid current is 3.70% in Figure 1.5, and it is within the stipulated IEEE-519 limits, i.e., less than 5%, and THD obtained in load current is 24.51%.

1.4.2   DYNAMIC STATE PERFORMANCE OF PV-DSTATCOM IN MODE-1

During dynamic conditions, the proposed system is tested with sudden load variation during t = 0.8 s to t = 1.2s and variation in solar irradiance also. Solar irradiancies varied from a minimum value of 200 W/m² to a maximum of 700 W/m². In mode-I, i.e., during the daytime, real-time irradiation has been taken into consideration for testing. The real-time data of solar irradiation taken [18] from morning 06:00 am to evening 06:00 pm is shown in Figure 1.8. It is clearly observed that the obtained data have dynamic and nonlinear characteristics. Therefore, the average reading of irradiation obtained is taken from the morning to evening and is divided into four equal intervals of three-hour duration each in the simulation. Table 1.2 shows the average irradiance reading. The real-time solar irradiance data is shown in Figure 1.8, and the average variation taken is shown in Figure 1.9.

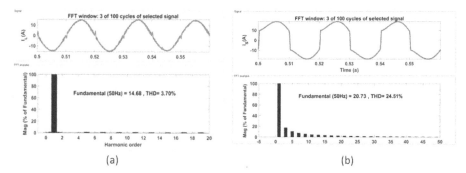

FIGURE 1.5   Harmonic spectrum of grid current and load current under steady state.

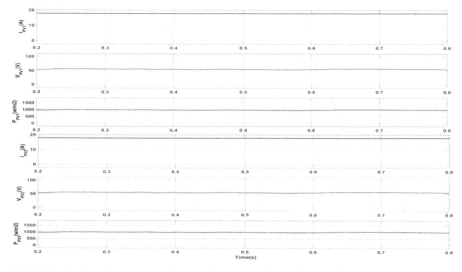

FIGURE 1.6   Steady state waveforms of currents (I_{PV1} & I_{PV2}), voltages (V_{PV1} & V_{PV2}) and power (P_{PV1} & P_{PV2}).

Figure 1.10 shows the waveforms of grid voltage V_g(V), grid current I_g(A), load current I_L(A), and total DC link voltage V_{DC}(V) and the PV array parameters such as currents (I_{PV1} & I_{PV2}), voltages (V_{PV1} & V_{PV2}) and power

$(P_{PV1}$ & $P_{PV2})$ are presented in Figure 1.11. Furthermore, the active and reactive power of grid $P_g(W)$, reactive power $Q_g(var)$, and the load active and reactive power, i.e., $P_L(W)$ and $Q_L(W)$, are represented in Figure 1.12.

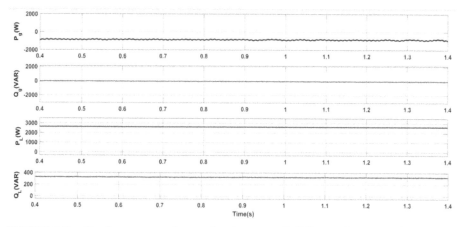

FIGURE 1.7 Steady-state waveforms of active power $P_g(W)$, reactive power $Q_g(VAR)$ of the grid, and the load active and reactive power, i.e., $P_L(W)$ and $Q_L(W)$.

TABLE 1.2   Average Value of Real-Time Data of Solar Irradiation

SL. No.	Duration of Interval	Average Reading Obtained	Used in Simulation
1.	06:00 am to 09:00 am	196.15 W/m²	200 W/m²
2.	09:00 am to 12:00 noon	584.77 W/m²	600 W/m²
3.	12:00 noon to 03:00 pm	711.55 W/m²	700 W/m²
4.	03:00 pm to 06:00 pm	274.22 W/m²	300 W/m²

The irradiance variation used in the simulation is shown in Figure 1.8.

The performance is satisfactory during dynamic load and solar irradiance conditions. The total DC link voltage V_{DC} rapidly achieves to reference value of 200 V due to the fast action of the PI controller and the proposed ISOGI-Q algorithm able to mitigate the harmonics generated by the load current and respond quickly under these sudden dynamic variations in load and solar irradiance. The grid current is almost sinusoidal but out of phase with the grid voltage. During daytime operation, the proposed system is able to feed the real power to the grid, and another advantage is the enhanced utilization of two PV arrays due to cascaded connections of inverters. However, in the literature, most of the authors have implemented single PV array-based 2-Level inverters.

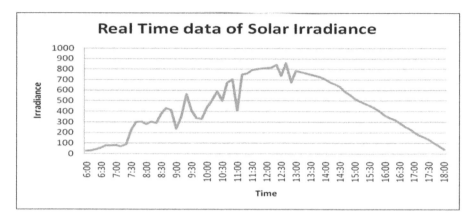

FIGURE 1.8 Real-time solar radiation obtained in Delhi /NCR region.

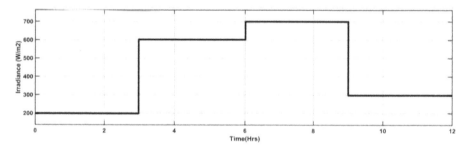

FIGURE 1.9 Average irradiance is taken for the simulation study.

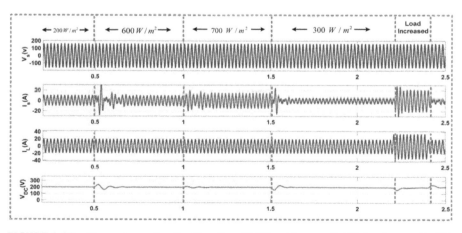

FIGURE 1.10 Dynamic results of grid voltage $V_g(V)$, grid current $I_g(A)$, load current $I_L(A)$, and total DC link voltage $V_{DC}(V)$ under solar irradiance at t = 0.5 s, 1 s, and 1.5 s and load variation during t = 2.2 s to 2.4 s.

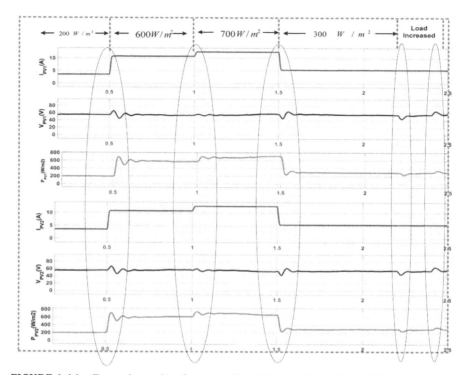

FIGURE 1.11 Dynamic results of currents (I_{PV1} & I_{PV2}), voltages (V_{PV1} & V_{PV2}), and power (P_{PV1} & P_{PV2}) under solar irradiance variation at t = 0.5 s, 1 s, and 1.5 s and load variation during t = 2.2 s to 2.4 s.

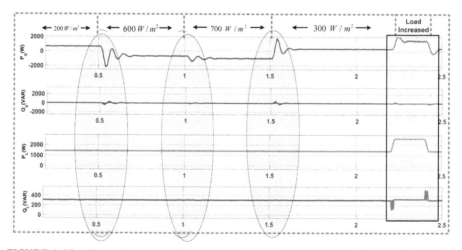

FIGURE 1.12 Dynamic results of active power P_g(W), reactive power Q_g(VAR) of the grid, and the load active and reactive power, i.e., P_L(W) and Q_L(W) under solar irradiance variation at t = 0.5 s, 1 s, and 1.5 s and load variation during t = 2.2 s to 2.4 s.

1.4.3 STEADY STATE PERFORMANCE OF PV-DSTATCOM IN MODE-2

In Mode-2, i.e., during nighttime, the PV array is disconnected, and the proposed system is acting as a DSTATCOM. Figure 1.13 shows the steady state wave-forms of grid voltage V_g(V), grid current I_g(A), load current I_L(A), and total DC link voltage V_{DC}(V). The active and reactive power of grid P_g(w), reactive power Q_g(VAR), and the load active and reactive power, i.e., P_L(W) and Q_L(W), are presented in Figure 1.14. In steady-state waveforms, the grid current and grid voltage is in a phase, which means the grid is supplying reactive power to the load.

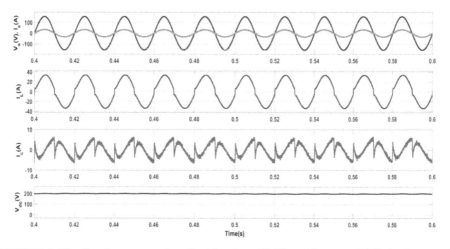

FIGURE 1.13 Steady-state results of grid voltage V_g(V), grid current I_g(A), load current I_L(A), and total DC link voltage V_{DC}(V).

1.4.4 DYNAMIC STATE PERFORMANCE OF PV-DSTATCOM IN MODE-2

The dynamic waveforms of the proposed system under load variations are shown in Figure 1.15. The grid is not feeding any reactive power to the load, i.e., the entire reactive power is supplied from DSTATCOM, making the source current in phase with grid voltage and unity power factor operation. The effective compensation has been achieved under varying load conditions.

It is observed from Figures 1.15 and 1.16 that during sudden load variation at t=0.5, there is a small variation in load reactive power (Q_L ↓). However, the load active power requirement is increased from 1,500 W to 2,600 W. The active power of the grid (Pg↑), but there is a slight variation in the reactive power requirement of the grid that means almost the unity power factor is maintained on the grid side.

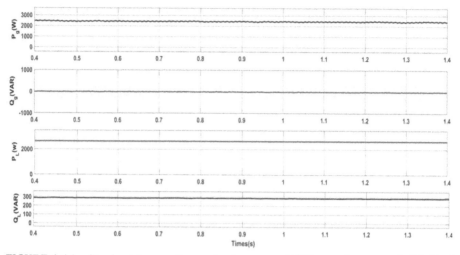

FIGURE 1.14   Steady-state waveforms of active power P_g(W), reactive power Q_g(VAR) of the grid, and the load active and reactive power, i.e., P_L(W) and Q_L(W).

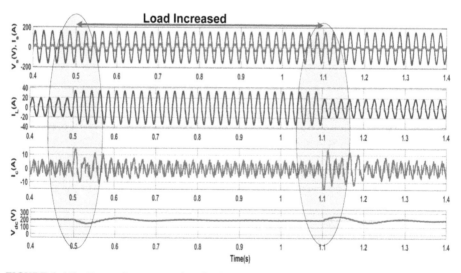

FIGURE 1.15   Dynamic state results of grid voltage V_g(V), grid current I_g(A), load current I_L(A), and total DC link voltage V_{DC}(V) during t = 0.5 to 1.1 s.

The performance of the proposed system is compared with conventional SOGI and SRFT algorithms in terms of fundamental weight extraction and error. Figures 1.17 and 1.18 shows the fundamental estimated component of weight and error, i.e., the difference between the actual load current and fundamental estimated load current, respectively. The load is varied at t = 0.3 s and t = 0.5

s, and it was observed that the SRFT algorithm takes more time to converge as compared to SOGI. ISOGI-Q and the conventional SOGI converge in almost the same duration, but sustained oscillations are observed in SOGI.

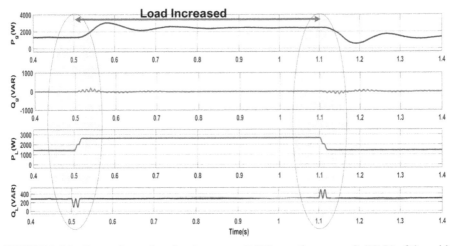

FIGURE 1.16 Dynamic results of active power P_g(W), reactive power Q_g(VAR) of the grid, and the load active and reactive power, i.e., P_L(W) and Q_L(W) under sudden load variation during t = 0.5 s to 1.1 s.

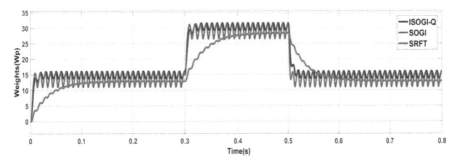

FIGURE 1.17 Weight estimation under dynamic load conditions.

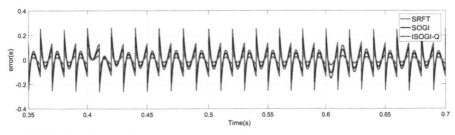

FIGURE 1.18 Error estimation.

1.5   CONCLUSION

PV-DSTATCOM-based single-stage grid is implemented in the proposed system, and the system is tested to operate in two separate modes, viz. daytime operation, and nighttime operation. The ISOGI-Q algorithm has been modeled in MATLAB/Simulink environment and tested extensively under a wide variation of solar irradiation and load disturbances. During MODE-I operation, i.e., during the day, the PV-DSTATCOM feeds real power to the grid and reactive power to the load, and during MODE-II operation, i.e., at night, the PV-DSTATCOM feeds reactive power to the load and improves power quality. The effective control makes the system free of harmonics on the grid side, allowing the system to achieve a power factor of approximately unity. In addition, the ISOGI-Q technique is used to extract the fundamental current from the nonlinear load current. Because of the CHB-MLI, the designed system operates at a low switching frequency and results in a low THD in the grid, currently meeting IEEE-519 norms. The performance of the proposed system under variable solar irradiation and load variations is satisfactory. Additionally, the performance of the proposed algorithm is better as compared to the conventional SOGI and SRFT algorithms in terms of fundamental weight estimation and reduced error under dynamic load variations.

KEYWORDS

- capacitors
- control algorithm
- DC link voltage reference
- multilevel inverter
- nonlinear load
- PV array
- PV-DSTATCOM
- steady-state performance
- unit vector template

REFERENCES

1. Koraki, D., & Strunz, K., (2018). Wind and solar power integration in electricity markets and distribution networks through service-centric virtual power plants. In: *IEEE Transactions on Power Systems* (Vol. 33, No. 1, pp. 473–485).
2. Yan, G., Cai, Y., Jia, Q., & Liang, S., (2019). Stability analysis of grid-connected PV generation with an adapted reactive power control strategy. In: *The Journal of Engineering* (pp. 2980–2985).
3. Shugar, D. S., & Ramon, S., (1990). Photovoltaic in the utility distribution system: The evaluation of system and distributed benefits. In: *Proc. IEEE Photovoltaic Specialist Conference* (Vol. 2, pp. 836–843).
4. Singh, B., Al-Haddad, K., & Chandra, A., (1999). A review of active filters for power quality improvement. *IEEE Trans. Ind. Electron., 46*(5), 960–971.
5. Ogunjuyigbe, A. S. O., Ayodele, T. R., Idika, V. E., & Ojo, O., (2017). Effect of lamp technologies on the power quality of electrical distribution network. In: *Proc. IEEE Power Energy System* (pp. 159–163).
6. Wu, T. F., Chang, C. H., Lin, L. C., & Kuo, C. L., (2011). Power loss comparison of single–and two-stage grid-connected photovoltaic systems. *IEEE Transactions Energy Conversion, 26*(2), 707–715.
7. Leon, J. I., Vazquez, S., & Franquelo, L. G. (2017). Multilevel Converters: Control and Modulation Techniques for Their Operation and Industrial Applications. *Proceedings of the IEEE, 105*(11), 2066–2081.
8. Chitra, S., & Valluvan, K. R., (2020). Design and implementation of cascaded H-bridge multilevel inverter using FPGA with multiple carrier phase disposition modulation scheme. *Microprocessors and Microsystems, 76*, 103–108.
9. Golestan, S., Monfared, M., Freijedo, F. D., & Guerrero, J. M., (2013). Dynamics assessment of advanced single-phase PLL structures. *IEEE Transactions on Industrial Electronics, 60*(6), 2167–2177.
10. Panda, A., Pathak, M. K., & Srivastava, S. P., (2016). Enhanced power quality based single phase photovoltaic distributed generation system. *International Journal of Electronics, 103*(8), 1262–1278.
11. Tripathi, R. N., Singh A., (2013). SRF theory-based interconnected solar photovoltaic (SPV) system with improved power quality. In: *Proc. IEEE Emerging Trends in Communication, Control, Signal Processing & Computing Application* (pp. 1–6).
12. Rodriguez, P., Luna, A., Candela, I., Mujal, R., Teodorescu, R., & Blaabjerg, F., (2016). Multiresonant frequency-locked loop for grid synchronization of power converters under distorted grid conditions. *IEEE Transactions on Industrial Electronics, 58*(1), 127–138.
13. Hemant, S., Alka, S., & Rai, J. N., (2016). Design and analysis of different PLLs as load compensation techniques in 1-Ø grid-tied PV system. *International Journal of Electronics, 106*(11), 1632–1659,
14. Soumyadeep, R., Nitin, G., & Ram, A. G., (2016). A comprehensive review on cascaded H-bridge inverter-based large-scale grid-connected photovoltaic. *IETE Technical Review, 34*(5), 463–477.

15. Ray, S., Gupta, N., & Gupta, R. A., (2017). Improved single-phase SRF algorithm for CHB inverter-based shunt active power filter under non-ideal supply conditions. *IEEE PES Asia-Pacific Power and Energy Engineering Conference, 2017 APPEEC '09.* Bangalore.
16. Saxena V., Kumar N., Singh, B., & Panigrahi, B. K., (2020). An enhanced multilayer GI-based control for grid-integrated solar PV system. *IEEE International Conference on Power Electronics, Drives and Energy Systems (PEDES)*, 1–6.
17. Bansal, P., & Singh, A., (2021). Nonlinear adaptive normalized Huber control algorithm for 5-level distribution static compensator. *Electrical Engineering.*
18. Saurabh, K. R., & Dharmendra, K. D., (2021). Mathematical modeling and experimental validation for the impact of high solar cell temperature on transformer loading and life. *Renewable Energy Focus, 39,* 27–36. ISSN 1755-0084.

CHAPTER 2

Thermal Efficiency Investigation of a Solar Parabolic Trough Collector System Using Two Dissimilar Reflectors

S. N. VIJAYAN,[1] S. SENDHIL KUMAR,[2] and S. KARTHIK[3]

[1]Assistant Professor, Department of Mechanical Engineering, Karpagam Institute of Technology, Coimbatore, Tamil Nadu, India

[2]Professor, Department of Aeronautical Engineering, Annasaheb Dange College of Engineering and Technology, Ashta, Sangli, Maharashtra, India

[3]Assistant Professor, Department of Mechanical Engineering, Sri Krishna College of Engineering and Technology, Coimbatore, Tamil Nadu, India

ABSTRACT

Solar radiation consists of energy in the form of electromagnetic radiation with wavelengths ranging from infrared to ultraviolet. Solar thermal collectors gather heat energy from the sun and transfer it to the working medium to generate energy. Solar collectors are devices that capture energy from the sun and are utilized as alternative energy sources (ES) in industrial and domestic applications. A parabolic trough is a type of solar thermal collector with a polished metal mirror that is straight in one dimension and has a parabola curve in the other. In this research, two types of reflective materials, such as stainless steel 303 and aluminum A011, are employed to conduct experiments in the western zone of Tamil Nadu with two different flow rates (1.139 kg/s and 2.174 kg/s). The results are validated to select suitable reflector material with highest efficiency for the application of industrial and domestic usage. A maximum temperature of 65°C was achieved using stainless steel 303 as a

The Internet of Energy: A Pragmatic Approach Towards Sustainable Development. Sheila Mahapatra, Mohan Krishna S., B. Chandra Sekhar, & Saurav Raj (Eds.)

reflector, whereas 70.1°C was achieved when aluminum A011 was employed as a reflector, which yields 5% of higher output temperature and 28.67% of higher thermal efficiency than stainless steel 303 reflective materials. Finally, it is investigated that the highest thermal efficiency can be obtained by employing aluminum A011 reflective material with a minimum flow rate (1.139 kg/s) of working medium.

2.1   INTRODUCTION

Inappropriate usage of available fossil fuels leads to a shortage of fossil fuels, which creates a negative imbalance in the environment. Hence need arises to search for new renewable energy resources to fulfill the requirements and save the earth from toxic gases. Solar energy is the primary renewable source of energy on the planet. Emitted energy from the sun is utilized to generate power by the different types of solar energy collector systems [1–3]. Heat energy is transferred from the sun to any working medium, which captures the heat and circulates it [4, 5]. Figure 2.1 represents the various stages of the heat energy transfer process. High-intensity rays fall on the reflector, and it reflects the heat energy contained in the intensity rays to the receiver tube mounted on the focal line. Heat transfer fluid is passed through the receiver pipe, which receives the heat energy from the solar rays through the pipe.

FIGURE 2.1   Heat energy transfer process.

Atmospheric temperature working medium (fluid) is given as an inlet and converted into high-temperature fluid by absorbing heat energy from the reflector, which can be used directly for the applications or for later usage saved in another mode. The performance of the system depends on the irradiation level of the sun. Irradiance is the flux received by a surface per unit space. Direct normal and global horizontal irradiance (GHI) are the two levels of the irradiance. Direct normal irradiance (DNI) is the quantity of radiation received per unit space by a surface that is constantly held perpendicular (or normal) to the rays from the direction of the sun. The average value of DNI is 5.35 kWh/m² for a day. GHI is the total quantity of shortwave radiation received from the source to the ground; the annual average GHI for a day is 5.86 kWh/m². The irradiation is measured at the latitude of 11.05° and longitude of 76.95°.

2.2 LITERATURE REVIEW

Stainless steel reflectors are used for the analysis of parabolic trough collectors with GI absorber tubes [1], and the collector thermal efficiency depends on the flow rate and temperature of the working medium. Obtained thermal efficiency of the collector is 65% for the optimum values of input parameters [4]. Concentrated collectors are used in solar power production for producing high-temperature output with high thermal efficiency using a minimum collecting area [6, 7]. A Parabolic Trough Collector (PTC) consists of a solar collector (reflector) to receive the intensity rays from the sun and reflect it to the receiver tube. The receiver comprises an absorber tube made of stainless steel or copper with a selective coating to absorb the solar radiation. Circulated the working medium through the absorber tubes to collect the heat energy from solar radiation and transfer the heat to the steam generator or to the heat storage system [8, 9].

A modified absorber has been used to enhance the efficiency of solar collector by 42.1%, while in conventional absorber, it is only 26.7% [10]. Performed the investigation by changing the parameters (aperture area, diameter of the receiver, and working medium) to optimize the design, thereby increasing the efficiency. To maintain the equilibrium between the aperture area and optical losses for minimizing the thermal losses [11, 12].

Different modes of heat transfer occur while transferring heat from the radiation to the working medium through the pipe. During the process, heat losses occur due to the conduction and convection modes of transfer on the surface of the receiver pipe and the fluctuation of wind speed [13, 14]. Assessment of year-round performance considering aluminum sheet as a reflector in a parabolic trough collector for different climatic conditions is conducted. And found that maximum water temperature was achieved during the month of April with the mass flow rate of 0.010 kg/s for the aperture area of 1.34 m² [15]. The impact of the failure of the absorber tube was analyzed through experimental and numerical approaches for three different types of tubes such as vacuum tube, lost vacuum (air), and broken glass (bare) tube. Heat loss is significantly reduced in a vacuum tube when compared with a broken tube and a lost vacuum tube [16].

PTC is used for cooking applications due to its ability to produce a high-temperature outlet with 36 m² of reflective area, which can produce 84,566

Kcal/day from 6 parabolic troughs with 60% collector efficiency. This energy is enough to prepare meals three times for 250 people in a day [17]. Thermal performance analysis of the parabolic trough collector is analyzed for the application of heating water by using copper and aluminum as absorber tubes with covered glass to arrest the emitted rays. A copper tube with a glass cover gives a maximum outlet temperature of 68.7°C, whereas an aluminum tube with a glass cover produces a maximum temperature of 62.4°C [18]. The solid plug is inserted in the focal line of the absorber tube to reduce the volume of the flow rate and increase the heat transfer rate from the sun by increasing flow velocity. Evacuated space prevents heat loss by arresting the emitted radiation from the tube [19, 20].

In this work, two types of reflective materials were used to inspect the thermal performance of solar parabolic trough collectors for the applications of industrial and domestic purposes.

2.3 CONSTRUCTION OF PTC

The main components of the solar collector are:

- Collector (receives the heat energy from the sun).
- Receiver (glass tubes which allow the working medium to circulate inside and gain the heat from the collector and transfer it to the heat storage device).
- Working medium or heat transfer medium (organic fluids, water, or any other fluids which transfer heat effectively).
- Tracking devices (E-W or N-S tracking or both) to increase the efficiency of the parabolic trough collector.
- Storage tank (used to store the heat energy for later use).
- Temperature sensor (contact type LCD digital thermometer sensor).

2.3.1 DESIGN CALCULATION

The calculation for designing the parabola curve and focal length for the collector is decided based on the following equations:

$$y = \frac{1}{5}x^2 \tag{1}$$

$$y = x^2/4f, \text{ where f is its focal length.} \tag{2}$$

$$f = \frac{x^2}{4y} 125 \text{ mm} \tag{3}$$

Figure 2.2 illustrates a parabola curve drawn using coordinates generated from the Eqns. (1) and (2) [21] and Eqn. (3) is used to calculate the focal length of the curve to fix the absorber tube for better absorption of radiation. The obtained value of the focal length of the system is 125 mm for the reflector size of 400 mm in the X direction and 320 mm in the Y direction. Three-dimensional modeling of the parabolic trough collector is made using solid works software; it is illustrated in Figure 2.3. The mass flow rate of the working medium is calculated as 2.17 kg/s using Eqn. (3) and the volume of the absorber tube is calculated (0.2393 m³) using Eqn. (4). The performance of the solar parabolic trough collector is based on the heat transfer rate of the absorber. The thermal efficiency of the parabolic trough collector was calculated using Eqn. (5) for both reflective materials. In the above equations, m is mass flow rate, Ao means the length of the parabolic curve (0.33 m³), Gb is a constant value of solar beam radiation (1,360 w/m²), cp is the specific heat of water (4.184), and Qu is useful to heat energy delivered.

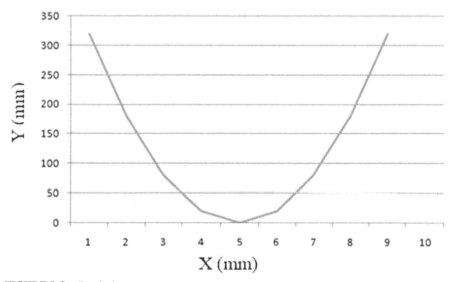

FIGURE 2.2 Parabola curve.

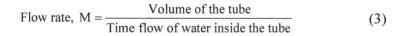

$$\text{Flow rate, } M = \frac{\text{Volume of the tube}}{\text{Time flow of water inside the tube}} \tag{3}$$

$$\text{Volume of the absorber tube} = \pi \, r^2 h = \pi \times (0.25)^2 \times 1.219 \qquad (4)$$

$$V = 0.2393 \text{ m}^3$$

$$\text{Thermal Efficiency} = \frac{Qu}{Qa} \qquad (5)$$

$$Q_u = mcp \, (T_{out} - T_{in}) \qquad (6)$$

$$Q_a = A_0 \times G_b \qquad (7)$$

From the above equations, the thermal efficiency of stainless steel 303 and aluminum was calculated.

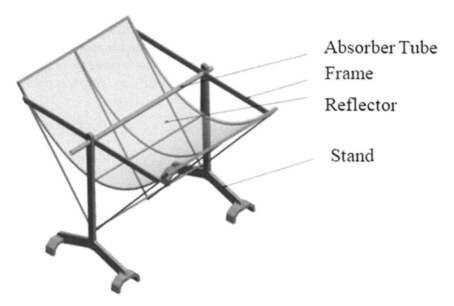

Absorber Tube

Frame

Reflector

Stand

FIGURE 2.3   Three-dimensional view of parabolic trough collector.

For stainless steel,

$$Q_u = mcp \, (T_{out} - T_{in}) = 2.175 \times 4.184 \times (55.2 - 29.1) = 0.2375 \text{ Kw}$$

$$Q_a = A_0 \times G_b,$$

where; $A_0 = 0.33$ m³

$$G_b = 1360 \text{ w/m}^2$$

$$Q_a = 0.33 \times 1360 = 448.8 \text{ W} = 0.448 \text{ Kw}$$

Efficiency of Stainless Steel 303 sheet

$$= \frac{Q_u}{Q_a} \times 100 = \frac{0.2375}{0.4488} \times 100 = 0.529 \times 100 = 52.9\%$$

For Aluminum 011 Sheet,

$$Q_u = mcp(T_{out} - T_{in}) = 2.175 \times 4.184 \times (62.1 - 28.5)$$
$$= 305.54 \text{ w} = 0.3055 \text{ Kw}$$

Efficiency for aluminum A011 sheet

$$= \frac{Q_u}{Q_a} \times 100 = \frac{0.3055}{0.4488} \times 100 = 0.6807 \times 100 = 68.07\%$$

The calculated efficiency of a parabolic trough collector with stainless steel as reflective material is 52.9%, and for the aluminum sheet is 68.07%. The maximum efficiency is obtained when the aluminum sheet is employed as reflective material when compared to stainless steel. The detailed specification of components used in the parabolic trough collector is tabulated in Table 2.1.

TABLE 2.1 Specifications of PTC

SL. No.	Description	Unit	Dimensions
1.	Focal length	mm	125
2.	Reflector sheet length	mm	1,219
3.	Reflector sheet width	mm	1,075
4.	Frame width	mm	800
5.	Receiver tube diameter (Do)	mm	25
6.	Receiver tube diameter (Di)	mm	22
7.	Receiver tube length	mm	1,219
8.	Volume of tube	m³	0.239
9.	Length of parabolic curve	m³	0.33
10.	Solar beam radiation	w/m²	1,360
11.	Specific heat of water	joules	4.184

2.4 EXPERIMENTAL INVESTIGATION

Two types of reflective materials, such as stainless steel 303 and aluminum A011, were selected for the investigation of the thermal performance of solar parabolic trough collectors. Water is used as a heat transfer medium, and the copper tube is used as the absorber tube. The frame and supporting structures are fabricated using mild steel material. Experimental analysis was carried out in the latitude of 11.05°N and longitude of 76.95°E during the month of March 2017 for both selected reflective materials under similar atmospheric conditions. The operating parameters of minimum and maximum flow rates are calculated as 1.139 kg/s and 2.174 kg/s, and the experimental setup of the parabolic trough collector is illustrated in Figure 2.4. The reflective capacity of stainless steel 303 sheets and aluminum A011 sheets is shown in Figures 2.5 and 2.6. Stainless steel material possesses a minimum amount of reflectivity. However, aluminum A011 gives a massive amount of radiation to the absorber tube. During the test, temperature measurement error is assumed to be of negligible quantity.

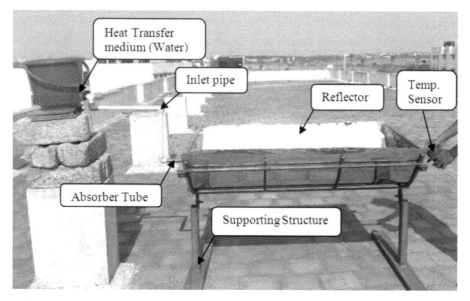

FIGURE 2.4   Experimental setup of parabolic trough collector.

At first, the stainless steel 303 sheet was fixed in the collector as a reflector to perform the investigation at a minimum flow rate of the working medium with constant time intervals (30 minutes) from 10.00 am to 3.00 pm. Initially, water is allowed to flow from the tank through the absorber tube. During the process, inlet and outlet temperatures were noted using a temperature sensor. The same process is continued for a maximum flow rate of the working medium and calculated the thermal efficiency of the collector. Instead of stainless steel 303 sheets, aluminum A011 reflective material was replaced, and the same procedure was followed to determine the reflective capacity and thermal efficiency of the solar parabolic trough collector.

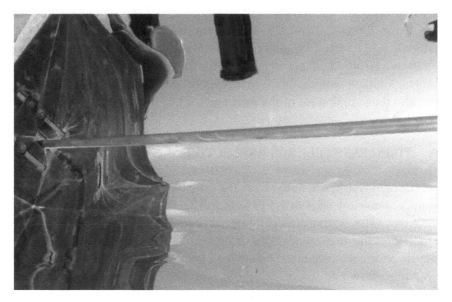

FIGURE 2.5   The focal point of stainless steel 303 sheets.

FIGURE 2.6   The focal point of aluminum A011 sheet.

2.5 RESULTS AND DISCUSSION

Experiments were conducted for both reflector materials for 10 days with the same operating parameters and conditions, and the output results were tabulated in Tables 2.2 and 2.3. Output temperatures for stainless steel and aluminum A011 reflective material for minimum and maximum flow rates are observed from 10.00 am to 3.00 pm for the respective time interval of 30 minutes. Ambient temperature is increased till noon and decreased gradually. It decides the temperature of the inlet and outlet.

2.6 STAINLESS STEEL 303 REFLECTIVE MATERIAL

Experimental results are noted from the time of 10.00 to 15.00 for minimum and maximum flow rates of 1.139 kg/s and 2.174 kg/s with ambient climatic conditions. The output temperature is varied with respect to ambient temperature; the highest level of ambient temperature is reached at noon time; therefore, maximum efficiency can be reached at the time of noon for an optimum flow rate.

TABLE 2.2   Experimental Results for Stainless Steel 303 Reflective Materials

SL. No.	Time	Temperature (°C)					
		Minimum Flow (1.139 kg/s)			Maximum Flow (2.174 kg/s)		
		Ambient	Inlet (T1)	Outlet (T2)	Ambient	Inlet (T1)	Outlet (T2)
1.	10.00	34	28.1	60.2	34	29.4	48.4
2.	10.30	35.7	29.2	62.4	35.7	28.2	49.5
3.	11.00	36.2	28.9	64.5	36.2	28.5	50.2
4.	11.30	38.5	30.1	63.8	38.5	29.6	51.3
5.	12.00	40.3	29.4	65	40.3	28.9	55.2
6.	12.30	39.4	28.5	62.1	39.4	29.1	50.1
7.	13.00	38	29.2	63.4	38	28.5	51.5
8.	13.30	37.6	30.2	64.1	37.6	28.8	52.3
9.	14.00	36.1	27.5	62.5	36.1	30.1	49.6
10.	14.30	36.2	28.8	59.2	36.2	29.4	48.4
11.	15.00	35	29.3	55.4	35	28.2	43.5

The output temperature gradually increases up to noon time, and further, it reduces due to the variation of ambient temperature. For the minimum flow rate of the working medium highest output temperature was obtained (65°C) at noon time (12.00), while for the maximum flow rate of the working medium highest output temperature occurred (55.2°C) at noon time (12.00). This means that the highest heat transfer rate takes place for a minimum flow rate of the working medium.

2.7 ALUMINUM A011 REFLECTIVE MATERIAL

Reflective material is replaced with Aluminum A011 to conduct an investigation with similar ambient climatic conditions. Experiments were carried out by maintaining a constant minimum and maximum flow rate of 1.139 kg/s and 2.174 kg/s. Output temperature is gradually increased up to noon time and further decreased due to the variation of ambient temperature. The highest output temperature obtained for the minimum and maximum flow rates is 70.1°C and 62.1°C at noon time (12.00). The lowest heat transfer rate takes place during low ambient temperatures, and the highest heat transfer rate takes place during the highest ambient temperature.

TABLE 2.3 Experimental Results for Aluminum A011 Reflective Material

SL. No.	Time	Temperature (°C)					
		Minimum Flow (1.139 kg/s)			Maximum Flow (2.174 kg/s)		
		Ambient	Inlet (T1)	Outlet (T2)	Ambient	Inlet (T1)	Outlet (T2)
1.	10.00	34	28.8	65.1	34	27.3	50.1
2.	10.30	35.7	28.4	66.5	35.7	27.5	52.4
3.	11.00	36.2	29.1	67.5	36.2	28.1	52.5
4.	11.30	38.5	29.5	68.3	38.5	28.5	55.8
5.	12.00	40.3	28.9	70.1	40.3	29.3	62.1
6.	12.30	39.4	28.8	68.8	39.4	28.2	56.1
7.	13.00	38	29.1	68.1	38	27.9	53.1
8.	13.30	37.6	28.8	67.9	37.6	28.3	54.5
9.	14.00	36.1	29.2	65.4	36.1	28.8	51.9
10.	14.30	36.2	29.3	65.2	36.2	29.5	49.8
11.	15.00	35	28.4	61.4	35	28.9	49.5

The graphical representation of output temperature variation is illustrated in Figures 2.7 and 2.8 for minimum and maximum flow rate of working medium for stainless steel 303 and aluminum A011 reflective material. The output temperature is gradually increased from 10.00 am and decreased after noon time (12.00) due to lack of intensity in radiation. At noon time (12.00), it can reach the highest temperature because of higher radiation emission. The highest temperature obtained for the minimum and maximum flow rate is taken into consideration for the discussion of parabolic trough collector performance.

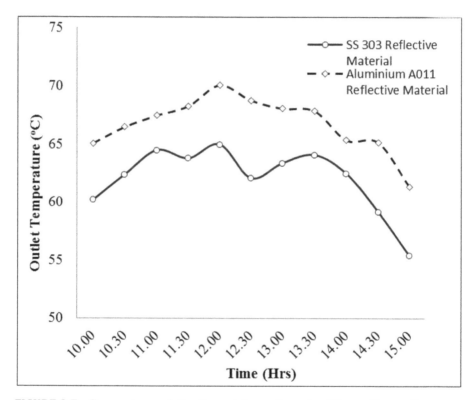

FIGURE 2.7 Temperature variation for a minimum flow rate of the working medium.

The highest temperatures of 55.2°C and 62.1°C were obtained at noon time (12.00) when stainless steel 303 and aluminum A011 were employed as reflective materials for a maximum flow rate of the working medium. Similarly, the highest temperatures of 65°C and 70.1°C were observed at noon time (12.00) for a minimum flow rate of the working medium. The thermal efficiency of the parabolic trough collector system is calculated using the analytical methods for stainless steel 303 and aluminum A011 reflective material as 52.9% and 68.07%, respectively. The highest thermal efficiency was obtained for aluminum A011 reflective material in the climatic condition of the western area of Tamil Nadu, India. The highest output temperature and thermal efficiency of parabolic trough collector aluminum A011 reflective material indicate that the reflective material has the highest reflectivity and transfers more heat energy to the absorber tube when compared to stainless steel 303 reflective materials. The highest thermal efficiency of a parabolic trough collector can be achieved while giving an optimum flow rate of the working medium as an inlet with aluminum A011 reflective material. Higher output temperature is produced by the stainless-steel reflector when compared with the previous work performed by Syed Mohd. Yasir Arsalan et al. [22] and using aluminum reflector material by Macedo-Valencia et al. [23].

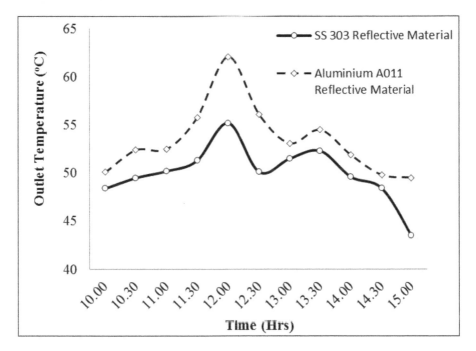

FIGURE 2.8 Temperature variation for a maximum flow rate of the working medium.

2.8 CONCLUSION

The parabolic trough collector was investigated in the western zone of Tamil Nadu. Thermal efficiency and maximum reflectivity of two dissimilar reflective materials (stainless steel 303 and aluminum A011) were analyzed. Experiments were conducted for minimum and maximum flow rates of heat transfer medium with constant time intervals of 30 minutes from 10.00 to 15.00. When stainless steel 303 is employed as reflective material, the highest output temperature is obtained as 65°C at 12.00 with minimum flow rate, and it was 70.1°C for aluminum A011. To gain the maximum utilization of reflective heat energy, the parabolic trough collector should be run with an optimal flow rate of the heat transfer medium. From the result, it is clearly evident that maximum output temperature and highest thermal efficiency can be obtained while using aluminum A011 as reflective material. Aluminum A011 reflective material gives 5% of higher output temperature and 28.67% of higher thermal efficiency than stainless steel 303 reflective materials. Due to the highest reflectivity of aluminum A011 reflective material, it is best suited for parabolic trough collector, which is used for the purpose of industrial and domestic applications.

KEYWORDS

- **domestic applications**
- **heat transfer**
- **parabolic trough collector**
- **reflector**
- **solar energy**
- **stainless steel reflector**
- **thermal efficiency**
- **thermal performance**

REFERENCES

1. Syed, M. et al., (2016). Performance evaluation of solar parabolic trough collector with stainless steel sheet as a reflector. *International Journal of Engineering Research & Science, 2*(6), 22–28.
2. Vijayan, S. N., & Sendhil, K. S., (2020). Numerical and analytical investigation of heat transfer enhancement in flat plate solar collector using internal fins in absorber tube and dissimilar working medium. *Journal of Xidian University, 14*(5), 35–45.
3. Vijayan, S. N., & Sendhil, K. S., (2017). Theoretical review on influencing factors in the design of parabolic trough collector. *World Academy of Science, Engineering and Technology International Journal of Mechanical and Materials Engineering 11*(11).

4. Roy, B. et al., (2016). Parametric study of parabolic trough collector – a case study for the climatic conditions of Silchar, Assam, India. *ISESCO Journal of Science and Technology 12*(21), 24–29.

5. Vijayan, S. N., et al., (2017). Performance analysis of non-concentrating solar collector: A review. *International Journal of Advanced Chemical Science and Applications 5*(2), 39–43.

6. Christos, T., & Evangelos, B., (2016). The use of parabolic trough collectors for solar cooling: A case study for Athens climate. *Case Studies in Thermal Engineering, 8,* 403–413.

7. Avadhesh, Y., et al., (2013). Experimental study and analysis of parabolic trough collector with various reflectors. *International Journal of Mathematical, Computational, Physical, Electrical and Computer Engineering, 7*(12), 1659–1663.

8. Amit, K. P., et al., (2014). Design, fabrication, and performance evaluation of a low-cost parabolic trough collector with copper receiver. *International Journal of Mechanical Engineering and Technology, 5*(5), 122–129.

9. Akdeniz, C., (2016). Design of a parabolic trough solar collector using a concentrator with high reflectivity. *Proceedings of the 2nd World Congress on Mechanical, Chemical and Material Engineering* (pp. 1–5). Budapest, Hungary.

10. Kajavali, A., et al., (2014). Investigation of heat transfer enhancement in a parabolic trough collector with a modified absorber. *International Energy Journal, 14,* 177–188.

11. Fauziah, S., et al., (2012). A simulated design and analysis of a solar thermal parabolic trough concentrator. *International Journal of Environmental, Chemical, Ecological, Geological, and Geophysical Engineering, 6*(12), 739–743.

12. Sendhilkumar, S., & Vijyan, S. N., (2016). Solar tracking system using a refrigerant as a working medium for solar energy conversion. *International Journal of Mechanical Aerospace Industrial Mechatronic and Manufacturing Engineering, 10*(8), 1548–1553.

13. Donald, J. G., et al., (2012). *Design, Construction, and Test of a Miniature Parabolic Trough Solar Concentrator* (pp. 1–73). Project Report Fall Quarter.

14. Vijayan, S. N., et al., (2013). Performance analysis of solar-assisted drying system. *Elixir International Journal, 63,* 18107–18109.

15. Devander, K., & Sudhir, K., (2015). Year-round performance assessment of a solar parabolic trough collector under the climatic condition of Bhiwani, India: A case study. *Energy Conversion and Management, 106,* 224–234.

16. Yaghoubi, M., et al., (2013). Analysis of heat losses of absorber tubes of parabolic through the collector of Shiraz (Iran) solar power plant. *Journal of Clean Energy Technologies, 1*(1), 33–37.

17. Shubham, G., & Mishra, R. S., (2016). Design and performance analysis of solar parabolic trough for cooking application. *International Journal of Research and Scientific Innovation, 3*(5), 136–139.

18. Mayank, V., et al., (2014). Thermal performance analysis of water heating system for a parabolic solar concentrator: An experimental model-based design. *International Journal of Current Engineering and Technology, 4*(5), 3649–3654.

19. Kaloudis, E., et al., (2016). Numerical simulations of a parabolic trough solar collector with nanofluid using a two-phase model. *Renewable Energy, 97,* 218–229.

20. Yu, Q., et al., (2017). Thermal performance analysis of a parabolic trough solar collector using supercritical CO_2 as heat transfer fluid under non-uniform solar flux. *Applied Thermal Engineering, 115*(25), 1255–1265.

21. Valan, A. A., & Sornakumar, T., (2007). Design, manufacture, and testing of fiberglass reinforced parabola trough for parabolic trough solar collectors. *Solar Energy, 81*(10), 1273–1279.

22. Syed, M. et al., (2016). To analyze the performance of solar parabolic trough concentrators with two different reflector materials. *International Journal of Scientific Engineering and Research, 4*(5), 65–69.

23. Macedo-Valencia, J., et al., (2014). Design, construction, and evaluation of parabolic trough collector as a demonstrative prototype. *ScienceDirect, Energy Procedia, 57*, 989–998.

CHAPTER 3

Smart Buildings Using IoT and Green Engineering

AMIT KUMAR SINGH,[1] ANSHUL GAUR,[2] and PRADEEP KUMAR[3]

[1]*Biomedical Engineering Department, VSB Engineering College, Karur, Tamil Nadu, India*

[2]*Electronics and Communication Department, Uttarakhand Technical University, Uttarakhand, India*

[3]*Electrical Engineering Department, Ambalika Institute of Technology and Management, Lucknow, Uttar Pradesh, India*

ABSTRACT

Residential and commercial buildings are rapidly evolving, and Internet of Things (IoT) technologies are helping to shape their future. Researchers have recently exploited the Internet of Things (IoT) in various applications and situations to change traditional buildings into smart, efficient, and secure structures. While feasible IoT techniques have been created, there is still a need to fully develop IoT applications and operations to realize the technology's promise. This can be accomplished by filling gaps in current research and laying a foundation for future studies. This chapter thoroughly assesses current technologies and IoT applications in residential and commercial buildings. First, we systematically investigated the BEMS using an IoT approach and how different segments in smart buildings are connected. The role of sensors and actuators in smart building systems was then discussed. Then, we demonstrated the role of IoT in green energy development for smart buildings and how to use it in smart building grid management. Solar and biomass energy have generated green energy for

The Internet of Energy: A Pragmatic Approach Towards Sustainable Development. Sheila Mahapatra, Mohan Krishna S., B. Chandra Sekhar, & Saurav Raj (Eds.)

smart buildings. Finally, the trends, existing advantages, hazards, and future obstacles of IoT deployment inbuilt settings are recognized and analyzed based on an analysis of each category. The most significant barriers to deployment have been identified as integrating multiple IoT technologies with varying capabilities, data storage and processing, and privacy and security concerns. We have also compared different wireless protocols used in the smart buildings and showed which protocol could be used for the smart building IoT infrastructure. This chapter assists IoT developers and researchers in defining their work boundaries and contributions by proposing future research directions.

3.1 INTRODUCTION

The growing human population is increasing the energy demand of the world. Uncontrolled energy consumption will undoubtedly exacerbate environmental damage, worsening global warming. International Energy Agency report data suggests that the world's energy demand will double its current level by 2035. Buildings consume the most energy, accounting for a rough 33% of global energy usage [1]. According to Shah et al., researchers were more concerned with predicting and optimizing electricity use in residential buildings to develop IoT-based smart building systems. Traditional energy managing methods used statistical analysis and machine learning techniques to gather data from electricity meters and were utilized for monitoring purposes [2]. Two approaches can be used to deal with the ongoing increase in demand. The first method is to increase energy production. The other way to reduce the need for power is to increase energy consumption efficiency, particularly in the building sector. As a result, renewable energy increases production capacity while decreasing demand by managing energy consumption. Energy management is the best practice for maximizing the energy efficacy parameter of smart buildings. Different authors' literature and research work show that the IoT can play a vital role in increasing the efficiency and monitoring of the smart building parameters to integrate them with the smart grid.

Because of technological advancements, the IoT has a significant role in the energy management of buildings, enabling them to become smart. Security, network architecture, sensors, health services, safety, and overall smart buildings are the features that make a building smart. The Internet of Things has a flexible architecture and necessary components to accomplish this. Sensors and actuators were used to monitor and regulate many smart building's energy consumption parameters. With the advancements in internet technology, smartphones can easily interface with sensor networks and actuators to control and monitor them and transfer the real-time data to the user using IoT. As a result, the Internet of Things (IoT) can significantly improve the energy efficiency of smart buildings. IoT was already being used to increase the production capacity of the solar-powered plant, which improves the efficiency of the existing infrastructure by monitoring and controlling the different parameters of the solar plant [3]. By properly monitoring and managing the biogas plant, the study reveals that biomass has a remarkable capacity to boost the electricity generation of rural/urban areas utilizing IoT [4].

Different researchers have used different algorithms to improve energy efficacy in building energy management system (BEMS), which uses the comfort-index and energy-consumption optimization by the fuzzy logic and considers the thermal, air quality comfort, and visual with the user's satisfaction at the top of the priority list. They also avoid overshooting and energy wastage while monitoring overall energy consumption to achieve the comfort index [2]. The IoT-based structure uses different methods to communicate with the various components of the smart buildings. Verma et al. demonstrated various gateways for transmitting data between different parts of the sensor and actuators networks, as well as other smart building components. Different protocols used for data transmission were Long-range (LoRa), Narrow Band-Internet of Things (NB-IoT), Long Term Evolution for Machines (LTE-M), Sigfox, Zig-Bee, wireless fidelity (Wi-Fi), Bluetooth, Ultra-wideband (UWB), and Wireless universal serial bus (USB), and comparing those technologies based on different parameters like the speed of the network, range, network topology, security, power consumption, complexity, cost and operating frequency of the network used [5, 53].

Figure 3.1 depicts the major components of the IoT-based smart building system. The smart building (residential/commercial) is at the heart of the system, with an IoT system and BEMS support. It gathers all the building information and sends it to a local server/web application via the internet gateway. The local server saves all of the needed data generated by the smart building and makes it available in real time to the building's end users. Solar and biomass green energy sources (ES) could be used to meet the smart building's energy requirements. When no other energy source is available, battery-based energy storage provides a vital point for power backup. There is also a waste disposal system, which produces biomass energy. Smart buildings use various sensors to monitor all environmental characteristics, such as temperature sensors for the building's ambient temperature and many different sensors for monitoring different parameters. The sensors and actuator networks established within the building premise control and monitored the electrical appliances installed at the smart facility. The installed security system improves the living standards of the persons living on the building premises. The space occupation system provides real-time data of human presence in a specific building area and, using various sensor and actuator networks, can turn on/off specific electrical appliances. This technology eliminates energy waste and boosts the energy management efficiency of smart buildings. Several studies suggest that net-zero energy buildings (nZEBs) can reduce building energy usage and make them self-sufficient by using efficient and sustainable methods.

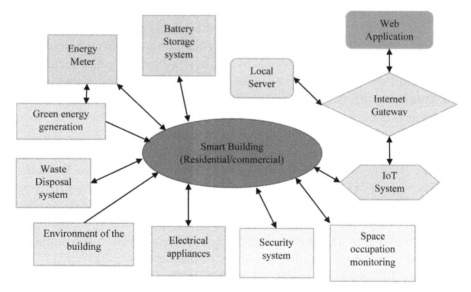

FIGURE 3.1   The block diagram of the smart building system (IoT-based).

3.2   ENERGY MANAGEMENT FOR SMART BUILDINGS

According to Nguyen et al., buildings consume a large amount of energy. The survey also found that buildings consume almost 40% of the total energy required on the planet, of which 60% of that energy was used solely for heating and cooling the atmosphere of the buildings [1]. Therefore, for the sustainable development of the world and controlling air pollution, there was the requirement of the proper monitoring and handling of the buildings' electrical equipment and meet the growing demand for the buildings. Ida et al. have demonstrated a two-way solution to the energy and environmental problems. To deal with the shortfall of energy, the two options are: (i) more energy production; and (ii) minimizing available energy resource consumption and reducing wastage. Energy production is costly for human society that takes more time and money, but it is possible to reduce energy usage by implementing straightforward preventative actions. The ISO 50001 energy management standard was developed by the industry sector to provide a framework for businesses to integrate energy efficiency into their operations and put it into practice in the commercial building sector [6].

So, Nguyen et al. have shown that a BEMS is an efficient technology to monitor and control building energy requirements. To solve the present problems of the current BEMS, such as the large energy data of BEMS, energy data loss, and overload problems, a BEMS based on the Internet of Energy (IoE) is currently being used. The reference energy model can assess the building's energy efficiency and establish critical setpoints. With the methodological framework given, the BEMS can improve smart buildings' energy efficiency and savings when considering energy consumption and cost-benefit analysis [1]. According to a study conducted by Doukas et al., numerous metrics were used to determine the user's comfort level by ensuring the desired living quality standards in all of the building's rooms and the need for energy savings [7].

Furthermore, the demands for essential thermal comfort, indoor air quality, and visual comfort are increasing, particularly in the present environment of price changes, the rapid growth of population, and technological advancement. The BEMS were typically used to control active systems, such as heating, ventilation, and air-conditioning (HVAC) systems, determining their operating times in addition. The major recent advances in BEMS have followed the developments achieved in computer technology, telecommunications, and information technology. Major BEMS technologies were implemented on residential and commercial buildings, transforming them into "intelligent buildings" or smart buildings [7].

Figure 3.2 shows the block diagram of the BEMS. The block diagram depicts the flow and the control of the energy common in residential and commercial buildings. It shows the power source of the building that can be coming from the external source or the green energy generated within the building premises itself. The load gets its power through the smart meter, which stores the energy profile of the building by monitoring the peak loads and other factors of the energy demand. The building monitoring and control system keeps the data and uses an IoT framework for data transmission among the different components of the smart buildings.

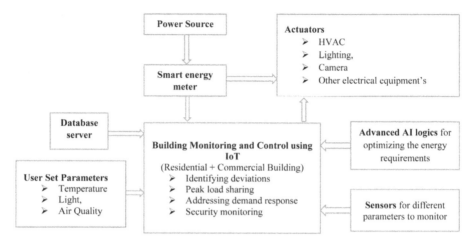

FIGURE 3.2   Block diagram of the BEMS.

The BEMS analyzes the gathered information per the user requirement, which has entered its energy demand as per user comfort. At the appropriate space and time, the user can enter input factors such as their room's temperature, lighting on and off timings, intensity control, and air quality. So, the feedback provided by the consumer and the energy demand manage the actuators controlling different electrical appliances to maintain the environment of the smart building and by on/off the air conditioner, heater, lights, and other electrical types of equipment as per the demand. The BEMS uses advanced artificial intelligence (AI) with different fuzzy logic techniques to optimize the energy requirement and control the actuators accordingly. Various sensors were deployed at different locations and for different purposes of the smart buildings, providing real-time data for the different ambient parameters of the indoor and outdoor environmental conditions. So, the complete system also uses the local server or cloud-based servers to store the enormous data generated by the facilities and helps further optimize energy efficiency and real-time monitoring of the smart buildings for its better utilization of the resources in terms of money, space, and time. The system makes the building smart, and IoT gets converted into IoE. The central building monitoring system identifies the deviations of the peak load demand. It takes corrective action to address the demand response and provides the consumer's security by monitoring the building in real-time with the help of an intelligent security system installed at the building premises.

3.2.1 ROLE OF SENSORS AND ACTUATORS USED IN SMART BUILDING MANAGEMENT

The sensors and actuators are one of the prime blocks of a smart building. The IoT-enabled sensors increase the efficiency of building management manifold. Buildings are becoming smarter as sensors and actuators, information and communication, and Internet of Things (IoT) technologies progress. The building management and information system (BMIS), together with the IoT, is called a "Building Internet of Things" [5]. Without accurate data, IoT would be useless. As a result, sensors and actuators are critical components of a smart building. According to different parameters inside a smart building, there are different sensors. Some essential types of sensors are environmental, optical, touch, level, leak, magnetic, electricity, acoustic, object presence, motion, gyroscope, and chemical.

Figure 3.3 shows different sensors and actuators used in smart buildings with IoT infrastructure support. The environmental sensors used in smart buildings are temperature, humidity, gas, smoke, and others. In addition, the temperature sensors provide thermal conditions of air inside the building, working environment (thermal comfort), early fire detection, and others [8]. The temperature sensors used in smart buildings are of the following types: Resistance temperature detectors (RTD), thermocouples, thermistors, thermopiles, infrared (IR) temperature sensors, semiconductor junctions, and others. These primarily used temperature sensors are negative temperature coefficient (NTC) thermistors and IR sensors in smart buildings. The NTC thermistor was a low-cost, nonlinear sensor with good temperature sensitivity. Through proper signal conditioning, they interface with IoT hardware devices such as ESP8266. IR temperature sensors are becoming popular due to their non-contact temperature-sensing feature. These are useful for providing occupant body temperature conditions and electrical wire heating status for safe operation. Temperature sensors are also used in HVAC and occupancy detection systems.

Another critical environmental parameter inside buildings is humidity. The vaporized water content in the air is humidity, generally expressed in relative humidity. Therefore, humidity provides indoor environment quality (IEQ) information in a smart building. Gulnizkij et al. have presented a hydrogel-based bistable sensor for relative humidity measurement inside a building [9]. Moisture is a broad term measured by different technologies like hair tension moisture sensors and psychrometers. The humidity sensors are used in manufacturing plants to ensure long shelf life and optimum performance. For example, in museums, hospitals create a healthy environment to prevent articles from corrosion. In greenhouses, the humidity is maintained at the optimum value for better plant growth. These were used in weather stations to transmit humidity-related information to the public.

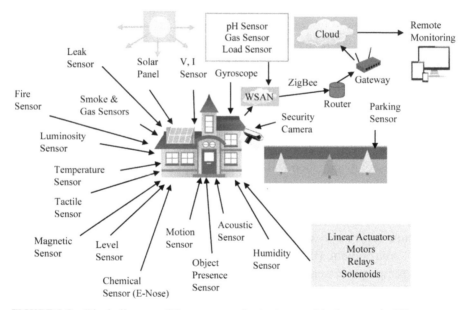

FIGURE 3.3 Block diagram of the sensors and actuators used in the smart building.

The next environmental parameters monitored inside buildings are the smoke and gases. The smoke could be due to a fire hazard inside a building. Most of the time, it occurs earlier as compared to flames. The broad classification of smoke sensors is nonvisual and visual smoke sensors. The nonvisual sensors are particle-based, and they suffer particle delay effects. Hence their response time was poor, whereas the accuracy was good. The visual sensors are volume sensors. They have a good response time but suffer false outputs. Machine learning techniques, particularly deep learning techniques, are used to improve performance by reducing false positives of visual smoke sensors [10]. Gas sensors used in smart buildings provide IEQ monitoring, occupancy detection, combustible or air pollution from gas leakage detection, and others. The significant gases measured inside a smart building are carbon dioxide (CO_2), carbon monoxide (CO), oxygen, hydrogen, nitrogen oxide, and air pollution gases. Bad air quality inside a building develops health problems such as tiredness, headaches, and carbon monoxide could cause the deaths of building occupants [11].

Another category of sensors used in smart buildings is optical sensors. A camera (visible and infrared range), proximity, photoconductive, photovoltaic, and other sensors fall under this group. The chief application of these sensors is to monitor the building space.

Cameras are used to surveillance the building, having the night vision feature. Different image processing and machine learning algorithms detect any suspicious activity near and inside the building space, and the fire was also detected. Optical sensors help adjust the buildings' surrounding light and make them smart. In smart buildings, touch (tactile) sensors sense touch or proximity without physical contact. The main challenge of these sensors was the lack of moving parts. Table 3.1 shows the types of sensors and their usage in a smart building. Therefore, they can replace the mechanical buttons used in several electronic devices inside a smart building. According to different regulations, touch sensors use Surface Acoustic Wave (SAW), IR, wire resistive, and capacitive techniques to detect objects. In addition, level sensors are used to monitor the water level in the tanks of the buildings, including float, ultrasonic, resistive, capacitive, optical, and other sensors. Another type of sensor used in the smart building is the leak sensor. These sensors are of the types such as rope style, spot type, and others are used to detect leakages in smart buildings. Magnetic sensors were used to provide a magnetic map of corridors of buildings. These maps are helpful under low-lighting conditions. These are mainly of the Hall Effect type. Electricity types of sensors are used in buildings to monitor the electrical load conditions, which help provide a cost estimate of electricity usage. These are also useful to measure the electric current flowing through the wires or cables. This information helped find the possibility of a fire hazard. Because when the electrical current becomes more prominent, the wire or cable heats up and may lead to a fire hazard. Acoustic, object presence, motion, gyroscope, and chemical sensors were also used in buildings.

3.2.2 NETWORK PROTOCOLS USED IN SMART BUILDINGS

The core of smart buildings is smart BEMS. Different communication networks aid in exchanging data between sensors and control devices. Commercial and

TABLE 3.1 Sensors Used in Smart Buildings

SL. No.	Sensor Type	Sensors Used in Smart Buildings	Usage	References
1.	Environmental	Temperature (RTD, thermistor, thermocouple, thermopile, IR temperature sensor, silicon diode, change of state sensors, and others.	HVAC, fire detection, health parameters, occupancy detection	[5, 10]
2.	Environment	Humidity (resistive, thermal subtypes, capacitive), psychrometer, and Hair tension moisture sensor.	Moisture content in a smart building environment for comfort and extending appliance's life.	[12]
3.	Environment	Smoke, gas (CO, CO_2, nitrogen oxides), photoelectric, ionization, and aspirating.	Gas and smoke sensing for a cleaner environment, fire detection, gas leakage detection for safety	[13]
4.	Optical	Camera (IR, visible), proximity sensor, charge-coupled devices, photoconductive, and photovoltaic.	Surveillance, fire detection, occupancy detection, adjustment of screen brightness of televisions, mobile phones, personal computers, and others depending on ambient light.	[14]
5.	Touch (tactile)	SAW sensor, IR sensor, wire resistive sensor, and capacitive sensor.	Remote control devices, as an alternative to mechanical buttons, detect touch or proximity.	[15, 16]
6.	Level	Float, ultrasonic, resistive, capacitive, and optical.	Flood alarms in buildings, water level detection in building tanks, and others.	[17]
7.	Leak	Spot leak detectors, rope style, under carpet leak detectors, and hydroscopic tape-based sensors.	Detection of leakages in buildings.	[18]
8.	Magnetic	Hall effect, and others.	Magnetic maps inside buildings are suitable in poor lighting conditions.	[19]
9.	Electricity	Current sensors and voltage sensors.	Monitoring the electricity parameters inside a building.	[20]
10.	Acoustic	Hydrophone and geophone.	For picking sound in liquids, to pick up ground vibrations (seismography).	[21]

TABLE 3.1 (*Continued*)

SL. No.	Sensor Type	Sensors Used in Smart Buildings	Usage	References
11.	Object presence	Doppler radar and occupancy sensors.	Manufacturing machines, robotics, smart parking.	[22]
12.	Motion	Ultrasonic, IR, PIR (passive IR), and radar (active and passive).	Surveillance, alarm systems, automatic lighting, and ventilation systems.	[23]
13.	Gyroscope	Accelerometer (mechanical and optical)	Measures rotation and angular velocity, which is beneficial for athletes for measuring accurate body movements.	[24]
14.	Chemical	E-Nose and others.	Industrial security systems, environment protection.	[25]

residential buildings use different protocols based on their energy, distance, and data load requirements. Wired and wireless are two open protocols. Both protocols have their benefits and shortcomings. The wireless protocols used for data transfer in existing buildings and the comparison of different wireless protocols used in the smart buildings are shown in Table 3.2. The protocols compared in Table 3.2 are described below:

1. **802.15.3 UWB:** 802.15.3 is the Institute of Electrical and Electronics Engineers (IEEE) standard for a high-data-rate wireless personal area network. The protocol design provides proper quality of service for the real-time distribution of video and music content. Therefore, it ideally suits a home multimedia wireless network.

2. **Bluetooth:** It is a communication protocol used for short-range communication. The protocol was commenced by Bluetooth's special interest group to exchange the data over short distances between fixed and mobile devices. Numerous products are compatible with Bluetooth automation. A low-energy variant of the Bluetooth network protocol is Bluetooth low energy (BLE) is a part of the Bluetooth v4.0 and the recent v4.2 stack. BLE is a global wireless communication protocol for transferring small data pieces. Bluetooth network protocol has a limited range.

3. **Long-Term Evolution (LTE):** This wireless network protocol uses fast wireless data communication. It is based on GSM/EDGE and UMTS/HSPA network technologies. It is commonly called 4G LTE. Nevertheless, LTE has been going into service in mobile phone communication as it is capable of providing multicasting and broadcasting service [26].

4. **LoRaWAN:** It refers to a Long-Range Wide Area Network that senses signals under the noise level over a long distance and low strength. LoRaWAN uses the internet-based connection of battery-powered items. The use of LoRaWAN includes smart cities having private or global networks.

5. **NB-IoT:** It is a low-power, wide-area network radio technology standard developed by 3GPP for cellular devices and services. The protocol utilizes a subset of the LTE standard and limits the bandwidth to a narrow band of 200 kHz. In addition, it uses orthogonal frequency-division multiplexing (OFDM) modulation and SC-FDMA for downlink and uplink communication, respectively.

6. **Sigfox:** It was established in 2021 by a French global network operator that initially built wireless networks to interconnect

TABLE 3.2 Comparison between Different Wireless Technologies Used in IoT-based Smart Buildings

Parameters	LoRa	Bluetooth	Zigbee	NB-IoT	802.11 (Wi-Fi)	LTE-M	802.15.3 UWB	Sigfox	Wireless USB
Networking topology	Star or mesh, point to point	Ad Hoc, small NWs	Ad Hoc, peer to peer, star, or mesh	Star	Point to hub	Star	Point to point	Star	Point to point
Security	Low	Low	Low	Low	Good	Moderate	Low	Moderate	Moderate
Complexity	Moderate	High	Low	Low	High	Moderate	Low	Moderate	Moderate
Range	5 km (Urban) and 15 km (Rural)	10 m	10 m	15 km	100 m	11 km	Less than 10 m	50 km	10 m
Data rate	0.3–38.4 kbps	1 Mbps	250 kbps	158.5 kbps (UL) 106 kbps (DL)	6.93 Mbps	1 Mbps	100–500 Mbps	100 bps (UL) 600 bps (DL)	62.5 kbps
Power usage	Very low	Low	Very low	High	High	Low	Low	Low	Low
Operating frequency	779–787 MHz (China), 863–870 MHz (EU), 902–928 (US)	2.4 GHz	2.4 GHz (worldwide), 868 MHz (Europe), and 900–928 MHz (NA)	Licensed (700–900 MHz)	2.4 and 5 GHz	Licensed (700–900 MHz)	3.1–10.6 GHz	Sub-GHz ISM: EU 868 MHz, US 902 MHz	2.4 GHz
Applications	Smart buildings (smart lighting)	Mobile phones Computers, digital devices	Home/industrial automation, health care, smart metering in renewable energies	Smart grid communication	Thermostats, intelligent devices, and access in smart buildings	Smart meters in buildings, Agricultural Monitoring	Localization, identification	Low power applications, Smart buildings (electric plugs)	Game controllers, digital cameras.

low-power objects. Example of these is energy meters and smart watches, which continuously send small amounts of data for communication.

7.   **Wi-Fi:** It stands for Wireless Fidelity. This wireless network protocol refers to IEEE 802.11 standard for Wireless Local Networks. Wi-Fi is the registered trademark of the non-profit Wi-Fi Allianz. Wi-Fi is the cost-effective and easily accessible wireless technology in commercial buildings that connects IoT devices. Data transferred faster when Wi-Fi protocol was used in smaller connected devices. Furthermore, Wi-Fi may be a great choice in smart buildings powered by cloud-based software applications, i.e., the Chariot platform. Wi-Fi application areas include thermostats, lighting, intelligent devices, and broadband internet access in smart buildings.

8.   **Wireless USB:** WUSB stands for Wireless USB (WUSB), a Universal Serial Bus (USB) protocol. It is a high-bandwidth and short-range wireless radio communication protocol. The protocol was originally developed by the Wireless USB Promoter Group and preserved by the WiMedia Alliance. It uses radiofrequency (RF) links rather than cables to provide the interfaces between a computer and peripherals, for example, external drives, monitors, printers, headsets, MP3 players and digital cameras.

9.   **Zigbee:** It is a shorter-range wireless communication standard specifically for commercial use. It is mostly used in home and building automation. It uses a mesh network to create long ranges and fast communications through radiofrequency. The main benefit of the mesh network is that for any link breakage, devices can search through the mesh network to find a new route. Thus, the addition of a new device is simple and less costly. More than 1,200 products were compatible with the Zigbee protocol. This wireless communication protocol is best suited for large buildings and campuses. The Zigbee technology has widespread applications in home automation, consumer electronics, healthcare, industrial automation, material tracking, and many others. The latest version of Zigbee is v3.0 is a single unified ZigBee standard.

3.2.3   *BENEFITS AND ROLE OF IoT IN SMART BUILDINGS*

IoT has a major role in transforming residential and commercial buildings into smart residential and commercial buildings. Smart buildings have all the systems connected through IoT and network protocols. With the help of IoT technology, smart buildings, regardless of their usage, started to upgrade, innovate, be efficient in terms of energy, and provide a healthy, dynamic, and tech-savvy smart society. The multiple profits of smart buildings originate from the data-generating systems that empower them. For example, it is much easier to recognize facilities when there is data to inform how people use them. Here we have discussed some of the benefits of smart buildings:

1. **Reduced Energy Consumption:** By using IoT in connecting and controlling different devices of smart buildings, various researchers try to reduce energy consumption. This controlled system using lesser energy costs benefits to commerce and the environment.

2. **Automation Opportunities:** The IoT initiates a robust automation system for many applications, including links between physical and digital management systems. For example, it helps implement floor sensors for occupancy, motion-sensitive lights, beacons to gauge workspace utilization, and many others.

3. **Improved Performance Management:** Remote monitoring permits administrations to integrate information about many assets and developments, regardless of time and the location of the resources. BEMS can instantly merge information about the body's performance while covering key performance indicators similar to overall equipment efficiency and total carbon (CO_2) emissions.

4. **Quantifiable Building Insights:** The energy data points provided by the workplace, which provide information on the energy consumption patterns of people living and working in and around buildings, aid in quantifying the building's energy consumption pattern. Therefore, each data point created by the IoT is a calculable part of the tangible workplace which adds the data points to find trends and actionable insights.

5. **Predictive Maintenance:** Technology has a more insightful overview of the building's operations through sensors attached to the IoT. The sensor data aids in the timely upkeep of the equipment installed for the building functionality at the right time, thereby increasing the devices' shelf life. The collected data also helps reduce the expenses of replacing the costly equipment used in the buildings and making the devices smart.

6. **Better Resource Utilization:** IoT helps better manage the workplace resources, space, and human resources of the smart buildings, especially in commercial buildings. Smart buildings make and

measure them within broader facilities. The result shows how people use those resources and information more efficiently.

7. **Reduced Operational Costs:** With better insights, one can quickly reduce the buildings' operating costs, especially for the commercial domains, by reducing energy wastage. There the IoT helps by controlling the wasteful use and thereby reducing the operational cost of the commercial buildings specially and saving the paramount money for the human. Moreover, the connection of sanitation devices with sensors also reduces the costs of smart buildings.

8. **New Workplace Opportunities:** With the growing technology, there is a new age of growing work styles and changes in the workplace. Due to the pandemic, flexible work and agile workspaces have become the new standard and the demand for systems to accomplish them. The intelligent IoT networks and strategies that produce data to support the new workplace can easily handle this oversight.

9. **Increase Productivity:** Smart buildings make humans more productive by continually monitoring and controlling the building use and ensuring the better facility of the workspace or residence. In addition, smart buildings provide healthier, productive, satisfied workers and a comfortable workspace for the people staying in with advanced climatic and lighting controls using IoT technology. Furthermore, a better ventilation system in the buildings also improves the indoor environment's air quality and reduces the sick building syndrome symptoms.

10. **Physical Security:** Due to the financial gap in society, there is an upswing in the community crime rate. So, IoT can provide better security for the persons residing in the buildings. Moreover, the advanced sensors and cameras with the IoT interface help provide safety and security in any building, and the automatic systems of the camera also reduce the security expenses of the buildings.

11. **Smart Buildings Give Smart Data:** Smart buildings provide real-time electricity, water consumption, hourly space utilization, and other data. It helps the residents to take corrective action about their constructive utilization.

12. **Smart Buildings Result in Increasing the Asset Value:** Buildings are becoming smart buildings when IoT and smart data are used, increasing the asset value of the system. In addition, smart energy efficiency measures ensure that the building is well maintained and hence does not depreciate over some time.

13. **Space Allocation Management:** With the help of IoT, smart building managers can take advantage of remote occupancy monitoring to allocate shared spaces tenants allocations management, thus improving the efficiency of the space use and increasing the person's income.

14. **Reduction of Manual Labor:** Remote monitoring of the systems enables remote inspections and reduces the need for hands-on work. Furthermore, it also helps the employees to utilize the remaining time to pursue more creative work.

Table 3.2 shows Sigfox protocol has got the highest range for data transfer while the 802.15.3 UWB protocol has the shortest range. 802.15.3 UWB has the highest data transfer rate, while the LoRa protocol was the slowest one. NB-IoT and 802.11 (Wi-Fi) have the highest power consumption, while the LoRa protocol has the lowest. 802.15.3 UWB protocol uses the highest bandwidth, and LoRa uses the lowest bandwidth frequency. LoRa has the lowest cost, while the 802.15.3 UWB protocol was the costliest. The best networking topology used by LoRa, and ZigBee is Peer to Peer, Star, or Mesh, which helps in the fastest and most reliable data transfer. Regarding data security over network communication, the LoRa was the worst, while the 802.11 has the best security. In terms of the complexity of the hardware, Bluetooth and 802.11 are highly complex, while the NbIoT is the least complex one. Hence, in terms of the features like security, good speed, cost-effectiveness, range, and lesser energy consumption, LoRa, and 802.11 protocols are found to be most suitable for data communication between different segments of BEMS.

3.3 ROLE OF IoT IN GRID MANAGEMENT

The traditional power grid connects generators, transformers, transmission lines, distribution systems, and various loads. Moreover, the traditional grid system is a one-way communication network. Power generation dependency is huge on fossil fuels in this system, and there is a burden on these resources due to an increase in electricity demand. Therefore, renewable energy resources have attracted power utilities in the last few decades. The major challenge in the power industry is to provide dependable and decent-quality power to consumers. With new technological advancements, IoT applications offer boundless solutions to power utility by giving two-way communication, which helps in transforming traditional power grids into Smart Grid [SGs].

A smart grid is an energy network consisting of control, automation, and computers, enabling two-way communication between power utilities and consumers. Figure 3.4 shows the block diagram of the layout of the smart grid. The application of IoT in smart grids allows the smart grid to share information between all components in the grid. As a result, the smart grid provides many great solutions to the challenges in the power industry. In addition, IoT can support technologies in SG. The applications of IoT in different parts of the smart grid are as follows:

- IoT is capable of monitoring electricity generation of different kinds of power plants (such as coal, solar, wind, biomass), energy storage, gas emissions, energy consumption, and forecasting necessary power to supply consumers.

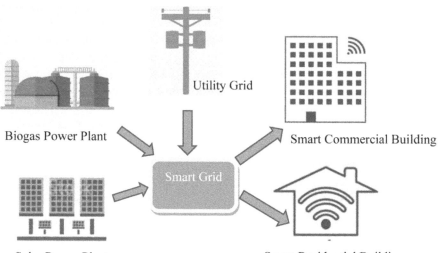

FIGURE 3.4   A layout of the smart grid used in the smart building systems.

- IoT uses to gain electricity consumption, monitor, dispatch, and protect transmission lines, substations, and towers, along with management of control equipment.
- IoT can be used on the customer side in smart meters to measure various parameters. These are intelligent power consumption, interoperability between various networks, charging and discharging electric vehicles (EVs), and managing energy efficiency for power demand.

3.4   ROLE OF IoT IN GREEN ENERGY

A smart building has continuous energy demand in various forms, such as heat energy, mechanical energy, light energy, and electrical energy. A significant share of this requirement is electrical power, which the electricity department fulfills. However, electricity generation mostly comes from fossil fuels, decreasing faster. So, the answer lies in the use of non-conventional sources based on solar energy, hydraulic energy, tidal energy, wind energy, geothermal energy, biogas, and others.

World bank report shows that around 93% of the universal population lives in countries with an average daily solar PV capacity between 3.0 and 5.0 kWh/kWp [27]. Where there are humans, there is biomass, and hence it covers 100% population of the world for the power source. Animals and plants also provide biomass for ES. In the ARENA 2017 report, geothermal power plants generated nearly 80.9 TWh in 2015 or approximately 0.3% of global electricity generation [28], which means that the percentage availability of geothermal was still not available in all the places in the world and hence not easy to harvest the potential it has. According to the reports, wind power's stake in worldwide electricity usage at the end of 2018 was 4.8% [53]. Still, it was less as it was unavailable at all the world's places. Hydro and tidal energy contributed a higher percentage but were still not available at all places. Therefore, we have chosen only solar and biomass energy as the source for the power of IoT-based smart buildings and discussed their role. Most existing technologies and literature surveys show solar as the green energy source for smart buildings. Some literature surveys also show solar and wind using IoT for smart buildings.

However, as we know, only solar and biomass covers more than 90% of the world's share as green energy source. Hence, in this chapter, we have discussed the role of IoT with solar and biomass as the energy source in smart buildings. Also, we have discussed the BEMS to make the IoT-based system for buildings smart with the help of sensors and actuators. We have shown in Table 3.1 the use of different sensors and their functions in smart buildings. As the IoT requires enormous amounts of data to be transferred to different segments of the smart buildings, we have also compared the technical features of the different wireless protocols used in the BEMS and their pros and cons. However, we have also discussed how the IoT can make utility electricity accessible for smart buildings through the usage of the smart grid if and when necessary and supply more power back to the utility if the building has excess power of its own.

The current literature shows that the development of the latest technologies, such as IoT, improves the performance of these green ES. Batcha et al. [29] surveyed IoT-based utilization of renewable energy. They focused on solar power. According to them, AI-based techniques also improve the performance of current renewable energy-based systems. Adhya et al. discussed and proposed IoT-based remote monitoring and control of a solar power plant [30]. This technique helps provide a solar power plant's maintenance, real-time monitoring, and fault detection. Patil et al. proposed a display system for the real-time power usage of solar energy [31]. They have used a flask framework based on Raspberry Pi. Hanumanthaiah et al. presented an IoT-based remote monitoring system. It will disconnect the load when a short circuit or illegal tapping occurs [32]. Khan et al. focused on IoT-based real-time solar energy monitoring using voltage and current sensors [33]. It helps find out the anomalies arising in a solar power plant. They have used Thingspeak™, an open-source IoT cloud platform, to monitor the parameters. Phoolwani et al. have used the thermal camera, solar power meter, and photovoltaic (PV) analyzer to detect the reduction in efficiency of the solar panel due to partial shading, dust, delamination, and cracks [34].

Chieochan et al. have developed an IoT-based off-grid solar energy system [35]. Kumar et al. discussed the role of IoT in monitoring the solar energy system [36]. According to them, the benefits of IoT are remote monitoring of the plant site, a reduction in human-to-human and human-to-computer interaction, and better identification of faults. Choi et al. explored the LoRa network in a solar-powered home to create an IoT-based renewable energy monitoring system [37]. Figure 3.5 shows the use of green energy usage in smart buildings. The energy provided by solar panels or biogas generated within the smart building's premises makes the building energy self-sufficient, and when there is a power deficit, it can draw power from the utility grid. So, the smart grid system installed at the building site with the help of IoT helps fulfill all the needs of the smart building and meets the concept of nZEB.

3.4.1 SOLAR ENERGY

The electricity derived from the sun is known as solar energy. The various life activities on earth occur due to this massive energy source. Szabo et al. discuss the milestones achieved by humankind for the conversion and use of solar energy [38]. Solar energy has the most significant percentage of all ES. The conversion of solar power into electrical power is one of the significant achievements of humankind. This section focuses on various aspects of this energy and its use in the building system. The solar system helps reduce building's greenhouse gas generation and improve public health. The main advantage of solar-based renewable energy is that it requires little or no water to operate. It works better under heat wave or drought conditions. If the energy becomes surplus, it can also be shared through grids. According to one study, solar PV systems having storage cause a 14% reduction in smart buildings energy demand, further reduced by using intelligent inverters by 12% [39]. This energy was latitude dependent. Countries with plenty of average sunlight throughout the year have great potential for this energy.

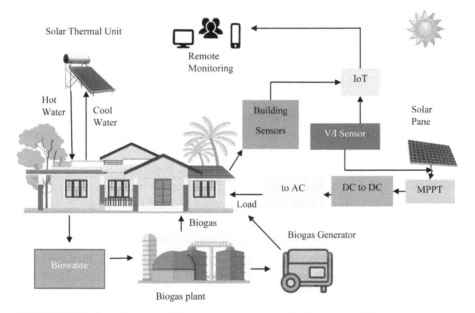

FIGURE 3.5 Block diagram of the green energy usage in the smart building.

The cloudy days produce 10% to 20% of energy compared to energy produced on a bright sunny day. The national renewable energy lab used light detection and ranging techniques, geospatial mapping, and large-scale simulations to detect America's (US) total rooftop PV energy potential. According to their analysis, 40% of the whole U.S. electricity generation was possible through it. Therefore, there is a need for a smart grid to tap this energy [40]. In addition, Teofilo et al. investigated the potential of large rooftop solar panels placed in significant space areas such as airports. These will help overcome the energy crisis in Australia [41].

This form of clean energy can reduce unemployment by creating jobs. It is possible to save money by reducing or eliminating energy bills, earning tax credits and rebates, and improving our environment with this clean energy source [42]. The other advantage of these renewable ES is the local heat and electricity generation and minimization of transmission losses. The other benefits of solar renewable energy systems are (i) the subsidies on the installation of such systems; (ii) an increase in overall property value; (iii) higher rental income; and (iv) income through selling the electricity through the grid, and hence non-requirement of batteries. Some drawbacks are also there, and these are (a) higher initial cost of installation; (b) requirement of ample space; (c) maintenance cost; and (d) architectural challenges of installations. The architecture should consider practical issues such as rainwater sealing protection from overheating. Two types of systems can tap solar energy. These are the solar thermal systems (STSs) and PVs. Residential, commercial, industrial, institutional, and agricultural buildings use PV power. The two PV system types are building-applied photovoltaic and building-integrated photovoltaic (BIPV) [43]. Kalogirov et al. also surveyed possible photovoltaic and solar thermal system integration solutions on building roofs and facades [44]. The solar components on the rooftop are often seen as foreign elements and need integration.

There are different aspects to be considered before installing STSs and PVs. These are mechanical firmness and structural integration; protection from rain, wind, snow, and hail; protection from fire and noise; cooling arrangement for excessive heat; calculation of the energy demand of the building.; availability of space; proper orientation of the solar panel to receive the maximum amount of solar energy; cost; size and type; off-grid connection with batteries or grid connection using inverters; count of meetings [45]. In addition, grid-connected solar PVs face the challenge of grid stability when considerable energy goes to the grid when the requirement is low [46].

The PV panels also provide shading to the rooftop of the building. It makes the roof relatively cooler. The additional functions of the PV panels are integrating them into the building envelope (Building-integrated PV (BIPV)), integrating heat collection purposes into the PV panel (building-integrated PV/thermal (BIPV/T), integrating light transmission functions into PV panels (Building-integrated PV/light (BIPV/L)). The PV can act as a roof or as a façade. Then there are the various sorts of solar collection technologies in use.

The IoT technology integration with solar systems increases efficiency and improves overall performance. Kumar et al. focused on the integration of

IoT technology along with solar energy in a smart building environment [36]. They performed experiments in a 12-story building. Moreover, in case of hardware disruptions or communication-related issues, they ensured reliable operation by IoT-enabled control techniques. Furthermore, they focused on solar tracking parameters. Terzio showed that solar energy is preferable to wind energy [47]. They proposed that the solar panel installation should be flexible. Some panels should fulfill the instant needs, whereas others are used to charge the batteries to achieve cost-effectiveness.

3.4.2 BIOGAS ENERGY

Biogas energy comes under the category of clean, renewable sources of energy and improves the quality of human life. Different countries are running programs related to biogas. For example, China and India installed larger plants for electricity and heat applications. The Indian Government was running programs such as "National Biogas and Manure Management Program," "Off-grid Biogas Power Generation Program," "Biogas based Distributed/Grid Power Generation Program," "Recovery of Energy from Industrial Wastes," and "Recovery of Energy from Urban Wastes." China runs the "Biogas Institute of the Ministry of Agriculture," part of the Chinese Academy of Agriculture Sciences. Other countries such as Italy, the US, and the United Kingdom run several programs. At the same time, some countries are not utilizing the full potential of biogas. One such country is South Africa, where commercial biogas plants are underutilized. Kemausuor et al. reviewed commercial biogas systems that treat organic waste from different sources [48]. According to them, the serious blockades to commercial biogas development are (i) the initial capital costs; (ii) weak environmental policies; (iii) poor institutional framework; (iv) poor infrastructure; and (v) a general lack of willpower to implement policies and targets.

Biogas is a mixture of methane (CH_4) (50%–70%), CO_2 (30%–40%), or other gases (hydrogen (5% to 10%), nitrogen (1% to 2%), water vapor (0.3%), hydrogen sulfide (traces) produced by process of anaerobic digestion of organic matter in an oxygen-free atmosphere [49]. Hence, an airtight container is the main requirement to ensure an oxygen-free environment. The airtight system where organic material (diluted in water) breaks by natural microorganisms is called a bio-digester. The whole process of biogas energy relies on generating a particular kind of bacteria in the proper amount. The carbon-nitrogen ratio for anaerobic bio-gasification plays an important role. For that purpose, sensing and IoT technology can play an important role. Some sources suggest it be between 20:1 and 25:1 ratio, whereas some suggest it be approximately 30:1 ratio [50]. The generation of the biogas depends on different chemical reactions. So, chemical sensors and IoT have an essential role in continuously monitoring the proper amount of these chemicals. The primary sensors used are the MQ4 sensor to measure the CH_4 level, the MG811 sensor to measure the CO_2 level, the temperature sensor, the pH sensor, the load sensor, and others. The Fourier transform infrared spectroscopy-based spectrometer helps detect chemical oxygen demand (COD), total organic carbon, and volatile fatty acids of the reactor. Some other feasible techniques are micro-gas chromatography (suitable to determine hydrogen, CH_4, hydrogen sulfide, nitrogen, and oxygen), membrane inlet mass spectroscopy (suitable to determine CH_4, CO_2, hydrogen sulfide, reduced organic sulfur compounds, and p-cresol) in the gas phase and with near IR spectroscopy, pH in the liquid phase [51].

The pH level of the digester slurry plays an important role. It has to be in the neutral or slightly base range. Different microorganisms require different pH values for better performance. According to three categories of the anaerobic digestion process, there was a requirement to maintain different temperature ranges. The temperature affects the enzyme activity, efficiency of the process, and microbial dynamics. It was also required to monitor inhibitors or toxins in the digester slurry that may lead to lesser production of CH_4. These are ammonia, sulfides, light metal ions such as chromium, iron, and other organic compounds.

Biogas energy not only helps in cooking, but the converted energy also helps generate the electric supply [52]. This kind of generated electricity comes under the category of green electricity. Power is generated in gas turbine power generation by using CH_4. Other biogas applications include lighting, food preparation, two/four-wheeler fuel, power generation, and running motors for pumping water. In addition, biological waste such as cow dung waste, nourishment waste, farming waste (crops and animal manure), and municipal waste generate biogas. Therefore, biogas generation leads to a decrease in waste transportation to the landfills and less emission of greenhouse gases. Biogas energy has another advantage: it is more demand-based energy than photovoltaic and wind power.

The sensor and actuator, IoT, and AI technologies help to improve the efficacy of a biogas power plant. The anaerobic digestion process needs improved monitoring. Conventional methods require an up-gradation. Correlation between different operational parameters is needed, and AI can fulfill this need. IoT will also help improve performance. Success monitoring of the biogas generation process is necessary to avoid any kind of crash. There should be a common control of feeding frequency. The technologies mentioned above will be helpful. These are also useful for detecting fire, leakage, explosions, poisoning at biogas plants, and alerting occupants. The control strategies used at the biogas plant are ON-OFF, PI/PID, adaptive, fuzzy, artificial neural networks (ANNs), and others.

3.5 CONCLUSION, CHALLENGES, AND FUTURE TRENDS

As the human population grows, so requires more energy. Most of the energy generated is presently being covered by fossil fuel burning, leading to greenhouse gas emissions that further change the earth's climate. This climate change has had an unprecedented effect on the human race. So, there should be sustainable growth in the energy sector. For sustainable development, energy production should have more percentage of generation from green energy, and the building sector should also reduce the wastage of energy, which consumes a significant portion of the generated power. In this chapter, we have shown the role of IoT and the BEMS in increasing the energy efficiency of the building and making them smart buildings. Several research studies show the approach to reaching nonzero energy buildings to optimize building energy consumption and self-dependability through efficient and sustainable ways. We have also shown how to implement IoT with solar and biogas-based energy integration with the BEMS to make smart buildings self-sustainable and reliable in terms of power, thus becoming smart. Different types of IoE-based BEMS, such as storage systems and materials, energy routers, renewable sources, and plug-and-play interfaces, are also discussed in the chapter. The use of enhanced public safety and surveillance, street lighting controls, infrastructure monitoring, meter reading, physical security, and optimization systems to make the city smart was also shown in the chapter. Different wireless technologies used for the considerable data transfer of the IoT-based BEMS system and different sensors and actuators used were also discussed. Hence in terms of security, fastest, cost-effective, range, and consumes less energy, LoRa and 802.11 protocols are suitable and can be used for the smart building IoT infrastructure for data communication.

There are still some challenges to deploying IoT in the energy sector, and it includes privacy and security. Although engineers and researchers are coming up with good solutions to these challenges, one is blockchain technology. Survey shows that energy policymakers, economists, and managers with an overview of IoT's role in optimizing energy systems can help the sustainable energy development of the world requirement. The cloud and fog computing platforms can ease the path for blockchain services in IoT. The energy ingesting of IoT devices is another critical challenge, especially in the large-scale deployment of these technologies. The IoT will require a significant amount of energy to run billions of devices. Many IoT devices will also yield a boundless deal of electronic waste. Hence, low-carbon and efficient communication networks are in need to tackle the challenges. Fortunately, these requirements have led to the appearance of the green Internet of Things (G-IoT). The key component of G-IoT is its power-efficient characteristics throughout the life cycle, that is, the design, deployment, production, and finally, clearance. As a result, smart applications in the energy sector convert buildings to smart buildings, converting towns to smart cities, which then proceed to smart states, leading to smart countries, and finally to a smart world. When human society reaches that level, the sustainable energy dream will come true.

KEYWORDS

- biogas
- building energy management system
- energy management
- green energy
- grid management
- IoT
- sensors and network protocols
- smart building
- solar energy

REFERENCES

1. Nguyen, V. T., Thanh, L. V., Nam, T. L., & Yeong, M. J., (2018). An overview of Internet of Energy (IoE) based building energy management system. In: *9t^{h} International Conference on Information and Communication Technology Convergence: ICT Convergence Powered by Smart Intelligence, ICTC 2018* (pp. 852–855). https://doi.org/10.1109/ICTC.2018.8539513.

2. Shah, A. S., Haidawati, N., Muhammad, F., Adidah, L., & Asadullah, S., (2019). A review on energy consumption optimization techniques in IoT-based smart building environments. *Information (Switzerland), 10*(3). https://doi.org/10.3390/info10030108.

3. Deshmukh, N. S., Bhuyar, D. L., & Jadhav, A T., (2018). *Review on IoT Based Smart Solar Photovoltaic Plant Remote Monitoring and Control Unit, 3*(3), 6–10.

4. Acharya, V., Vinay, V. H., Anjan, K., & Manoj, K. M., (2018). IoT (internet of things) based efficiency monitoring system for biogas plants. In: *2nd International Conference on Computational Systems and Information Technology for Sustainable Solutions, CSITSS 2017*, (pp. 113–117). https://doi.org/10.1109/CSITSS.2017.8447567.

5. Verma, A., Surya, P., Vishal, S., Anuj, K., & Subhas, C. M., (2019). Sensing, controlling, and IoT infrastructure in smart building: A review. *IEEE Sensors Journal, 19*(20), 9036–9046. https://doi.org/10.1109/JSEN.2019.2922409.

6. Ida, B. G. P., Nyoman, S. I., & Made, S., (2020). Application of IoT-based system for monitoring energy consumption. *International Journal of Engineering and Emerging Technology, 5*(2), 81–93. https://doi.org/10.24843/IJEET.2020.v05.i02.p014.

7. Doukas, H., Konstantinos, D. P., Konstantinos, I., & John, P., (2007). *Intelligent Building Energy Management System Using Rule Sets, 42*, 3562–3569. https://doi.org/10.1016/j.buildenv.2006.10.024.

8. Paniagua, E., Macazana, J., Lopez, J., & Tarrillo, J., (2019). IoT-based temperature monitoring for buildings thermal comfort analysis. In: *2019 IEEE XXVI International Conference on Electronics, Electrical Engineering and Computing (INTERCON)* (pp. 1–4). https://doi.org/10.1109/INTERCON.2019.8853608.

9. Gulnizkij, N., & Gerlach, G., (2019). Bistable hydrogel-based sensor switch for monitoring relative humidity. In: *2019 IEEE Sensors* (pp. 1–4). https://doi.org/10.1109/SENSORS43011.2019.8956498.

10. Gaur, A., Singh, A., Kumar, A., Kulkarni, K. S., Lala, S., Kapoor, K., Srivastava, V., et al., (2019). Fire sensing technologies: A review. *IEEE Sensors Journal, 19*(9), 3191–3202. https://doi.org/10.1109/JSEN.2019.2894665.

11. Liu, X., Sitian, C., Hong, L., Sha, H., Daqiang, Z., Huansheng, N., & Information Engineering, (2012). *A Survey on Gas Sensing Technology, 9635–9665.* https://doi.org/10.3390/s120709635.

12. Roveti, D. K., (2001). *Choosing a Humidity Sensor: A Review of Three Technologies.* Sensors Online. https://www.fierceelectronics.com/components/choosing-a-humidity-sensor-a-review-three-technologies (accessed on 12 June 2023).

13. Prajapati (Shekhar), C., Rohith, S., Rudraswamy, S. B., Nayak, M. M., & Navakanta, B., (2017). Single-chip gas sensor array for air quality monitoring. *Journal of Microelectromechanical Systems.* https://doi.org/10.1109/JMEMS.2017.2657788.

14. Singh, A., Balam, N. B., Kumar, A., & Kumar, A. (2016). An Intelligent Color Sensing system for building wall. *International Conference on Emerging Trends in Communication Technologies (ETCT)*, Dehradun, India, 1–4, doi: 10.1109/ETCT.2016.7882930.

15. Kwon, O., An, J., & Hong, S., (2018). Capacitive touch systems with styli for touch sensors: A review. *IEEE Sensors Journal, 18*(12), 4832–4846. https://doi.org/10.1109/JSEN.2018.2830660.

16. Hub, Admini of Electronics, (2019). *Introduction to Touch Sensors-Working, Capacitive and Resistive.* Electronics Hub. https://www.electronicshub.org/touch-sensors/ (accessed on 12 June 2023).

17. Sood, S., Rajinder, S., Karan, S., & Victor, C., (2017). IoT, big data, and HPC-based smart flood management framework. *Sustainable Computing: Informatics and Systems,* 20. https://doi.org/10.1016/j.suscom.2017.12.001.
18. Expert, NetworkTech., (2013). *Water Detection Sensors: Types and Applications.* Network Technologies Incorporated. https://www.networktechinc.com/blog/water-detection-sensors-types-and-applications/303/ (accessed on 12 June 2023).
19. Ashraf, I., Yousaf, B. Z., Soojung, H., & Yongwan, P. (2020). A comprehensive analysis of magnetic field based indoor positioning with smartphones: Opportunities, challenges and practical limitations. *IEEE Access,* 4. https://doi.org/10.1109/ACCESS.2020.3046288.
20. Bedi, G., (2018). *Internet of Things and Intelligent Technologies for Efficient Energy Management in a Smart Building Environment.* ProQuest Dissertations and Theses. https://search.proquest.com/docview/2239986856?accountid=14598 (accessed on 12 June 2023).
21. Fernandes, H., Alam, M., Ferreira, J., & Fonseca, J., (2017). Acoustic smart sensors based integrated system for smart homes. In: *2017 International Smart Cities Conference (ISC2),* (pp. 1–5). https://doi.org/10.1109/ISC2.2017.8090869.
22. Ehsan, Y., Victor, M. L., & Boric-Lubecke, O., (2013). True human presence detection with radar technology. *IEEE Life Sciences.* https://lifesciences.ieee.org/lifesciences-newsletter/2013/february-2013/true-human-presence-detection-with-radar-technology/ (accessed on 12 June 2023).
23. Luppe, C., & Amir, S., (2017). *Towards Reliable Intelligent Occupancy Detection for Smart Building Applications.* https://doi.org/10.1109/CCECE.2017.7946831.
24. Marcello, F., Virginia, P., & Daniele, G., (2019). Sensor-based early activity recognition inside buildings to support energy and comfort management systems. *Energies, 12*(13). https://doi.org/10.3390/en12132631.
25. Sánchez-Garrido, C., Javier, M., & González-Jiménez, J., (2014). A configurable smart e-nose for spatio-temporal olfactory analysis. *Proceedings of IEEE Sensors, 2014.* https://doi.org/10.1109/ICSENS.2014.6985418.
26. Jia, M., Ali, K., Yueren, W., & Ravi, S. S., (2019). Adopting Internet of Things for the development of smart buildings: A review of enabling technologies and applications. *Automation in Construction, 101,* 111–126. https://doi.org/https://doi.org/10.1016/j.autcon.2019.01.023.
27. Report, World Bank, (2020). *Solar Photovoltaic Power Potential by Country.* World Bank. https://www.worldbank.org/en/topic/energy/publication/solar-photovoltaic-power-potential-by-country (accessed on 12 June 2023).
28. Team ARENA, (2017). *Annual Report 2017–2018.*
29. Batcha, R. R., & Geetha, M. K., (2020). A survey on IoT based on renewable energy for efficient energy conservation using machine learning approaches. In: *2020 3ʳᵈ International Conference on Emerging Technologies in Computer Engineering: Machine Learning and Internet of Things (ICETCE)* (pp. 123–128). https://doi.org/10.1109/ICETCE48199.2020.9091737.
30. Adhya, S., Dipak, C. S., Abhijit, D., Joydip, J., & Hiranmay, S., (2016). An IoT-based smart solar photovoltaic remote monitoring and control unit. In: *2016 2ⁿᵈ International Conference on Control, Instrumentation, Energy & Communication (CIEC)* (pp. 432–436).
31. Patil, S. M., Vijayalashmi, M., & Tapaskar, R., (2017). IoT-based solar energy monitoring system. In: *2017 International Conference on Energy, Communication, Data Analytics and Soft Computing (ICECDS)* (pp. 1574–1579). https://doi.org/10.1109/ICECDS.2017.8389711.

32. Hanumanthaiah, A., Liya, M. L., Arun, C., & Aswathy, M. (2019). IoT-based solar power monitor & controller for village electrification. In: *2019 9th International Symposium on Embedded Computing and System Design (ISED)* (pp. 1–5). https://doi.org/10.1109/ISED48680.2019.9096221.

33. Khan, M. S., Sharma, H., & Haque, A., (2019). IoT enabled real-time energy monitoring for photovoltaic systems. In: *2019 International Conference on Machine Learning, Big Data, Cloud and Parallel Computing (COMITCon)* (pp. 323–327). https://doi.org/10.1109/COMITCon.2019.8862246.

34. Phoolwani, U. K., Sharma, T., Singh, A., & Gawre, S. K. (2020). IoT-based solar panel analysis using thermal imaging. In: *2020 IEEE International Students' Conference on Electrical, Electronics and Computer Science (SCEECS)* (pp. 1–5). https://doi.org/10.1109/SCEECS48394.2020.114.

35. Chieochan, O., Anukit, S., & Ekkarat, B., (2017). Internet of Things (IoT) for smart solar energy: A case study of the smart farm at Maejo University. In: *2017 International Conference on Control, Automation, and Information Sciences, ICCAIS 2017* (pp. 262–267). https://doi.org/10.1109/ICCAIS.2017.8217588.

36. Kumar, N. M., Karthik, A., & Sriteja, P., (2018). Internet of Things (IoT) in photovoltaic systems. In: *2018 National Power Engineering Conference, NPEC 2018.* https://doi.org/10.1109/NPEC.2018.8476807.

37. Choi, C., Jeong, J., Lee, I., & Park, W. (2018). LoRa-based renewable energy monitoring system with open IoT platform. In: *2018 International Conference on Electronics, Information, and Communication (ICEIC)* (pp. 1, 2). https://doi.org/10.23919/ELINFOCOM.2018.8330550.

38. Szabó, L., (2017). The history of using solar energy. *Proceedings – 2017 International Conference on Modern Power Systems, MPS 2017.* https://doi.org/10.1109/MPS.2017.7974451.

39. King, J., & Christopher, P., (2017). *Smart Buildings: Using Smart Technology to Save Energy in Existing Buildings.* American Council for an Energy-Efficient Economy. https://www.aceee.org/sites/default/files/publications/researchreports/a1701.pdf (accessed on 26 July 2023).

40. Trinastic, J., (2016). *The Potential of Rooftop Solar Energy: 40% of Total U.S. Electricity Generation is Possible.* Blog. https://www.nature.com/scitable/blog/eyes-on-environment/the_power_of_rooftop_solar (accessed on 12 June 2023).

41. Teofilo, A., Qian (Chayn), S., Nenad, R., Yaguang, T., Jerome, I., & Chengyang, L., (2021). Investigating potential rooftop solar energy generated by leased federal airports in Australia: Framework and implications. *Journal of Building Engineering, 41,* 102390. https://doi.org/https://doi.org/10.1016/j.jobe.2021.102390.

42. Holowka, T., (2017). *Top Four Benefits of Installing Solar Panels on Your Home.* US Green Building Council. https://www.usgbc.org/articles/top-four-benefits-installing-solar-panels-your-home (accessed on 12 June 2023).

43. Canada, Govt. of. (2020). *Solar Photovoltaic Energy in Buildings.* Govt. of Canada. https://www.nrcan.gc.ca/energy-efficiency/data-research-insights-energy-ef/buildings-innovation/solar-photovoltaic-energy-buildings/3907 (accessed on 12 June 2023).

44. Kalogirou, S., (2015). *Building Integration of Solar Renewable Energy Systems Towards Zero or Nearly Zero Energy Buildings. 10*(4), 379-385. https://doi.org/10.1093/ijlct/ctt071.

45. Ganguli, B., (2019). *Home Solar System: Renewable Energy Solutions for Residential Users.* The Economic Times.

46. Phung, M. D., (2019). *IoT-Enabled Dependable Control for Solar Energy Harvesting in Smart Buildings.* https://doi.org/10.1049/iet-smc.2019.0052.
47. Terzio, H., (2015). *A New Approach to the Installation of Solar Panels.* https://doi.org/10.1109/ICISCE.2015.133.
48. Kemausuor, F., & Muyiwa, S. A., (2018). *A Review of Commercial Biogas Systems and Lessons for Africa,* 1–21. https://doi.org/10.3390/en11112984.
49. Singh, A. K., Jegan, R., & Kumar, P., (2022). Review on the usage of green energy and IoT in smart buildings. In: *2022 6th International Conference on Computing Methodologies and Communication (ICCMC)* (pp. 452–458). https://doi.org/10.1109/ICCMC53470.2022.9754130.
50. Satyanarayan, S., Kaul, S. N., Badrinath, S. D., & Gadkari, S. K., (1987). *Biogas from Human Waste.* Consortium on Rural Technology. https://www.ircwash.org/sites/default/files/352.1-4961.pdf (accessed on 12 June 2023).
51. Cinar, S., Cinar, S. O., Wieczorek, N., Sohoo, I., & Kuchta, K., (2021). Integration of Artificial Intelligence into Biogas Plant Operation. *Processes.* 9(1), 85. https://doi.org/10.3390/pr9010085.
52. Anbarasu, V., Karthikeyan, P., & Anandaraj, S. P., (2020). Turning human and food waste into reusable energy in a multilevel apartment using IoT. In: *2020 6th International Conference on Advanced Computing and Communication Systems, ICACCS 2020* (pp. 440–444). https://doi.org/10.1109/ICACCS48705.2020.9074170.
53. Wind Power by Country – Wikipedia [Internet]. [cited 2023 Jul 20]. Available from: https://en.wikipedia.org/wiki/Wind_power_by_country#cite_note-9 (accessed on 28 July 2023).

CHAPTER 4

A Novel Electrical Load Forecasting Model Using a Deep Learning Approach

NEELAPALA ANIL KUMAR,[1] RAVURI DANIEL,[2] and
PRUDHVI KIRAN PASAM[3]

[1]Department of Electronics and Communication Engineering (ACED),
Alliance University, Bangalore, Karnataka, India

[2]Department of Computer Science and Engineering, Prasad V. Potluri
Siddhartha Institute of Technology, Vijayawada, Andhra Pradesh, India

[3]Department of Information Technology, SRKR Engineering College,
Bhimavaram, Andhra Pradesh, India

ABSTRACT

The estimate of electricity appeal in modernistic years is becoming progressively relevant thanks to market-free trade and, thus, the initiation of sustainable assets. To satisfy the demands, leading intelligent models are built to form sure explicit power forecasts for multi-time prospects. The load forecasting of electric Power is a crucial process in devising the electric industry and operating electric power systems. Short-term forecasts are adopted to program the power generation and transmission of electricity. Medium-term forecasts are meant to line up the fuel purchases. This necessitates the implementation of the productive determination of algorithms could be a fundamental feature of smart grids and an efficient tool for determining ambiguity for better cost and energy ability decisions like slate the origination, authenticity, power escalation of the system, and monetary smart grid activities. This work introduces a model for the evaluation of the utilization of electricity, which can accurately forecast

The Internet of Energy: A Pragmatic Approach Towards Sustainable Development. Sheila Mahapatra,
Mohan Krishna S., B. Chandra Sekhar, & Saurav Raj (Eds.)

subsequently estimated from minimum to maximum duration with significant improvement in the accuracy of forecasting through advanced deep learning techniques. The analyzes or findings also can provide interesting results for energy consumption with parameters like forecasting efficiency and error with duration of data monitoring algorithms namely (LSTM)–long short-term memory (RNN) – recurrent neural networks and multi-layer perceptron algorithms (MLP). These algorithms furnish the most interesting results with respective to the duration of data. Mainly, MLP and RNN proved to produce favorable results for 24-hour data. Similarly, LSTM has proved better for 15-day data and monthly data with consistency in terms of errors, squared, and mean square. To anticipate data ranging from day to month, the minimal Forecasting error was attained by adopting MLP with R^2 (0.91). On hour-based data, R^2 of LSTM holds effective for half-monthly and monthly data with (0.88 and 0.93), RMSE (89.54 and 84.98), MAPE (3.51 and 2.47). RNN has been proven to attain the moderate outputs comparatively. MLP for half-monthly and monthly in terms of R^2 (0.81 and 0.92), RMSE (90.72 and 85.78) and MAPE (4.25 and 4.01). The result of LSTM acknowledges the enhanced attainment and substantial achievements of electrical load forecasting.

4.1 INTRODUCTION

Major aspects of advanced technology are moving around the consideration of power. Traditional electrical power generation and distribution systems getting obsolete due to energy loss and less effective affairs. The major reason for the claim could be outdated structure. On the other hand, reduction of natural resources and pollution which are used in the power generation process. Utmost of these problems were addressed by replacing the traditional grids with smart grids. Smart grids give the installation of monitoring, optimization, distribution, and consumption of electricity. In the present payload existence service, at most important and attainable aspects to satisfy the forthcoming payload appeals mainly by gratuitous loading, knockouts. This highly requires a measure of precautions and instantaneous monitoring of power stations with respect to the payload, starting from minimal level observation from hours to the maximal observations of the month. From the practical observations, flutters arrive in a short span of intervals, which demands hourly base payload monitoring [1]. Forecasting of the payload is a problem based on probability rather than being inconsistent. This demand is for forecasting stochastic data rather than static data analysis for safeguarding the power plant [2]. In this aspect, most data scientists followed the statistical approach-based data analysis algorithms to forecast the payload of power plants [3] and hierarchical forecast [4]. The load for the casting of power plants demands the dynamic duration of data retrieval and analysis. This has a common requirement that the extensive approaches have been monitored for a short time to long time data series intervals in the process of forecasting data for identifying the failures to safeguard the power plant and environment. The short-term payload forecasting turn into a severe problem for the operating and delivering of power networks to intrude pivotal issues associated with failures in power systems. Taking the considerations of maintenance cost of the power plant in the present scenario, it is made important to automate the process of monitoring load. This necessitates the typical deep learning intelligent systems for effective and economically safeguarding monitoring process of payloads in a power plant. It is also an accessory for optimizing the energy conservation while reducing emigration. Effective electric payload forecasting can protect electricity while using minimal energy coffers and allowing for easy distribution. Contrarily, if payload forecasting is done duly and straight, helps in managing the unborn electric payload demands with efficient consignment planning [5].

4.1.1 ASPECTS OF ELECTRICAL LOAD

Traditional methods incorporated in electrical systems for the operation might be definite from others, and a single electrical device is used most frequently. This type of adaption in the existing systems leads to the problem of variation in separate loads by casting up with major installation weights. The various aspects affect the electric load involves climate, time aspects, and arbitrary aspects.

1. **Climate:** It is one of the considerable impacting aspects of power forecasting. The climate is an extruded important rudiment among the capacity cast script. Generally, load pay models are erected and tested, with the considerations of factual rainfall reading. The rainfall cast obviously companion to the declination and instill in the achievement of the model. Forecasting of Weather, speed of wind, and climatic conditions attain a major part in short-term load forecasting (STLF) by varying the chart. The weather has major betters on ST electrical distribution forecasting of power systems [6]. Sensible climatic consignments have a vast effect on scaled-down

artificial energy architectures. Climate rudiments impact hourly power payload forecasting. In times of maximum humidity climates, the cooling outfit can be utilized for lengthy duty patterns for removal of non-significant condensation of conditioned air and holds the capability of temperature reduction in the air, which impacts the minimization of bracing loads.

2. **Time Aspects:** Power load fluctuations are highly related to the working timings for the non-working day power load cast is the same as on weekends. During non-working days, the power loads are lower compared to working days due to hugeness. Power load cast schemes for day-to-day daily cycle is general all through day and week. While seasonal issues describe the modifications in long-term improvement and pattern of rainfall.

3. **Random Aspects:** Arbitrary conditions which can agitate energy load definition include required loads that had not been planned and the absenteeism of workers making the forecast delicate. Load forecast has been given more importance in terms of recent times smart energy operation systems. Load forecasting has been added with users on an enduring base. Substantially, STLF is incorporated to maintain payload forwarding the transfer energy control schedule to 30 glitters for a single day. Thus, any improvement gained along the operational perfection of STLF tends to reduce the reduction of electrical charges of OS along with the effective energy network enhancement [7]. STLF facilitates the rearmost forecast information of rainfall, new payload forecasting strategy, and power system arbitrary behavior [8]. The load cast is getting more significant because of its adding importance towards renewable sources of energy trends, smart grids, and micro-grids. Several strategies have been imposed for operation and consignment, similarly as seasonal bus-accumulative, bus-accumulative integrated moving average models, retrogression, precipitously integrated moving average. In recent days, alternative ways were found to improve the effectiveness of load cast operation of power systems and the elegancy similar to artificial intelligence (AI) and fuzzy sense. These results were effectively calculated on the technologies of AI to clarify forecasting load issues.

Ranaweera et al. [9] presented acquainted rules of fuzzy developed by applying a literacy algorithm type to introduce load operation and data of literal rainfall. He et al. [8] suggested an advanced system for the quantifying the stringed unstable power load to gain information subsequently for awaited load with neural structure quintile regression construction frame in the development of statistical ways for castings. Later different styles of AI were implemented for different electrical payloads, including structures of smart grids [10].

In mere future payload demands, distributed system load forecasting and artificial neural network (ANN) exogenous auto-regressive vector inputs. In addition to these AI-based fuzzy logistics specialized network structures, neural systems based on Bayesian neural networks and SVM are extensively used for handling the issues of load forecasting of power loads. STLF advance in getting a claim because of its non-stationary load data in addition to durable dependencies, carrying related data for forecasting. This is a reason for thinking about the application of long-short-term memory, which is an intermittent distinct type of (RNN) structure [11], to overcome the issue of STLF. LSTM operates adequately about horizon anticipation of long duration comparatively. For the remaining AI methods derived with diversified payload for series time outgrowth connections.

4.2 MATERIALS AND METHODOLOGY

In this section, the detailed materials and methodology are presented. Neural network development is an extended process that requires a lot of thought behind the architecture and a whole group of nuances that make up the system.

4.2.1 MATERIALS

In this section, the details of the data set and tools used to develop the proposed work are presented. Some of the tools used in this work are Scikit Learn, TensorFlow, and Kera's. The tools are proven very efficient to build the neural network models.

1. **Dataset:** The data was chosen from a confluent to fulfill the maximum payload conditions for analysis. The data was supported with a Kaggle database from one of its electricity systems based on the hourly load demands of one grid as utilized in Ref. [12]. The load time series was generated for min duration starting from hours to month as short, medium, and maximum durational data intervals.

The data set of the electricity load collected from Kaggle is shown in Figure 4.1.

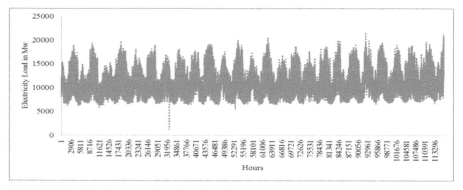

FIGURE 4.1   The data set of the electricity load.

2. **Tools:** The major tools are used in this work Scikit Learn, Tensor-Flow, and Keras. The Scikit Learn is a simple and efficient tool for foretelling data analysis. The TensorFlow is a deep learning and open-source library used for the smooth running of deep learning applications. The Keras is a high-level neural network API and is developed mainly to enable faster experimentation.

4.2.2 METHODOLOGY

Figure 4.2 depicts the schematic block diagram of the methodology, which interprets the forecasting system for electric load for different durations ranging from day to month data with several innovative forecasting models. This model is incorporated with STLF [13] and MTLF used for the planning of the power systems ranging from a single day to a single month. Once after introducing the data pre-processing, we progress with robust deep neural network proficiency models, namely MPL, RNN, and LSTM. The performance was determined based on the ground test set standard error performance criteria as RMSE, MAPE, and R-squared. Eventually, load forecasting for short to medium durations was performed in terms of durational data for forecasting load demands.

1. **Multilayer Perceptron (MLP):** It can be trained with local learning procedures. The training process is performed with a few specimens selected from the neighborhood of the area of interest. From the data

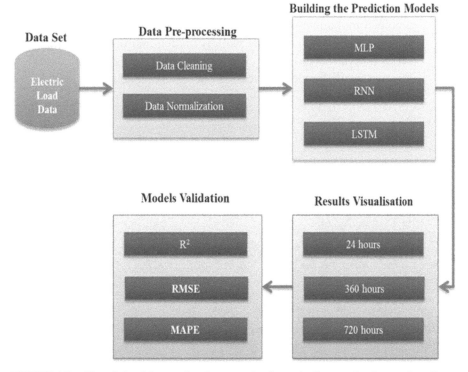

FIGURE 4.2   Electric load forecasting framework schematic diagram for short and medium-term load forecasting.

observations complication of the local target function is less than the global complexity target function. This feature facilitates the MLP design with minimum neurons to train and learn the system faster. In Ref. [14], the authors addressed that the utility of a one-neuron sigmoid acceleration function delivers productive output comparatively to the other system, functioning across several neurons within the concealed surface. MLP can be worked with a backpropagation algorithm which is claimed to be acquisition monitored algorithm for handling the problems of large-scale prediction implementation. Few significantly, in annual forecasting of gross electricity demand Indicator established on climatic and socio-economic parameters. Closest neighbors' identification is mainly performed with hyper-parameter by utilizing the LLP method. Levenberg-Marquardt algorithm with Bayesian regularization for MLP absorption [15] minimizes the squared miscalculations consolidation of weights to avoid overfitting. The effective utility of MLP is represented in the algorithm shown in Figure 4.3, which depicts the MLP algorithm architecture and working, which consists of several normalized time series input loads appeals as well as hidden neurons and actual time series activation functions. The final error metrics are calculated to forecast the difference between the actual load and the predicted load requirements.

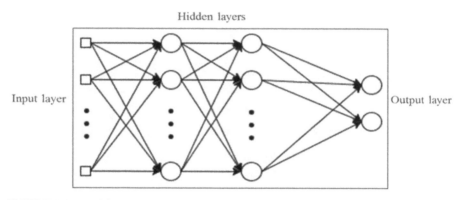

FIGURE 4.3   Architecture of multi-layer perceptron (MPL).

Deep learning is a subdivision of AI, works on par with ANNs and machine learning (ML) performance. AI functions as a mimicking pattern of the human brain. ANN utilizes informational data to process the required information using neuron groups which originate as layers [16]. These neurons deliver information from one neuron source to another. Where a small quantity of output information is fed back to the preceding layer until it has retrieved the expected output. Information is represented in the form of regression at the output layer or classification. DL proved to be effective for forecasting the complex problems [17] with feature extraction and automatic data method improvements.

2. **Recurrent Neural Network (RNN):** The most promising and powerful robust type of neural network, with its single internal memory of use, can be claimed as RNN. This algorithm is mainly preferred for its sequence data modeling. The behavior of this algorithm mimics the brain functionality; it keeps generating forecasting responses. This algorithm is relatively ancient, identified in the 1980s but came into existence recently to forecast and process the time series data [18]. RNN is also one of the deep learning algorithms

used to solve the huge computational data with a short series of time intervals. In the 1990s, the massive utility of RNN came into existence because of its internal memory considerations. As this algorithm can only focus on the significant things about the received input. Which facilitates them to be most particular in forecasting the next outcome. This feature placed RNN in higher demand, specifically for its time series sequential data widely utilized for several applications. RNN placed the highest priority for its most in-depth understanding of the sequence of data compared to any other algorithms. The understanding of RNN rest in a feed-forward neural network with time series data [19]. It does not have any notation of time order. If a decision must make, it investigates the current input and previously received inputs from what it learned. From the above details, it is observed that the limitations of RNN regarding the computation speed are observed to be low, and it is difficult to access the long-time go information as it cannot consider any future input for the current state.

Figure 4.4 shows the RNN model's construction. Forecasting 1-dimensional load time sequence as input for RNN model, which was refined in various RNN model layers, and the output is the error or variance between actual and forecast load values. RNN holds better for the constant model size and input, and computation takes consideration of historical information shared across the time.

Time Series Input Data

Output

FIGURE 4.4   Load forecasting RNN architecture.

3. **Long Short-Term Memory (LSTM):** It is designed to recognize sequences of time series data patterns of deep learning applications [20]. It is represented as artificial recurrent neural network (RNN) architecture. LSTM is recognized as a famous model expert [21] in forecasting time sequences that can handle both long and short-term data dependents effectively. RNN architecture issues for motivation

are mostly designed with LSTM to control disappearing gradients. Especially for long-term dependent data, acceptable to form the short- and long-term neural network. Mostly, Recursive Neural Network architecture is a framework that introduces the forgetting gate, input barrier, and output gateway neurons. This new approach to development can efficiently guide the issue of vanishing gradient [22]. This addons make LSTM architecture most preferred for long-term data problem dependencies, which makes LSTM to be most widely used for forecasting the time series. Most of the applications pertaining to neuron-cognitive performance use RNN's methods from LSTM. The LSTM solved the well-known problems of machine learning, like gradient learning with the back-flow error problems. It also proves to be most appropriate for compressible input sequence, noise, and efficient adaptive learning, in the absence of short time lag capabilities losses. It is of utmost importance to solve lag tasks for complicated problems, which are difficult in conventional machine learning. The LSTM has linear hidden layers, but the gradient flow through long sequences is permitted by self-loop memory blocks. The recurrent blocks of LSTM, called memory blocks, comprise multiplicative units of three, majorly forget gates, input, output [23], and recurrent memory cells. Memory sections saving and connect information for a maximum duration for solving vanishing problems can be granted by this cell. The LSTM is additionally admitted for improving the functionality by memory cell resetting [24]. Figure 4.5 represents the general LSTM model architecture.

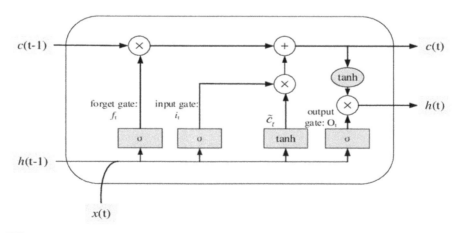

FIGURE 4.5   Architecture of LSTM model.

As a major innovation, the LSTM memory cell is used to limit the data-storing state. The forget gate used in the first step will eliminate the unnecessary information. The application is then followed by a sigmoid function to measure the accelerated state of forget f_g.

$$f_g = \sigma(w_i[h_{t-1}, x_g] + b_i) \tag{1}$$

The second step involves locating the necessary data that must be saved within the cell state. The "input gate layer," another sigmoid layer, is used to update information. Then, a vector $\tilde{c}t$ is created using the tanh function, and new values that must be updated based on the current state.

$$i_g = (w_j[h_{t-1}, x_g] + b_i) \tag{2}$$

$$\tilde{C}_g = \tanh(w_j[h_{t-1}, x_g] + b_i) \tag{3}$$

The old cell state c_{g-1} of the second step is changed to the new cell state cg. Multiply c_{g-1} by f_g to remove the information content from the ancient cell. Then add $i_t * \tilde{c}_g$. The updated data provided is represented by the new candidate values.

$$C_g = f_g * C_{g-1} + i_g * C_t \tag{4}$$

Finally, the output must be analyzed in several steps, with the sigmoid function serving as an output barrier for emphasizing the cell state. In addition, the achieved cell state is transferred across tanh (.) to the output, which is the product used to calculate the required information.

$$o_g = (w_1[h_{t-1}, x_g] + b_1) \tag{5}$$

$$h_g = o_g * \tanh(C_g) \tag{6}$$

In Eqns. (1)–(5), w_i, w_j, w_k, and w_1 represent the weight matrices along with representation of bias vectors b_i, b_j, b_k, and b_1.

4.2.3 METRICS FOR EVALUATING PERFORMANCE

The quantitative casting accuracy was measured in terms of determining coefficient errors by squared error (R^2), root mean squared (RMSE), and mean absolute percentage (MAPE) errors. The usage of the detailed metrics for the following renowned error forecasting is summarized [25].

1. **Co-Efficient of Determination (R^2):** The following function shows the computed R^2.

$$R^2 = \frac{\sum\limits_{i=1}^{n} (x_i - y_i)^2}{\sum\limits_{i=1}^{n} (x_i - y_i)^2} \qquad (7)$$

Here y denotes the average values of the entire samples.

2. **Root Mean Squared Error (RMSE):** The evaluation parameter to test the forecasted quality requires a measure for quantitative accuracy metric. In the prevalent study, a volume termed RMSE is denoted with the given formula.

$$RMSE = \sqrt{\sum\limits_{i=1}^{n} (x_i - y_i)} \qquad (8)$$

The measured and forecast values of the ith sample are denoted by x_i and y_i, and 'n' represents the total number of the training dataset samples. The RMSE smaller value specifies the better-selected set descriptors.

3. **Mean Absolute Percentage Error (MAPE):** It is a measure of accuracy for forecasting systems where the accuracy is measured in terms of percentage and also can calculate the average absolute percentage error for every instant of time period minus actual values divided by actual values The MAPE is computed using the following formula.

$$\text{MAPE} = \frac{100}{n} \sum\limits_{i=1}^{n} \frac{(y_i - x_i)}{n} \qquad (9)$$

4.3 RESULTS

As a process of forecasting of electricity load at regular instants of time ranging from day load to monthly load had been performed by various deep learning algorithms, equipped, and analyzed, with parameters namely MLP, LSTM, and RNN applied for load forecasting time series data with a load of day (24-hours) to a 1 month. The determination of performance in terms of R^2, MAPE, and RMSE for forecasting load demands of electricity. The input data of the different timing data has been collected from the Kaggle database. The computed difference between obtained and forecast values suggests that it is smaller and unbiased, indicating that it is the best fit for the given data. Statistically, the residual plots which measure the better fit reveal undesirable residual patterns indicating that misleading results are more effective than numbers. This metric is known as the coefficient of determination, and it indicates the degree to which data is correlated.

4.3.1   24 HOURS ELECTRIC LOAD FORECASTING

Table 4.1 represents the 24-hour electrical load forecasting based on the error parameters R^2, RMSE, and MAPE. The following tabular content depicts the R^2 maximum for the MLP with a value of 0.91, moderate for RNN as 0.75, and comparatively least for LSTM with a value of 0.37. The parameters RMSE and MAPE holds better for MLP, with a comparative minimum value of 62.46 and 1.74. This result reveals that MLP is more effective than other algorithms for the short-term duration (24 hours).

Figure 4.6(a)–(c) shows a pictorial depiction of the results of three different procedures for obtaining 24-hour electrical load forecasts (MLP, LSTM, and RNN). The MLP is the closest to the real load, according to extracted graphs, followed by LSTM and RNN.

TABLE 4.1   Forecasting of One-Day Electricity Load Forecasting

Method	R^2	RMSE	MAPE
MLP	0.91	62.46	1.74
RNN	0.75	91.45	2.53
LSTM	0.37	286.59	5.04

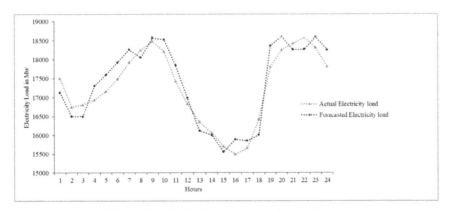

(a) One-day electricity load forecasting using MLP.

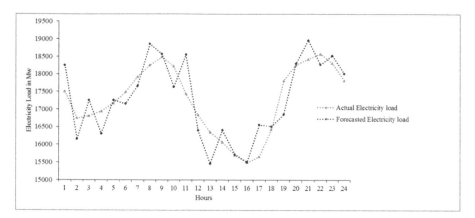

(b) One-day electricity load forecasting using RNN.

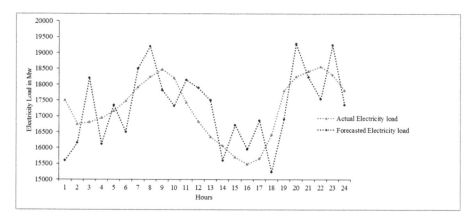

(c) One-day electricity load forecasting using LSTM.

FIGURE 4.6   One-day load forecasting using (a) MLP; (b) RNN; and (c) LSTM.

4.3.2   15-DAYS ELECTRICITY LOAD FORECASTING

Table 4.2 reveals the forecasting of 15 days electrical load forecasting. Focused on R^2, RMSE, and MAPE with the input database of the electricity load. The following tabular content depicts the maximum as 0.88 for the LSTM, moderate for RNN as 0.81, and comparatively least for MLP as 0.78. The parameters RMSE and MAPE hold better for LSTM with a comparative minimum value of 89.54 and 3.51. This result reveals that LSTM outperforms comparatively for the mid-term duration (15 days).

TABLE 4.2 Forecasting of the 15 Days Electricity Load Forecasting

Method	R²	RMSE	MAPE
MLP	0.78	120.422	5.44
RNN	0.81	90.72	4.25
LSTM	0.88	89.54	3.51

Figure 4.7(a)–(c) presents the results of 15 days electrical forecasting payload for the 3 different methods (MLP, LSTM, and RNN) with graphical representation. From the extracted waveforms, it is evident that LSTM data values are nearest to the real load curve accompanied by RNN and MLP.

4.3.3 30-DAYS ELECTRICITY LOAD FORECASTING

Table 4.3 represents the forecasting of load forecasting for one month. In comparison to the R^2-squared error method, the LSTM provides the best forecasting with R^2 0.93, accompanied by RNN as R^2 0.92 and MLP as R^2 0.65. According to MAPE, the longest duration of one month of forecasting was obtained as 2.47 for LSTM, 4.01 for RNN, and 5.89 for MLP. Likewise, with the consideration of RMSE, effective values have been stated as 84.98 for LSTM, 85.78 for RNN, and 155.65 for MLP. From the data forecast, it is evident that LSTM performance will be more effective for the max duration time series data in providing maximum squared error and minimal quantity of RMSE and MAPE.

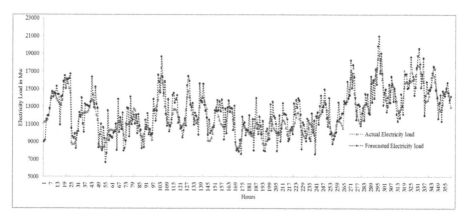

(a) 15-Day's electricity load forecasting using MLP.

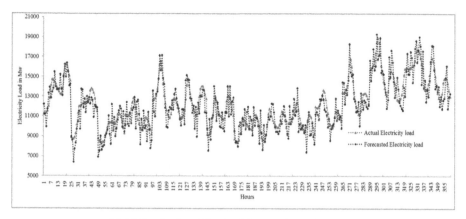

(b) 15-Day's electricity load forecasting using RNN.

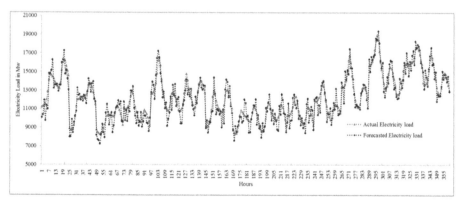

(c) 15-Day's electricity load forecasting using LSTM.

FIGURE 4.7   15-day electricity load forecasting using (a) MLP; (b) RNN; and (c) LSTM.

TABLE 4.3   Forecasting of 30 Days Electricity Load Forecasting

Method	R²	RMSE	MAPE
MLP	0.65	155.65	5.89
RNN	0.92	85.78	4.01
LSTM	0.93	84.98	2.47

Figure 4.8(a)–(c) presents the results of 30 days (one month) payload forecasts acquired utilizing three different techniques (MLP, LSTM, and RNN). According to the extracted waveforms, the LSTM is the nearest to the real load profile, accompanied by RNN and MLP.

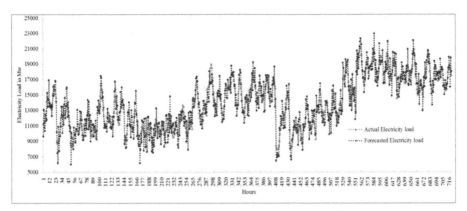

(a) 30-days electricity load forecasting using MLP.

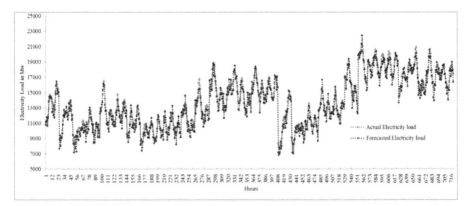

(b) 30 days electricity load forecasting using RNN.

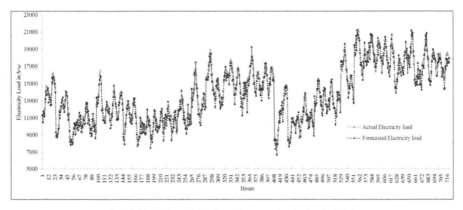

(c) 30 days electricity load forecasting using LSTM.

FIGURE 4.8 30-day load forecasting using (a) MLP; (b) RNN; and (c) LSTM.

4.3.4   PERFORMANCE OF ALGORITHMS USING R^2, RMSE, AND MAPE

Figure 4.9 depicts the comparative performance among the parameter used for load forecasting of electrical payloads with different time duration of the input data with the parameters, namely Squared error R^2, root mean square error, MSE, Moving average absolute percentage error MAPE. Figure 4.9(a) reveals the performance of algorithms respective to R^2 and reveals the fact that for the day data, MLP yields maximum output, while LSTM proves its effective capability for 15 days data and month data. Figure 4.9(b) marks the performance of comparative algorithms with the parameter RMSE. RMSE is proven more effective with LSTM for the day data and MLP for 15 days data and finally well proved for both CNN and LSTM. Figure 4.9(c) exhibits the effective performance with the MAPE parameter. Where MLP proved to be more effective for both the Medium and maximum data, i.e., for 15 days and one month, and LSTM holds better for small duration data for day data. As a common view in all the parameter performance, RNN is provided with the most stable results maintaining the substantial results for only the short time series data. This result shows the scope for mixed algorithm requirements, which proved the LSTM more effective with desired results for all the parameters with diversified durations.

4.3.5   DISCUSSIONS

Authentic load forecasting can manually mitigate the effect of sustainable power admittance to the network, expedite the electricity flora to organize unit protection, and motivates the electricity provider organizations to expand a possible strength invoice deal. Many electricity gadgets consisting of the protection of generator scheduling, renewable power integration, and electricity grids rely on the burden forecasting LSTM is substantially utilized in forecasting. LSTM can apprehend sample traits in conjunction with scale-invariant traits whilst the nearby facts have stable relationships with one some other. The layout of the regionally set route of load informational records in nearby hours will be extricated via way of means of LSTM. In some other load forecasting layout makes use of LSTM infrastructure and make an evaluation of the relaxation of the neural systems. The examinations exhibit the truth approximately; LSTM infrastructure is an important concern in load forecasting. Considering the above description, LSTM has nicely proven to offer excessive exactness forecast in medium and most due to their wonderful function to seize hid traits.

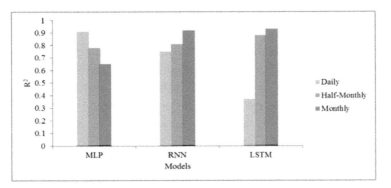

(a) R² performance in comparison with MLP, RNN, and LSTM for one day, half-month, and one month.

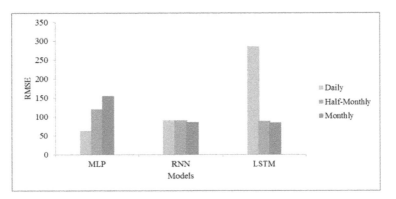

(b) RMSE performance in comparison with MLP, RNN, and LSTM for one day, half-month, and one month.

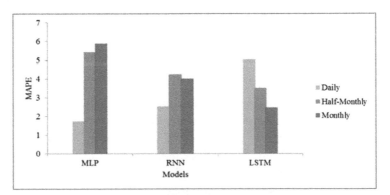

(c) MAPE performance in comparison with MLP, RNN, and LSTM for one day, half-month, and one month.

FIGURE 4.9   Parameter performance of algorithms (a) R²; (b) RMSE; and (c) MAPE.

Along those lines, a hybrid neural logical shape is far essential to capture and contain such unique, unseen developments in order to provide powerful execution. The LSTM, RNN, and featured fusion modules are all included in this package. The forget gate in the LSTM module, as well as the memory mobility in the RNN module, have been used to eliminate designs of nearby styles and similar layouts that show up in a variety of places. The included fusion aggregate module is used to keep track of those unanticipated requirements and generate the most up-to-date forecast. To predict a real phase electric-powered load collection of time, the suggested RNN-LSTM framework was constructed and used. In addition, some techniques have been updated to compare to our recommended version. The RNN and LSTM modules have been tracked individually to demonstrate the validity of the counseled version. Furthermore, the records file was divided into segments to test the effectiveness of the advised version. In summary, this chapter presents a deep mastering shape that can capture and contain the hidden qualities of RNN and LSTM to execute some distance with greater accuracy and power. We used MLP, LSTM, and RNN to compute forecasts for 1 day, 1/2 month, and one month in this study. R2, MAPE, and RMSE were used to calculate the computational overall effectiveness. We enhanced the metrics of those methodologies in order to improve overall effectiveness and reap the previously mentioned STLF and MTLF. The results of our comparison with previous findings in terms of RMSE and MAPE for 15 days and one month were identical. This well-known demonstrates that our recommended fashions with specifications enhancement provide the best beforehand detection overall performance when combined with the combined set of rules layout with LSTM and RNN for STLF and MTLF.

4.4 CONCLUSION

Electrical load forecasting for different duration of time ranging from min of 24 hours (day) to one month has been verified with different machine learning and deep learning algorithms. For this study, the database has been utilized from the Kaggle database feeder, and load forecasting for 24 hours (one day), 15 days, and one month has been computed. To estimate and improve the forecasting performance of all types of load demands, various AI algorithms (MLP, LSTM, RNN) have been utilized. Forecasting performance is analyzed R^2 error, MAPE, and RMSE are examples of robust error metrics. The smallest error value observed for the 24 hours was 84.98, 2.47. For 15 days and one month, load forecast data indicated the best fit. The smallest the value of the R^2 statistical, the more closely the data is fitted. From the practical observation with sample input data, the following conclusions were made w.r.t to parameters and several machine and deep learning algorithms. The data states that the R^2 yields the lowest error at 85.78 and 84.98 in all these cases except the LSTM, observed by MAPE and RMSE. A better forecast is obtained with STLF using MLP. In return, a similar forecast is obtained for MTLF using LSTM and RNN. It indicates that, for higher data demands, the deep learning models provide better forecasting with a greater neuron in the hidden layer and improved activation functions. We can deduce from this experimental validation that power systems with highly complicated expansion, power production consumption, and power planning can be better forecasted using this approach.

4.4.1 FUTURE DIRECTIONS

There are various factors which affect the load demand growth. This must be enhanced with real-time data, and comparison must be drawn with a practical grid along with each parameter for understanding the apparent capability and range. Hence this analysis helps in the practical estimation of load demands for suitable areas.

KEYWORDS

- deep learning
- electricity load forecasting
- load demand growth
- Long short-term load forecasting (LSTM)
- multi-layer perceptron (MLP)
- real-time data
- recursive neural networks (RNN)
- smart grids

REFERENCES

1. Singla, M. K., & Hans, S., (2018). Load forecasting using fuzzy logic toolbox. *Global Research and Development Journal for Engineering, 38,* 12–19.
2. Hong, T., Pinson, P., Fan, S., Zareipour, H., Troccoli, A., & Hyndman, R. J. (2016). Probabilistic energy forecasting: Global energy forecasting competition 2014 and beyond. *Int. J. Forecast., 32,* 896–913.
3. Hong, T., Pinson, P., & Fan, S., (2014). Global energy forecasting competition 2012. *Int. J. Forecast., 30,* 357–363.
4. Hussain, L., Nadeem, M. S., & Shah, S. A. A., (2014). Short-term load forecasting system based on support vector kernel methods. *Int. J. Comput. Sci. Inf. Technol., 6,* 93–102.
5. Jamaaluddin, J., Hadidjaja, D., Sulistiyowati, I., Suprayitno, E. A., Anshory, I., & Syahrorini, S., (2018). Very short-term load forecasting peak load time using fuzzy logic. *IOP Conference Series: Materials Science and Engineering, 403,* 012070.
6. Hossain, E., Khan, I., Un-Noor, F., Sikander, S. S., & Sunny, M. S. H., (2019). Application of big data and machine learning in smart grid, and associated security concerns: A review. *IEEE Access, 7,* 13960–13988.
7. Raza, M. Q., & Khosravi, A., (2015). A review on artificial intelligence-based load demand forecasting techniques for smart grid and buildings. *Renew. Sustain. Energy Rev., 50,* 1352–1372.
8. He, Y., Xu, Q., Wan, J., & Yang, S., (2016). Short-term power load probability density forecasting based on quantile regression neural network and triangle kernel function. *Energy, 114,* 498–512.
9. Laouafi, J., Mordjaoui, M., & Boukelia, T. E., (2018). An adaptive neuro-fuzzy inference system-based approach for daily load curve prediction. *Journal of Energy Systems,* 115–126.
10. Velasco, L. C. P., Villezas, C. R., Palahang, P. N. C., & Dagaang, J. A. A., (2015). Next-day electric load forecasting using artificial neural networks. In: *2015 Int. Conf. Humanoid, Nanotechnology, Inf. Technol. Control. Environ. Manag.* (pp. 1–6). IEEE.
11. Williams, R. J., & Zipser, D., (1989). A learning algorithm for continually running fully recurrent neural networks. *Neural Comput., 1,* 270–280.
12. Buitrago, J., & Asfour, S., (2017). Short-term forecasting of electric loads using nonlinear autoregressive artificial neural networks with exogenous vector inputs. *Energies, 10,* 40.
13. Kim, K. H., Park, J. K., Hwang, K. J., & Kim, S. H., (1995). Implementation of a hybrid short-term load forecasting system using artificial neural networks and fuzzy expert systems. *IEEE Trans. Power Syst., 10,* 1534–1539.
14. Lee, W. J., & Hong, J., (2015). A hybrid dynamic and fuzzy time series model for mid-term power load forecasting. *Int. J. Electr. Power Energy Syst., 64,* 1057–1062.
15. Niu, D. X., Shi, H. F., & Wu, D. D., (2012). Short-term load forecasting using Bayesian neural networks learned by hybrid Monte Carlo algorithm. *Appl. Soft Comput., 12,* 1822–1827.
16. Hernández, L., Baladrón, C., Aguiar, J. M., Calavia, L., Carro, B., Sánchez-Esguevillas, A., et al., (2014). Artificial neural network for short-term load forecasting in distribution systems. *Energies, 7,* 1576–1598.
17. Li, X., & Wu, X., (2015). Constructing long short-term memory-based deep recurrent neural networks for large vocabulary speech recognition. In: *2015 IEEE Int. Conf. Acoust. Speech Signal Process* (pp. 4520–4524). IEEE.

18. Zaremba, W., Sutskever, I., & Vinyals, O., (2014). *Recurrent Neural Network Regularization*. arXiv preprint arXiv: 1409.2329.

19. James, P. E., Kit, M. H., Vaithilingam, C. A., & Chiat, A. T. W., (2020). Recurrent neural network-based speech recognition using MATLAB. *International Journal of Intelligent Enterprise* (Vol. 7, pp. 56–66).

20. Jozefowicz, R., Zaremba, W., & Sutskever, I., (2015). An empirical exploration of recurrent network architectures. *International Conference on Machine Learning (ICML)*, 2342–2350.

21. Wang, F. F., Zhang, X., Liu, Y., Wei, L., & Shi, Y. (2019). LSTM-based short-term load forecasting for building electricity consumption. In: *2019 IEEE 28th Int. Symp. Ind. Electron* (pp. 1418–1423). IEEE.

22. Glorot, X., & Bengio, Y., (2010). Understanding the difficulty of training deep feedforward neural networks. In: *Proc. Thirteenth Int. Conf. Artif. Intell. Stat.* (pp. 249–256).

23. Zheng, H., Yuan, J., & Chen, L., (2017). Short-term load forecasting using EMD-LSTM neural networks with an Xgboost algorithm for feature importance evaluation. *Energies*, *10*, 1168.

24. Rafi, S. H., & Al-Masood, N., (2020). Highly efficient short-term load forecasting scheme using long short-term memory network, In *2020 8th Int. Electr. Eng. Congr.* (pp. 1–4). IEE.

25. Dudek, G., (2020). Multilayer perceptron for short-term load forecasting: From global to local approach. *Neural Comput. Appl.*, *32*, 3695–3707.

CHAPTER 5

Battery Plant Model Development for BMS Application

G. N. DHANYA and K. V. ABHINAND

Software Engineer, e-Powertrain KPIT, Bangalore, Karnataka, India

ABSTRACT

In order to develop an efficient BMS and serve as a key element of vehicle level simulation-studies for battery pack size and range estimation calculations, circuit-based models are preferred. Equivalent Circuit Models (ECM) have gained importance among EV designers for SOC and SOH estimation mainly due to the simpler mathematical and numerical methods, reducing the need for computationally intensive techniques. These models use passive and active electrical components, such as resistors and capacitors, to resemble the behavior of a real battery. In the proposed work, a simplified ECM for 48V battery system having cells with two RC pair network is modeled. Parameter estimation method is employed to generate the required parameters for the two RC pair ECM. The dependency of the parameters on the cell SOC and temperature is also considered.

5.1 INTRODUCTION

At present, with the development of new energy vehicles, batteries have become one of the main power sources of EVs. Among the various types of batteries, lithium-ion batteries are used in a diverse range of applications as they possess more advantages. But due to the fragile nature of these batteries, advanced monitoring is essential for safe operation when

The Internet of Energy: A Pragmatic Approach Towards Sustainable Development. Sheila Mahapatra, Mohan Krishna S., B. Chandra Sekhar, & Saurav Raj (Eds.)

compared to lead-acid or nickel-cadmium batteries. Thus, BMS becomes a major requirement in order to monitor and maintain the battery state, measure the secondary data, record the data, and control its environment and has functions to detect and notify the user in case of irregular conditions such as over-charging or over-discharging, over-heating and charge balancing between cells [1]. In particular, a battery management system (BMS) is used to ensure the safe and efficient use of batteries in electric vehicles.

In the past, a variety of models have been built to describe and simulate lithium cells. Lumped-parameter equivalent circuit designs have acquired a lot of popularity among EV designers for real-time battery state estimation and power management applications mainly due to the simpler mathematical and numerical methods, which reduce the need for computationally intensive techniques [3, 5]. These models use passive and active electrical components, such as resistors and capacitors, to resemble the behavior of a real battery [6]. It is essential to model the battery accurately in order to predict the battery performance more closely. One such approach is to estimate the battery parameters using the parameterization technique [5, 7].

Most of the existing models are done using a fixed-parameter approach [2]. But, in practice, these parameters vary with respect to state of charge, temperature, etc. [3, 8]. The accuracy of the fixed parameter model is not adequate, especially during the transient conditions. Hence the proposed work aims at developing a more accurate model which takes into account the dependency of parameters on SOC and temperature.

The objective of the proposed work is to develop a high-fidelity ECM that will closely describe the behavior of a real battery. A 48 V battery system having cells with two RC pair networks is modeled by considering the dependency of the parameters on the cell SOC and temperature, for which the parameter estimation method is employed to generate the required parameters for the two RC pair ECM.

5.2   METHODOLOGY

5.2.1   LI-ION CELL MODELING

A study of the battery model can be carried out by building up behavioral/ phenomenological analogs using common circuit elements. The end result of such a process will be an 'equivalent circuit model (ECM),' which will give a response similar to that of a real-life cell or battery [2].

The battery is a source of voltage. So, a voltage source is the closest component that can be used to represent a battery. This is a very oversimplified model, as the voltage remains constant irrespective of time or current drawn or any other parameters. It is evident from experiments carried out that when the battery is charged, it shows a higher OCV, and when it's discharged, it shows a lower SOC value. Figure 5.1 shows the change of OCV with respect to SOC. So, replacing the independent voltage source with a voltage source that depends on SOC value will increase the fidelity of the model.

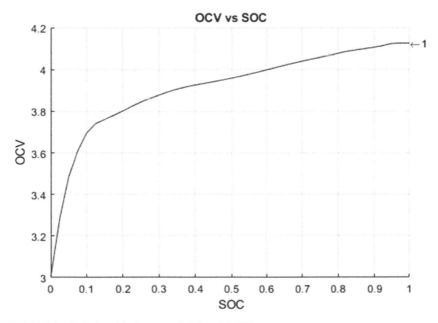

FIGURE 5.1 Relationship between SOC and OCV.

If we define the capacity as the total amount of charge removed from the cell when its SOC is removed from 100% to 0%, SOC can be represented as:

$$SOC(t) = SOC(0) + \frac{\int \eta * I(t)}{C * 3600} \tag{1}$$

where; I(t) is the current flow into or out from the battery; SOC(t) is determined by the previous SOC(to); C is the capacity in Ah and is the coulombic efficiency. The terminal voltage of a battery under a pulsed load is given in Figure 5.2. The voltage source depending on SOC is inadequate to describe this type of behavior under similar load conditions. It can be observed that when under load, the battery terminal voltage drops to a lower value. This voltage dip can be assumed due to the presence of a resistor within the cell, called the 'Internal resistance.' It can be modeled as,

$$V_0 = i\,(t).\,R_0 \tag{2}$$

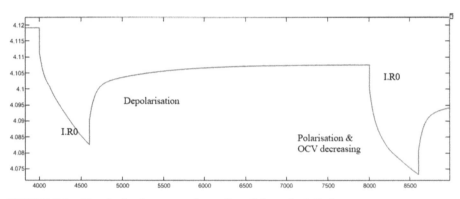

FIGURE 5.2   Terminal voltage waveform of a cell for pulsed discharge current.

The phenomenon is known as linear polarization. So, in the cell model, in addition to the voltage source, a series resistor is also connected. This model is good enough for many elementary electronic circuit systems but cannot be employed for sophisticated consumer electronics or EV applications. As seen in the pulse response waveform (Figure 5.2), we also observe a dynamic (non-instantaneous) response to a current pulse. When the cell is allowed to rest, the terminal voltage is not jumping back to OCV in an instant, but the process is gradual. This is due to the slow diffusion process within the cell, and hence it's called 'diffusion voltages.' This behavior can be approximated by connecting parallel connected resistor-capacitor pairs in series. As the number of RC pairs increases, the accuracy of the model will also increase, trading simplicity. Considering two RC pairs are sufficient enough for most of the applications. The voltage across the two RC pairs can be modeled as in the equation (Figure 5.3).

Let the current flowing through the resistor and capacitor be, respectively. Applying KCL to the circuit:

$$i(t) = i_R(t) + i_C(t) \tag{3}$$

where; $i_R(t)$ and $i_C(t)$ are the currents flowing through the resistor and capacitor of the RC network. We have voltage across the capacitor:

$$V_c(t) = \frac{1}{C} \int i_c(t).dt \qquad (4)$$

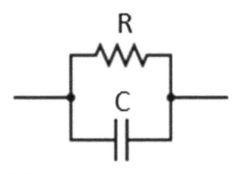

FIGURE 5.3   Parallel R-C branch.

Thus,

$$i_C(t) = C.\frac{dV_c(t)}{dt} \qquad (5)$$

i.e.,

$$i_{C1}(t) = C_1 \dot{V}_{c1} \qquad (6)$$

Substituting in Eqn. (3) we get,

$$i(t) = i_{R1}(t) + C_1 \dot{V}_{c1} \qquad (7)$$

$$i(t) = \frac{V_{R1}}{R_1} + C_1 \dot{V}_{c1} \qquad (8)$$

Therefore,

$$\dot{V}_{c1}(t) = \frac{-V_{R1}}{R_1 C_1} + \frac{i(t)}{C_1} \qquad (9)$$

Since $V_{R1} = V_{C1}$, we can re-write the equation as:

$$\dot{V}_{c1}(t) = \frac{-V_{C1}}{R_1 C_1} + \frac{i(t)}{C_1} \qquad (10)$$

Similarly,

$$\dot{V}_{c2}(t) = \frac{-V_{C2}}{R_2 C_2} + \frac{i(t)}{C_2} \qquad (11)$$

Therefore, the cell terminal voltage is determined by applying KVL to the equivalent circuit:

$$Vt = V_{oc} - V_0 - V_{c1} - V_{c2} \tag{12}$$

where; V_t is the cell terminal voltage; V_{oc} is the open circuit voltage; V_0 is the drop across the internal resistor R_0; V_{c1} and V_{c2} are assumed as the voltage across the two RC networks, respectively.

Summing up, all the above inferences are represented in Figure 5.4 as an equivalent circuit diagram.

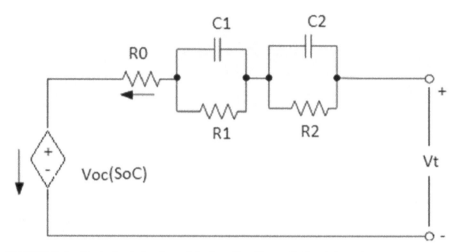

FIGURE 5.4   Equivalent circuit model of a cell with two R-C pair branches.

5.2.2   *PARAMETER ESTIMATION*

For battery systems, the accurate estimation of values representing the internal states and parameters of the cell is essential. Parameter estimation is one such technique which is commonly used to fit an ECM to a specific battery cell. In order to achieve a valid generalized fit between measured and simulated results, a numerical optimization algorithm using Simulink Parameter Estimation is used to estimate the values of the parameters [4, 5]. The parameter estimation method is employed to generate the required parameters for the two RC pair ECM.

In this work, a two RC equivalent cell is modeled to fit the real data of one RC model. For this, a pulse discharge test is performed. The pulse data of terminal voltage and the input current of the equivalent cell are considered for the pulse discharge test. The cell current and terminal voltage is recorded at three different temperatures (i.e., at 5°C, 25°C, and 45°C) and loaded to the parameterization algorithm. In the parameterization procedure, initially, the pulse data considered are loaded into the algorithm. Appropriate settings are chosen for the optimization algorithm. The algorithm facilitates pulse discharge sequences from 100% to 0% SOC. Then, the pulses and the relaxation periods are identified for different SOC values by the algorithm. Since the algorithm is automated, it estimates the parameters, sets initial values, optimizes, and generates the parameters for a two RC network which is then fed to a two RC ECM of a cell.

5.2.3   48 V LI-ION BATTERY PACK MODELING

To realize a specific voltage range on the battery pack level, cells must be connected in series, while parallel connections increase the capacity. In this work, the cells modeled by the equivalent cell model are connected in series and parallel as per the voltage and capacity required. The cell rating is 3.7 V, 2.5 Ah. The battery pack is modeled such that it consists of four modules in parallel, each module consisting of 13 cells connected in series. Thus, a total of 52 cells are arranged as a combination of series and parallel connections appropriately to form a complete battery pack of 48 V with 10 Ah capacity.

5.3   RESULTS AND DISCUSSION

5.3.1   EFFECT OF OPERATING PARAMETERS

5.3.1.1   PARAMETER GENERATION

The result of the pulse discharge test is shown in Figure 5.5. The waveform is of the pulse data of terminal voltage at three different temperatures showing the variation of terminal voltage with respect to temperature.

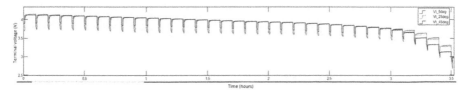

FIGURE 5.5   Pulse data of terminal voltage of the equivalent cell at three different temperatures while discharging.

After the considered pulse data are loaded into the algorithm, the pulses and the relaxation periods are identified for different SOC values by the algorithm, as shown in Figure 5.6.

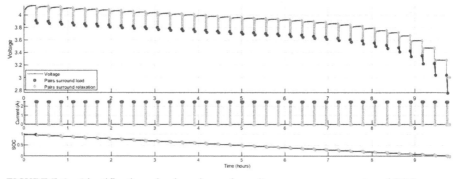

FIGURE 5.6   Identification of pulses shown for voltage response, current, and SOC.

The parameters generated after performing the parameter estimation technique at three different temperatures (i.e., at 5°C, 25°C, and 45°C) are as shown in Figure 5.7. It shows the results which provide the generated parameters, that is, OCV, R_0, R_1, R_2, C_1; and C_2, which are functions of SOC and temperature.

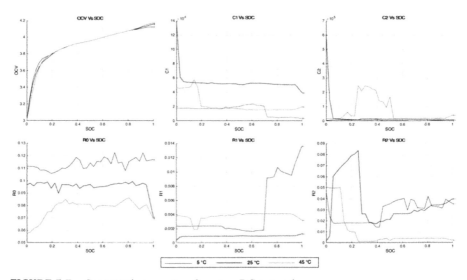

FIGURE 5.7   Generated parameters for a two RC network.

5.3.1.2 LI-ION CELL MODEL

The terminal voltage, current pulse, and state of charge during discharging of a two RC pair equivalent cell are obtained as shown in Figure 5.8. The SOC dips at every pulse of current, indicating the discharging operation.

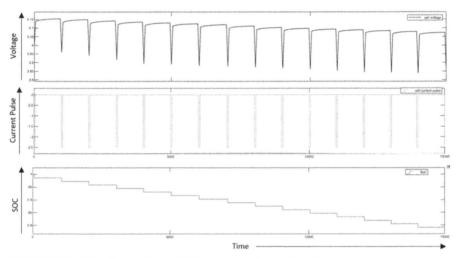

FIGURE 5.8   Waveforms of a two RC pair equivalent cell model.

5.3.1.3 BATTERY PACK MODEL

The battery pack voltage for a given current is shown in Figure 5.9. The battery is modeled for a nominal voltage of 48 V with 10 Ah capacity.

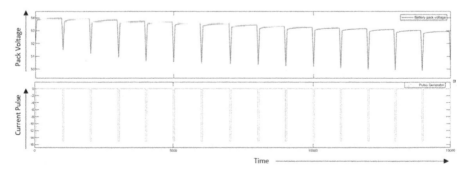

FIGURE 5.9   Battery pack waveforms.

5.4   CONCLUSION AND FUTURE SCOPE

A detailed understanding of the behavior and characteristics of the cell and equivalent cell modeling approach is described in this chapter. An approach to generate parameters for a two RC pair equivalent cell model using the Parameter estimation technique has been illustrated, and the corresponding equivalent cell is modeled wherein the parameters are a function of both SOC and temperature. From this, a 48 V, 10 Ah battery pack is being modeled. Thus, an electrical equivalent battery is modeled, and the results obtained for the cell and the battery show satisfactory characteristics with the data researched in the literature survey. Therefore, the battery system modeled on the Simulink environment can be utilized for virtual validation of BMS control algorithms in a real-time simulation environment.

In this model, the cell degradation effects are not considered. This is one of the areas which can be explored further to improve the fidelity of the battery plant model.

KEYWORDS

- **battery**
- **battery management system**
- **BMS**
- **electric vehicles**
- **equivalent circuit modeling**
- **EV**
- **parameter estimation**
- **SOC**
- **state of charge**

REFERENCES

1. Shen, M., & Qing, G. (2019). A review of the battery management system from the modeling efforts to its multiapplication and integration. *International Journal of Energy Research, 43*(10), 5042–5075.

2. Nemes, R., Sorina, C., Mircea, R., Horia, H., & Claudia, M., (2019). Modeling and simulation of first-order Li-ion battery cell with experimental validation. In: *2019 8th International Conference on Modern Power Systems (MPS)* (pp. 1–6). IEEE.

3. Lu, Z., Yu, X. L., Wei, L. C., Cao, F., Zhang, L. Y., Meng, X. Z., & Jin, L. W., (2019). A comprehensive experimental study on temperature-dependent performance of lithium-ion battery. *Applied Thermal Engineering, 158*, 113800.

4. Miniguano, H., Andrés, B., Antonio, L., Pablo, Z., & Cristina, F., (2019). General parameter identification procedure and comparative study of Li-ion battery models. *IEEE Transactions on Vehicular Technology, 69*(1), 235–245.

5. Ahmed, R., Javier, G., Simona, O., Saeid, H., Robyn, J., Kevin, R., Jimi, T., & LeSage, J., (2015). Model-based parameter identification of healthy and aged Li-ion batteries for electric vehicle applications. *SAE International Journal of Alternative Powertrains, 4*(2), 233–247.

6. Wang, Y., Jiaqiang, T., Zhendong, S., Li, W., Ruilong, X., Mince, L., & Zonghai, C., (2020). A comprehensive review of battery modeling and state estimation approaches for advanced battery management systems. *Renewable and Sustainable Energy Reviews, 131*, 110015.

7. Surya, S., Janamejaya, C., Shantanu, D. D., Abhay, S. J., & Ashita, V., (2020). Accurate battery modeling based on pulse charging using MATLAB/Simulink. In: *2020 IEEE International Conference on Power Electronics, Drives and Energy Systems (PEDES)* (pp. 1–3). IEEE.

8. Bai, B. (2020). Estimate the parameter and modeling of a battery energy storage system. In: *2020 Chinese Control and Decision Conference (CCDC)* (pp. 5444–5448). IEEE.

CHAPTER 6

Modeling of Constant Voltage Control in Synchronous Buck and Boost Converters Using MATLAB/Simulink for Point-of-Load Application

SUMUKH SURYA[1] and VINEETH PATIL[2]

[1]Senior Engineer, Bosch Global Software Technologies Private Limited, Bangalore, Karnataka, India

[2]Department of Electrical and Electronics Engineering, Manipal Institute of Technology, Manipal Academy of Higher Education, Manipal, Karnataka, India

ABSTRACT

DC-DC converters play a vital role in the design of Electric Vehicle (EV) charging systems, LED drivers and power supply units for critical loads like micro-processors. The converters show different behaviors during ideal and non-ideal scenarios. In the present work, synchronous buck and synchronous boost converters operating in Continuous Conduction Mode (CCM) considering the non-idealities were modeled using MATLAB/Simulink. The shift in the output voltage was observed when the Equivalent Series Resistance (ESR) values for the inductor and capacitor were varied. The mathematical models for the converters were derived using volt-sec and amp-sec balance equations for the ideal and non-ideal cases. It was observed that the practical synchronous buck converter showed stability. The practical synchronous boost converter showed instability due to the right half plane zero in its transfer function as a result of which phase reversal occurred at high frequencies.

The Internet of Energy: A Pragmatic Approach Towards Sustainable Development. Sheila Mahapatra, Mohan Krishna S., B. Chandra Sekhar, & Saurav Raj (Eds.)
© 2024 Apple Academic Press, Inc. Co-published with CRC Press (Taylor & Francis)

6.1   INTRODUCTION

The primary goal of any converter is to electrically transform power from one form to another. AC–AC, AC–DC, DC–DC, and DC–AC are the major converters used in modern power electronics. Switched-mode converters and linear regulators are more commonly used DC–DC converters. Since operating the load at the rated load current causes consequent I2R losses in the latter, the former is preferred. The primary role of DC-DC converters is to supply the DC power at required regulation to the load at high efficiency. The converters are designed such that the inductor shall never short-circuit, and the capacitor shall never be open-circuited. The practical converters show distinct behavior compared to the ideal converters in terms of duty cycle and inductor design. Hence, the modeling of practical converters is close to reality.

In Ref. [1], the modeling of steady state was shown using governing converter equations. The inductor and capacitor values were derived using the waveforms for continuous conduction mode (CCM) and discontinuous conduction mode (DCM). In Ref. [2], the steady-state modeling considering the inductor ESR was considered, and their effect on the output voltage was shown. Different methods to obtain the open-loop transfer functions during CCM and DCM operations were discussed.

In Refs. [3, 4], the mathematical modeling of non-isolated converters was carried out and analyzed using MATLAB/Simulink. The converter chosen did not consider any switch drops.

In Ref. [5], the modeling of isolated ideal DC-DC converters, namely Flyback and Forward converters were performed. The open-loop and closed-loop analyses were performed using MATLAB/Simulink. It was shown that in the closed-loop operation, the flyback converter showed a small over-shoot, and the time taken to reach the steady-state voltage was slower than the forward converter.

In Ref. [6], the steady-state and the average model for ideal and non-ideal boost converter operating in CCM and DCM operations were discussed. The effect of output voltage (V0), a ripple in the inductor current (ΔIL), and maximum and minimum inductor current (ILMax and ILMin) in ideal and non-ideal cases of CCM and DCM operations were discussed. The choice of selection of components was also discussed.

In traditional converters, controlled (MOSFET) and uncontrolled switches (diode) are used. However, in synchronous converters, the uncontrolled switch is replaced by a controlled switch. One of the many advantages is in regard to the conduction losses. The voltage drop provided by the diode would be typically 0.65 to 0.7 V. However, the voltage drop across a controlled device like MOSFET would be typically 0.3 to 0.35 V. In Ref. [7], a small signal model for a practical synchronous buck converter operating under CCM for light load conditions was discussed. It was shown that the resonant frequency drastically increased with the increase in the load resistance, R. Simulation study using MathCAD software and hardware implementation was performed. The developed model accurately predicted the AC behavior of the converter. This generic model can be used for lightly loaded synchronous converters during the critical conduction mode.

In Ref. [8], non-linear modeling of DC–DC converters using Hammerstein's Approach using black-box identification methods are presented for a boost converter. Controller design involves two major steps viz: (i) For each duty cycle value, a controller is designed to operate in a particular operative situation Gain Scheduling Technique is used; and (ii) to take parameter variations and unmodeled dynamics into account, a robust controller is designed assuming that the converter can be described by a set of models instead of only one model. DC–DC converters play a major role in photovoltaic systems. In Ref. [9], computationally efficient DC–DC converters for PV applications are presented. This approach shown is a model that is mathematically formed and can be added efficiently in a large simulation system. This model automatically detects the steady state, and lowered computational costs are incurred.

In this work, the mathematical modeling for the synchronous buck and boost converters was performed using governing equations. Synchronous converters are also referred to as Point of Load (POL) converters, as they are used as regulators in low-power circuits like a microprocessor. The developed model was analyzed using MATLAB/Simulink. A small-signal model for analyzing the control study was derived, and the investigation on their stability was made. It was noted that instability was observed for constant voltage (CV) operation in the synchronous buck converter. However, the synchronous buck converter was stable.

6.2 MATHEMATICAL MODELING OF AN IDEAL SYNCHRONOUS BUCK CONVERTER

Figure 6.1 shows an ideal converter with two switches S_1 and S_2. The switches S_1 and S_2 are typically MOSFETs. The main advantage of using a MOSFET at S_2 is that the drop across it is small. Hence, the drop at the output voltage can be minimized.

The operation of the converter is similar to that of the conventional boost converter except that the diode is replaced by a MOSFET (at S_2). Initially, the switch S_1 is closed due to which i_L gets charged. The supply current charges the capacitor. Later, switch S_2 is turned ON with S_1 in OFF condition. The inductor reverses its polarity and discharges the capacitor through the load.

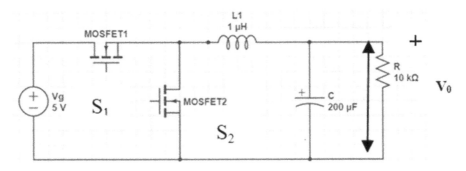

FIGURE 6.1   Circuit diagram of an ideal buck converter.

When S_1 ON and S_2 OFF,

$$V_L = V_g - V_0 \tag{1}$$

$$i_c = i_L - \frac{V_0}{R} \tag{2}$$

When S_2 ON and S_1 OFF,

$$V_L = -V_0 \tag{3}$$

$$i_c = i_L - \frac{V_0}{R} \tag{4}$$

Combining Eqns. (1) and (3) with (2) and (4)

$$V_L = L\frac{di_L}{dt} = (V_g - V_0)s + (1-s)(-V_0) \tag{5}$$

$$i_c = C\frac{dV_0}{dt} = (i_L - \frac{V_0}{R})s + (1-s)(i_L - \frac{V_0}{R}) \tag{6}$$

Eqns. (5) and (6) were solved using MATLAB/Simulink 2018a with a proper step size and solver.

6.3 MATHEMATICAL MODELING FOR SYNCHRONOUS BUCK CONVERTER

Figure 6.2 shows a practical converter. Considering the non-idealities, Eqns. (5) and (6) would be modified as Eqns. (7) and (8).

$$V_L = L\frac{di_L}{dt} = V_g D - i_L D(R_L - R_{on2}) - i_L(R_{on1} + R_{on2}) - V_0 \tag{7}$$

$$i_c = C\frac{dV_0}{dt} = (i_L - \frac{V_0}{R})s + (1-s)(i_L - \frac{V_0}{R}) \tag{8}$$

where; s is the instantaneous duty cycle.

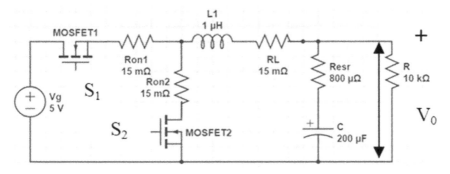

FIGURE 6.2 Circuit diagram of non-ideal synchronous buck converter.

6.4 SPECIFICATIONS OF THE CONVERTER (Table 6.1)

TABLE 6.1 The Specifications of the Synchronous Buck Converter

SL. No.	Specification	Value
1.	Input voltage (V_g)	5 V
2.	Output voltage (V_0)	1.8 V
3.	Output current (I_0)	0 to 5 A
4.	Inductor (L)	1 µH
5.	Inductor ESR (R_L)	15 mΩ
6.	MOSFET resistances (R_{on1} and R_{on2})	31 mΩ
7.	Duty cycle (D)	0.36
8.	Capacitor (C)	200 µH
9.	Capacitor ESR (R_{esr})	0.8 mΩ

6.5   SMALL-SIGNAL MODELING

In the present work, the transfer function for the practical converters is derived and their characteristics using bode plot are studied using MATLAB/Simulink 2018a.
Applying volt-sec balance to Figure 6.2,

$$V_L = L\frac{di_L}{dt} = D(V_g - i_L(R_L - R_{on2})) + (1-D)(-i_L(R_{on1} + R_{on2}) - V_0) \tag{9}$$

On Simplification,

$$V_L = V_g D - V_0 + i_L(R_{on2}(D-1) - DR_{on1} - R_L) \tag{10}$$

Perturbing and linearizing Eqn. (10)

$$V_L + \hat{v}_L = (V_g + \hat{v}_g)(D + \hat{d}) - (V_0 + \hat{v}_0) + (i_L + \hat{i}_L) \\ (-R_{on2}(1 - D - \hat{d}) - (D + \hat{d})R_{on1} - R_L) \tag{11}$$

On Simplification and eliminating the product of perturbed quantities,

$$V_L + \hat{v}_L = \hat{d}(V_g + I_L(R_{on2} - R_{on1})) + D(\hat{v}_g - \hat{i}_L R_{on1} + \hat{i}_L R_{on2}) - \hat{i}_L(R_L + R_{on2}) - \hat{v}_0 \tag{12}$$

where;

$$I_L = \frac{V_0}{R} \tag{13}$$

Applying amp-sec balance to Figure 6.2,

$$I_c + \hat{i}_c = D(I_L - \frac{V_0}{R}) + (1-D)(I_L - \frac{V_0}{R}) \tag{14}$$

$$I_c + \hat{i}_c = \hat{i}_c - \frac{\hat{v}_0}{R} \tag{15}$$

Equivalent circuit for Eqns. (12) and (15) are shown in Figures 6.3 and 6.4, respectively.

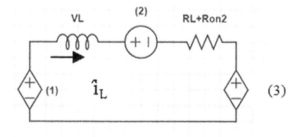

FIGURE 6.3   Equivalent circuit based on Eqn. (12).

where; (1) is '$D(\hat{v}_g - i_L R_{on1} + \hat{i}_L R_{on2})$', (2) is '$V_g + I_L(R_{on2} - R_{on1})$' and (3) is '$\hat{v}_0$'

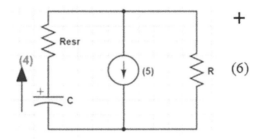

FIGURE 6.4 Equivalent circuit based on Eqn. (23).

where; Eqn. (4) is '$\hat{i}_c$,' Eqn. (5) is '$\hat{i}_L$' and Eqn. (6) is '$\hat{v}_0$'

The supply current i_g flows in the inductor when the switch S_1 is closed. Hence, it can be averaged as shown in Eqn. (24)

$$i_g = Di_L \qquad (16)$$

The small-signal model for Eqn. (24) can be obtained by perturbing and linearizing it.

$$I_g + \hat{i}_g = (D + \hat{d})(I_L + \hat{i}_L) \qquad (17)$$

$$\hat{i}_g = D\hat{i}_L + \hat{d}I_L \qquad (18)$$

The equivalent circuit for Eqn. (18) is shown in Figure 6.5.

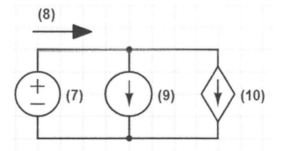

FIGURE 6.5 Equivalent circuit.

where; Eqn. (7) is '$\hat{v}_g$,' Eqn. (8) is '$\hat{i}_g$,' Eqn. (9) is $\hat{d}I_L$ and Eqn. (10) is '$D\hat{i}_L$'

Combining the circuits shown in Figures 6.3–6.6 Canonical model is obtained.

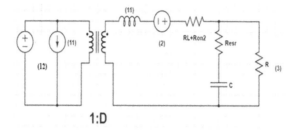

FIGURE 6.6   The canonical model for a synchronous buck converter.

where; Eqn. (12) is $V_g + \hat{v}_g$

For obtaining $G_{vd} = \dfrac{\hat{v}_0}{\hat{d}}$, the terms having $\hat{v}_g$ are set to zero. Hence, the circuit reduces to Figure 6.7.

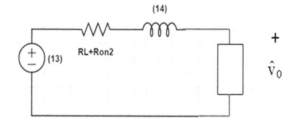

FIGURE 6.7   Reduced canonical model.

where; Eqn. (13) is $\hat{v}_g D$ and Eqn. (14) is sLD^2 with the voltage drop as $\hat{v}_L$.
   When the values shown in Table 6.1,

$$G_{vd} = \frac{4000(s + 6.25 * 10^6)}{s^2 + 3.08 * 10^4 s + 5 * 10^9} \qquad (19)$$

6.6   SMALL-SIGNAL MODELING OF SYNCHRONOUS BOOST CONVERTER

Applying the volt-sec and amp-sec balance to Eqns. (13) and (14) considering non-idealities in the circuit

$$V_L = V_g - i_L(DR_{on2} + R_L + R_{on2}(1-D)) - V_0(1-D) \tag{20}$$

$$i_c = i_L - Di_L - \frac{V_0}{R} \tag{21}$$

Perturbing and linearizing Eqn. (20)

$$\hat{v}_L = \hat{v}_g - \hat{d}(I_L R_{on1} - I_L R_{on2} - V_0) - D$$
$$(R_{on1}\hat{i}_L - R_{on2}\hat{i}_L - \hat{v}_0) - \hat{i}_L(R_L - R_{on2}) - \hat{v}_0 \tag{22}$$

Perturbing and linearizing Eqn. (21)

$$\hat{i}_c = -I_L \hat{d} + \hat{i}_L(1-D) - \frac{\hat{v}_0}{R} \tag{23}$$

where; $I_L = \dfrac{V_0}{R(1-D)}$

The equivalent circuit for Eqns. (22) and (23) are shown in Figures 6.8 and 6.9, respectively.

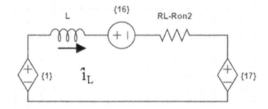

FIGURE 6.8   Reduced canonical model.

where; {1} is $\hat{v}_g$ {16} is $\hat{d}(I_L R_{on1} - I_L R_{on2} - V_0)$ and {17} is $D(\hat{i}_L R_{on1} - \hat{i}_L R_{on2} - \hat{v}_0)$.

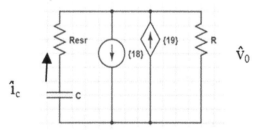

FIGURE 6.9   Equivalent circuit.

where; {18} is $-I_L \hat{d}$ and {19} is $(1-D)\hat{i}_L$

Combining Figures 6.8 and 6.9, the canonical model for the converter is shown in Figure 6.10.

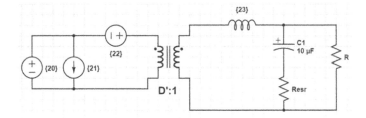

FIGURE 6.10   Canonical model of a synchronous boost converter.

where; $\{20\}$ is $\hat{v}_g$, $\{21\}$ is $\dfrac{i_L \hat{d}}{1-D}$, $\{22\}$ is

$$-\hat{d}\,(I_L R_{on1} - I_L R_{on2} - V_0 + \dfrac{I_L sL}{1-D} + \dfrac{I_L R_L}{1-D} + \dfrac{I_L R_{on2}}{1-D})\text{ and }\{23\}\text{ is }\dfrac{sL + R_L - R_{on2}}{(1-D)^2}$$

In order to obtain G_{vd}, $\hat{v}_g$ sources are zero (shorted) and the circuit reduces to Figure 6.11.

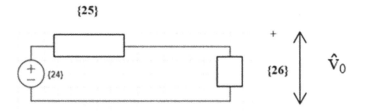

FIGURE 6.11   Reduced canonical model.

where; $\{24\}$ is, $\dfrac{-\hat{d}}{1-D}(I_L R_{on1} - I_L R_{on2} - V_0 + \dfrac{I_L sL}{1-D})$, $\{25\}$ is $\dfrac{sL + R_L - R_{on2}}{(1-D)^2}$ and $\{26\}$

is $\dfrac{RR_{esr}sC + R}{RR_{esr}sC + 1 + sRC}$

By plugging in values shown in Table 6.1,

$$G_{vd} = \dfrac{6.6856 * 10^5 (s + 1.327 * 10^5)}{(s + 1.714 * 10^4)(s + 7313)} \tag{24}$$

6.7   RESULTS AND CONCLUSION

Figure 6.12 shows the mathematical model for a practical synchronous buck converter operating in CCM. The inductor current (i_L) and output voltage (V_0) variations with time are observed in Figure 6.13.

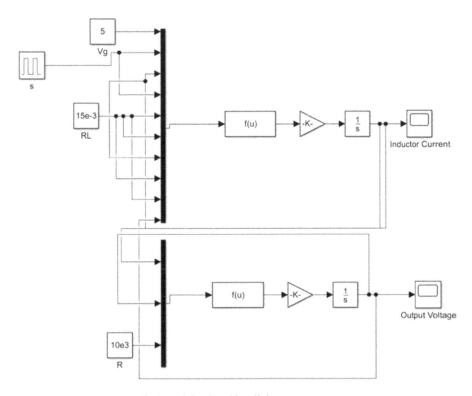

FIGURE 6.12   Mathematical model using Simulink.

It is observed that i_L initially showed a transient of about 20A and later settled at 0.526 A. Hence, the inductor must be designed to handle a peak current of 20 A. Similarly, V_0 showed a transient of about 2.8 V and later settled at 1.8 V. Hence, the voltage withstanding capacity of the capacitor at the output must be close to 3 V and not 1.8 V.

The variation of V_0 as a function of R_L is seen in Figure 6.14. A step-change in R_L from 15 mΩ to 17 mΩ in steps of 1 mΩ was applied. It is found that higher values of R_L produced higher V_0. From Figure 6.14, it was observed that $R_L = 17$ mΩ showed higher transients than others. It was observed that the output voltage and R_L are directly proportional when the DC transfer function was extracted [5].

$$\frac{V_0}{V_g} = \frac{DV_g + V_d(1-D)}{V_g} * \frac{R_{sw}D + R_d(1-D) + R_L + R(1-D)^2}{R(1-D)} \tag{25}$$

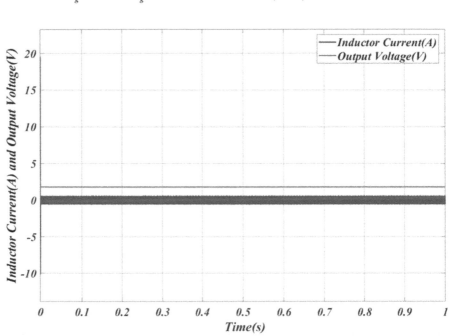

FIGURE 6.13   Inductor current (A) and output voltage (V) vs. time.

The variation of V_0 as a function of the R_{esr} is seen in Figure 6.15. A step-change in R_{esr} from 0.8 mΩ to 1 mΩ in steps of 0.1 m was applied. It was noted that the higher values of R_{esr} produced higher transient V_0. From Figure 6.15, it was observed that $R_{esr} = 1$ mΩ showed higher transients than others.

Figure 6.16 shows the variation of V_0 with respect to control voltage (V_c) proportional to D, varied from 0.2 to 0.9 in steps of 0.01. It is observed that V_0 showed linear variation with respect to the control voltage. As D increased, V_0 also increased.

Frequency response of G_{vd} as described in Eqn. (27) is shown in Figure 6.17. The plot initially had a DC offset of about 14 dB, showed a resonance at around 10^5 rad/s and later a decrease in magnitude due to the effect of second-order poles which were complex in nature. The bode plot shows the converter is stable in the open loop operation. However, for a DC–DC to be highly stable the Phase Margin (PM) should be greater than 42° [2].

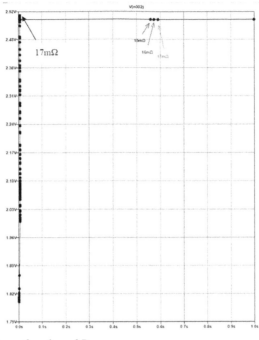

FIGURE 6.14 V_0 as a function of R_L.

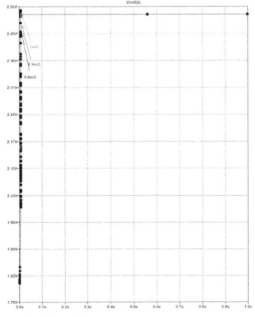

FIGURE 6.15 V_0 as a function of R_{esr}.

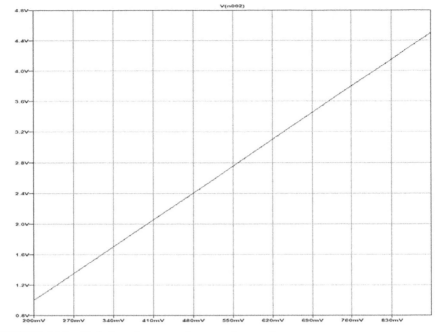

FIGURE 6.16   Variation of output voltage vs. duty cycle.

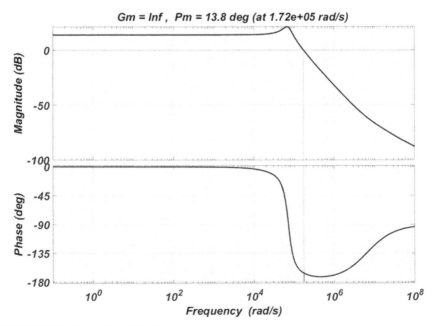

FIGURE 6.17   Bode plot of G_{vd} in open-loop.

Figure 6.18 shows the root locus of G_{vd}. It shows a real zero and complex poles.

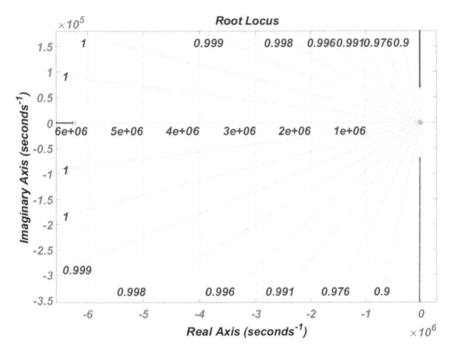

FIGURE 6.18   Root locus of G_{vd} in open-loop.

Figure 6.19 shows the mathematical model of a synchronous boost converter operating in CCM. Figure 6.20 shows i_L and V_0 variations. It is observed that V_0 showed an initial transient of about 16A and later settled around 2.5A at the steady state. Similarly, V_0 showed an initial transient of 18 V and a steady state voltage of around 12 V.

The variation of V_0 as a function of R_L is seen in Figure 6.21. The latter is varied from 71 mΩ to 91 mΩ in steps of 1 mΩ. In addition, D is varied from 0.62 to 0.65. It was found that the lowest R_L showed the highest V_0. As seen from Figure 6.21, 71 mΩ showed the largest V_0. This is due to the fact that D α V_0 as shown in Eqn. (33).

The variation of V_0 as a function of R_{esr} is seen in Figure 6.22. R_{esr} varied from 4 mΩ to 6 mΩ in steps of 1 mΩ. D is varied from 0.62 to 0.65. It was found that the R_{esr} showed no effect on V_0. As seen from Figure 6.22, all the values of R_{esr} converged to yield V_0 of around 14.65 V.

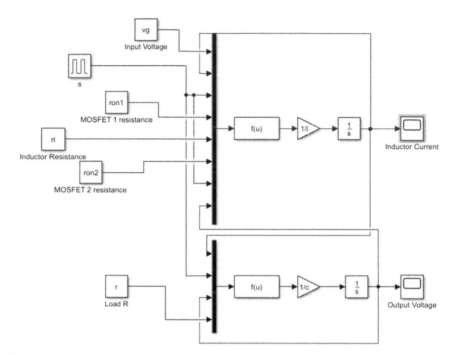

FIGURE 6.19   Mathematical modeling of synchronous boost converter.

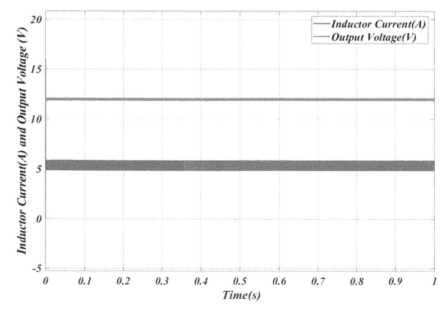

FIGURE 6.20 i_L and V_0 vs. time.

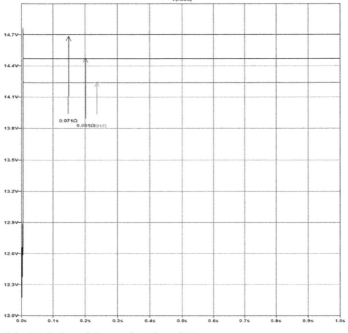

FIGURE 6.21 Variation of R_L as a function of V_0.

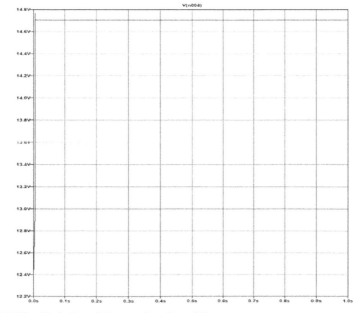

FIGURE 6.22 Variation of R_{esr} as a function of V_0.

The variation of V_0 as a function of V_c is seen in Figure 6.23. V_c proportional to D varied from 0.2 to 0.9 in steps of 0.01. In addition, R_L is varied from 71 mΩ to 91 mΩ. As seen from Figure 6.23, lower R_L provided higher V_0.

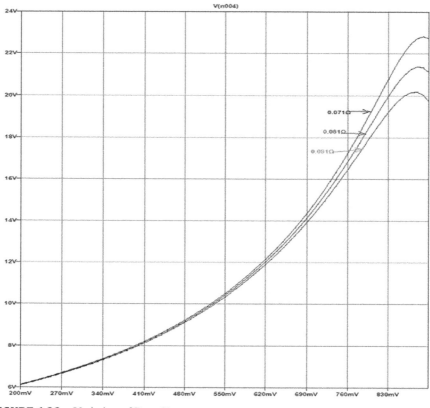

FIGURE 6.23   Variation of D on V_0.

Figure 6.24 shows the bode plot of G_{vd}. It can be observed that the system has a negative gain margin (GM) and phase margin. G_{vd} initially showed a DC offset of around 60 dB. Later, due to the effect of two poles, the system showed a – 40 dB/dec fall at high frequency. Root locus for G_{vd} is shown in Figure 6.25. It possesses a right-hand side zero, thus making an unstable system.

The variation of load resistance, R as a function of V_0 is shown in Figure 6.26. R is varied from 6 Ω to 8 Ω in steps of 2 Ω. It was observed that higher value of resistance results in larger value of resonance.

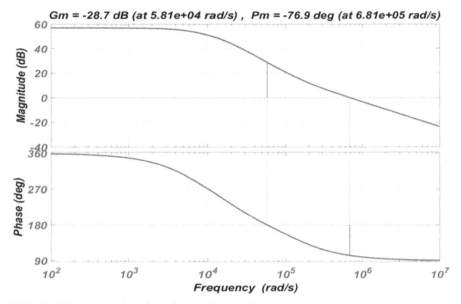

FIGURE 6.24 Bode plot of G_{vd} for synchronous boost converter.

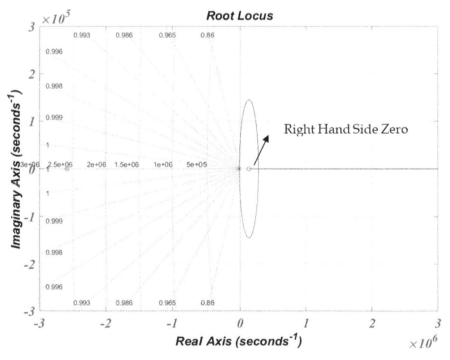

FIGURE 6.25 Root locus of G_{vd}

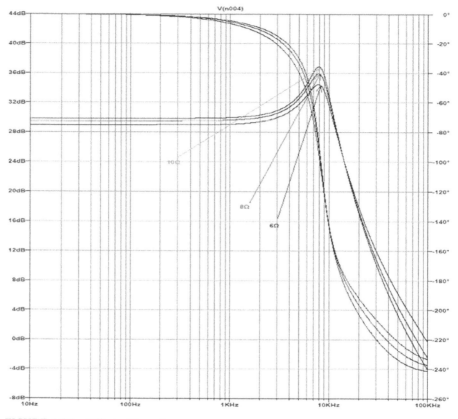

FIGURE 6.26 Effect of R on G_{vd}.

Figure 6.27 shows the physical origin of right half plane zero in the case of synchronous boost converter. D was varied from 0.628 to 0.828 in steps of 0.1, and i_L, and V_0 were plotted.

It is observed that V_0 decreased when the duty cycle was changed due to the load discharging the capacitor. As i_L builds up to a larger value (at steady state, $i_L = (\dfrac{V_0}{R(1-D)^2})$, V_0 regains its value.

The Right Half Plane (RHP) zero causes significant issues during the feedback design. The general form of a right half plane zero is shown in Eqn. (34).

$$G(s) = 1 - \frac{s}{\omega_0} \qquad\qquad (26)$$

At the low frequency, the transfer function for Eqn. (34) would be one or zero dB. However, at high frequency ($\omega_0 >> \omega$), there would be a phase reversal due to the negative sign.

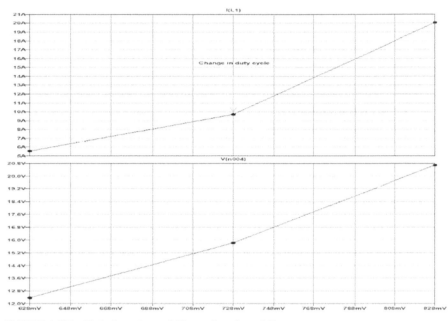

FIGURE 6.27 Variation of i_L and V_0 as a function of V_c.

6.8 CONCLUSION

In the present work, synchronous buck and synchronous boost converters operating in CCM considering the non-idealities are modeled using MATLAB/Simulink. The shift in the output voltage is observed when the ESR values for the inductor and capacitor were varied. The mathematical models for the converters are derived using governing equations for the ideal and non-ideal cases. It is observed that the practical synchronous buck converter showed high stability. The practical synchronous boost converter showed instability due to the RHP zero in its transfer function due to which phase reversal occurred at high frequencies.

The authors declare no conflict of interest and no data/information were taken from Bosch Global Software Technologies (BGSW).

KEYWORDS

- **CCM**
- **duty cycle**
- **instability**
- **mathematical modeling**
- **MATLAB/Simulink**
- **non-ideality**
- **small signal modeling**
- **synchronous converters**

REFERENCES

1. Hart, D. W., (2011). *Power Electronics*. Tata McGraw-Hill Education.
2. Erickson, R. W., & Dragan, M., (2007). *Fundamentals of Power Electronics*. Springer Science & Business Media.
3. Surya, S., & Arjun, M. N., (2021). Mathematical modeling of power electronic converters. *SN Computer Science, 2*(4), 1–9.
4. Hinov, N. L., (2018). Mathematical modeling of transformerless DC-DC converters. In: *2018 IEEE XXVII International Scientific Conference Electronics-ET*. IEEE.
5. Surya, S., & Srividya, R., (2021). Isolated converters as LED drivers. *Cognitive Informatics and Soft Computing* (pp. 167–179). Springer, Singapore.
6. Arjun, M., & Vineeth, P., (2015). Steady state and averaged state space modelling of non-ideal boost converter. *International Journal of Power Electronics, 7*(1, 2), 109–133.
7. Zhu, J. Y., (2005). Interpreting small signal behavior of the synchronous buck converter at light load. *IEEE Power Electronics Letters, 3*(4), 144–147.
8. Alonge, F., et al., (2007). Nonlinear modeling of DC/DC converters using the Hammerstein's approach. *IEEE Transactions on Power Electronics, 22*(4), 1210–1221.
9. Corti, F., et al., (2020). Computationally efficient modeling of DC-DC converters for PV applications. *Energies, 13*(19), 5100.

CHAPTER 7

Determination of Open Loop Responses of Switched DC–DC Converters Using Various Modeling Techniques

SUMUKH SURYA

Senior Engineer, Bosch Global Software Technologies (BGSW), Bangalore, Karnataka, India

ABSTRACT

In this chapter, transfer functions for open loop ideal second order converters like Buck, Boost and Buck – Boost converters operating in Continuous Conduction Model (CCM) are derived using different methods viz., (a) Small signal model and (b) State Space Averaging. Using State space averaging technique, transfer functions for duty cycle (G_{vd}) to output voltage and output to input voltage (G_{vg}) were derived. Circuit averaging technique provided the frequency response of G_{vd}. MATLAB/Simulink were used to study the open loop behavior using the principles of volt-sec and amp-sec balance equations. It was observed that the boost and buck-boost converters showed a right hand side zero, leading to instability in the open loop configuration. However, the buck converter showed stability with two poles on the left hand side of the s-plane.

7.1 INTRODUCTION

Switch mode power supplies (SMPS) have become fashionable due to the advent of Electric Vehicles (EVs). SMPS are considered superior to voltage regulators in terms of noise-free output, high efficiency, and less weight.

The Internet of Energy: A Pragmatic Approach Towards Sustainable Development. Sheila Mahapatra, Mohan Krishna S., B. Chandra Sekhar, & Saurav Raj (Eds.)

Some of the major applications of SMPS are (a) LED driver; and (b) battery charging and discharging. The main components of a SMPS are: (a) switches like MOSFET's and diodes; (b) inductor (L); (c) capacitor (C); and (d) a resistor (R). L and C behave as a low-pass filter, filtering out current and voltage harmonics which are caused due to fast switching.

It is important to maintain constant load voltages and currents in terms of critical loads. Hence, controlling the output voltage, duty cycle, and the inductor current irrespective of the Vg changes is extremely important.

In Ref. [1], a small signal model for an ideal buck-boost converter operating in Continuous Conduction Mode (CCM) was derived using a canonical model. The concept of State Space Averaging (SSA) was discussed. Circuit averaging (CA) for second-order converters operating in CCM and DCM (Discontinuous Conduction Mode) was shown. However, the CA for higher-order converters, including all the non-idealities, was not proposed.

In Ref. [2], the DC-DC converter is considered as a black box and is modeled using generalized Power Conservative POPI networks. It is one of the most effective methods to model the first-order power converter properties.

DC–DC converters and their behavior have been studied for many years. A large signal-averaged model for converters was proposed in Ref. [3]. An approach which is generalized is derived independent of the operation (CCM/DCM). The modeling and experimental results were presented for buck, boost, and SEPIC converters, $\mu DCM > \mu CCM$

In Ref. [4], a generalized procedure involving input voltage, output voltage, and duty cycle operating in the DCM for a buck-boost converter modeling was presented. In reality, increased conduction losses are observed due to equivalent series resistance (ESR).

In Ref. [5], a switch and an average model were modeled and analyzed using MATLAB/Simulink and validated using experiments. The transfer functions Gvd (constant output voltage) and Gvg (constant output voltage) were derived. Lower input current harmonics for SEPIC was the major conclusion drawn.

In Ref. [6], DCM operation for a SEPIC was demonstrated using State Space averaging technique. For frequencies below 10 kHz, a strong match was observed. For higher frequencies, differences were observed in LTspice when compared with MATLAB software. It was concluded that the deviation was due to approximation.

In Ref. [7], for power factor correction fourth order converters operating under DCM used. A single-phase rectifier was used to feed the converter. In

Ref. [8], an LTSpice-based averaged model of the converters operating in CCM was designed using a block named CCM1.

In Ref. [9], a generalized signal flow graph model of the fourth-order DC–DC converter topology is proposed. This technique is useful for generating unified models. Using these models, the steady-state models were derived and verified experimentally.

In Ref. [10], the CA for the Cuk converter was carried out using a Saber circuit simulator, and Gvd and Gvg were found out theoretically and verified using simulation. Gvd showed complex pole conjugates, and Gvg showed a Right Half Plane (RHP) zero. This chapter did not account for the diode drop (Vd) and Dynamic resistance (Rd).

In Ref. [8], CA using LTSpice for basic converters during ideal conditions was carried out. However, the simulation and analysis for non-ideal higher-order converters was not carried out. In Ref. [12], the modeling of switching DC-DC converters was shown using state space modeling, including parasitic values considering switching and conduction losses.

In the present work, the open loop behavior of second-order converters like Buck, Boost, and Buck Boost converters are studied in CCM operation using MATLAB/Simulink. Large signal and small signal models for the converters were derived using the 'SSA Technique.' Later, values of R, L, C, and duty cycle (D) were designed based on the switching frequency (fs). These values were used to study the open-loop behavior of the ideal converters using Root Locus and Bode techniques. Most of the papers concentrate on a specific control tool for the analysis of transfer function. In this work, three different control tools, namely, small signal model, SSA, and Generalized CA technique, are applied on the same converter operating in CCM. The software tools used are MATLAB and LTspice. It is shown that the bode plots of Gvd obtained for the converters using two different software tools closely match in terms of low-frequency gain, gain, and phase margins. Hence, extensive software validation is performed, which plays a vital role before the hardware development.

7.2 STATE SPACE MODELING

7.2.1 BUCK CONVERTER

Figures 7.1–7.3 shows an ideal buck converter operating in CCM with no lossy elements.

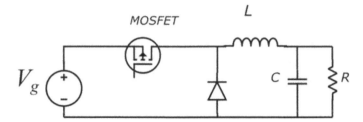

FIGURE 7.1   Schematic of the converter.

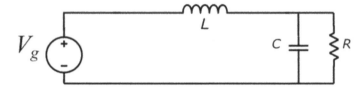

FIGURE 7.2   Switch closed.

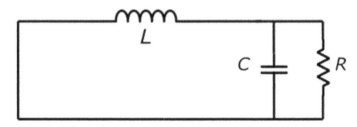

FIGURE 7.3   Switch opened.

State space averaging technique was used to find the large signal and small signal models for the converters.

The basic form of the state space equations is shown in Eqns. (1) and (2)

$$\overset{\circ}{X} = AX + BU \tag{1}$$

$$Y = CX + EU \tag{2}$$

When the switch is closed,

$$V_g - V_0 = V_L \tag{3}$$

$$i_c = i_L - \frac{V_0}{R} \tag{4}$$

When the switch is opened,

$$V_L = -V_0 \tag{5}$$

$$i_c = i_L - \frac{V_0}{R} \tag{6}$$

Reducing Eqns. (3), (4) and (5), (6) in the form of Eqns. (1) and (2)

Assuming $u = V_g$, $X = \begin{bmatrix} di_L / dt\ dV_0 / dt \end{bmatrix}^T$ and $Y = V_0$

$$A_1 = \begin{bmatrix} 0 & \dfrac{-1}{L} \\ \dfrac{1}{C} & \dfrac{1}{RC} \end{bmatrix} \tag{7}$$

$$B_1 = \begin{bmatrix} \dfrac{1}{L} \\ 0 \end{bmatrix} \tag{8}$$

$$A_2 = \begin{bmatrix} 0 & \dfrac{-1}{L} \\ \dfrac{1}{C} & \dfrac{1}{RC} \end{bmatrix} \tag{9}$$

$$B_2 = \begin{bmatrix} 0 \\ 0 \end{bmatrix} \tag{10}$$

$$C_1 = \begin{bmatrix} 0 & 1 \end{bmatrix} \tag{11}$$

$$C_2 = \begin{bmatrix} 0 & 1 \end{bmatrix} \tag{12}$$

Combining Eqns. (7), (8), ..., (12)

$$\begin{bmatrix} \dfrac{Ldi_L}{dt} \\ \dfrac{CdV_0}{dt} \end{bmatrix} = \begin{bmatrix} 0 & \dfrac{-1}{L} \\ \dfrac{1}{C} & \dfrac{-1}{RC} \end{bmatrix} \begin{bmatrix} i_L \\ V_0 \end{bmatrix} + \begin{bmatrix} \dfrac{1}{L} \\ 0 \end{bmatrix} \begin{bmatrix} V_g & 0 \end{bmatrix} \tag{13}$$

$$\begin{bmatrix} \dfrac{Ldi_L}{dt} \\ \dfrac{CdV_0}{dt} \end{bmatrix} = \begin{bmatrix} 0 & \dfrac{-1}{L} \\ \dfrac{1}{C} & \dfrac{-1}{RC} \end{bmatrix} \begin{bmatrix} i_L \\ V_0 \end{bmatrix} + \begin{bmatrix} 0 \\ 0 \end{bmatrix} \begin{bmatrix} V_g & 0 \end{bmatrix} \tag{14}$$

To obtain the large signal model,

$$A = A_1 D + A_2 (1 - D) \tag{15}$$

$$B = B_1 D + B_2 (1 - D) \tag{16}$$

$$C = C_1 D + C_2 (1 - D) \tag{17}$$

$$E = E_1 D + E_2 (1 - D) \tag{18}$$

$$A = \begin{bmatrix} 0 & \dfrac{-1}{L} \\ \dfrac{1}{C} & \dfrac{-1}{RC} \end{bmatrix}, \; B = \begin{bmatrix} \dfrac{D}{L} \\ 0 \end{bmatrix}$$

$$C = \begin{bmatrix} 0 & 1 \end{bmatrix} \text{ and } E = \begin{bmatrix} 0 \end{bmatrix}$$

Since, the term $\hat{X} = 0$ under steady state conditions,

$$X = -A^{-1}BU \tag{19}$$

$$\begin{bmatrix} i_L \\ V_0 \end{bmatrix} = -\left(\dfrac{1}{\Delta}\right) \begin{bmatrix} \dfrac{-1}{RC} & \dfrac{1}{L} \\ \dfrac{-1}{C} & \dfrac{-1}{RC} \end{bmatrix} \begin{bmatrix} \dfrac{D}{L} \\ 0 \end{bmatrix} \begin{bmatrix} V_g & 0 \end{bmatrix}$$

Hence, upon simplification

$$\frac{i_L}{V_g} = \frac{D}{R} \tag{20}$$

$$\frac{V_0}{V_g} = D \tag{21}$$

where;

$$\Delta = \frac{1}{LC} \tag{22}$$

Eqns. (22) and (23) describe the small signal model for the converter.

$$\hat{X}(s) = [sI - A]^{-1}\{B\hat{U}(s) + \hat{d}[[A_1 - A_2]X + [B_1 - B_2]U]\} \tag{23}$$

To find $\dfrac{\hat{X}}{\hat{d}}$, $\hat{U}(s)$ should be made zero, as per superposition theorem.

$$A_1 - A_2 = [0] \;\&\; B_1 - B_2 = \begin{bmatrix} \dfrac{1}{L} \\ 0 \end{bmatrix}$$

$$\frac{\hat{X}}{\hat{d}} = \left(\frac{1}{\Delta}\right) \begin{bmatrix} (\dfrac{Vg}{L}) * (s + \dfrac{1}{RC}) \\ \dfrac{V_g}{LC} \end{bmatrix}$$

where;

$$\Delta = s^2 + \frac{s}{RC} + \frac{1}{LC}$$ (24)

$$\frac{\hat{V}_0}{\hat{d}} = \frac{1}{\Delta} * (\frac{V_g}{LC})$$ (25)

$$\frac{\hat{i}_L}{\hat{d}} = \frac{(\frac{V_g}{L}) * (s + \frac{1}{RC})}{\Delta}$$ (26)

7.2.2 BOOST CONVERTER

Figure 7.4 depicts a schematic of boost converter operating in CCM.

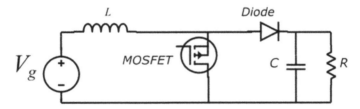

FIGURE 7.4 Schematic of a circuit.

When the switch is closed (Figure 7.5):

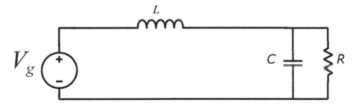

FIGURE 7.5 Switch closed.

$$V_g = V_L$$ (27)

$$i_c = -\frac{V_0}{R}$$ (28)

When the switch is opened (Figure 7.6).

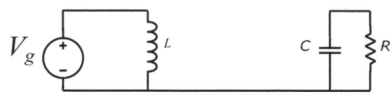

FIGURE 7.6   Switch opened.

$$V_L = V_g - V_0 \tag{29}$$

$$i_c = i_L - \frac{V_0}{R} \tag{30}$$

Assuming $u = V_g$, $X = [di_L/dt \, dV_0/dt]^T$ and $Y = V_0$ & reducing it to the form shown in Eqns. (1) and (2),

$$A_1 = \begin{bmatrix} 0 & 0 \\ 0 & \dfrac{-1}{RC} \end{bmatrix} \tag{31}$$

$$B_1 = \begin{bmatrix} \dfrac{1}{L} \\ 0 \end{bmatrix} \tag{32}$$

$$A_2 = \begin{bmatrix} 0 & \dfrac{-1}{L} \\ \dfrac{1}{C} & \dfrac{-1}{RC} \end{bmatrix} \tag{33}$$

$$B_2 = \begin{bmatrix} \dfrac{1}{L} \\ 0 \end{bmatrix} \tag{34}$$

$$C_1 = \begin{bmatrix} 0 & 1 \end{bmatrix} \tag{35}$$

$$C_2 = \begin{bmatrix} 0 & 1 \end{bmatrix} \tag{36}$$

To obtain the large signal model, Eqns. (15)–(18) are used.

$$A = \begin{bmatrix} 0 & \dfrac{-(1-D)}{L} \\ \dfrac{1-D}{C} & \dfrac{-1}{RC} \end{bmatrix}$$

$$B = \begin{bmatrix} \frac{1}{L} \\ 0 \end{bmatrix}$$

$$C = \begin{bmatrix} 0 & 1 \end{bmatrix} \&$$

$$E = [0]$$

Since, the term $\hat{X} = 0$ under steady state conditions, $X = -A^{-1} BU$

$$\begin{bmatrix} i_L \\ V_0 \end{bmatrix} = \left(\frac{-1}{\Delta} \right) \begin{bmatrix} \dfrac{-1}{RC} & \dfrac{-D'}{L} \\ \dfrac{-D'}{C} & 0 \end{bmatrix} \begin{bmatrix} V_g & 0 \end{bmatrix}$$

Upon simplification,

$$\Delta = \frac{(1-D)^2}{LC} \tag{37}$$

$$\frac{i_L}{V_g} = \frac{1}{RLC\Delta} \tag{38}$$

$$\frac{V_0}{V_g} = \frac{1}{1-D} \tag{39}$$

To find the small signal model for the converter, Eqn. (23) is considered.

$$\begin{bmatrix} i_L \\ V_0 \end{bmatrix} = \left(\frac{-1}{\Delta} \right) \begin{bmatrix} s + \dfrac{1}{RC} & \dfrac{-D'}{L} \\ \dfrac{-D'}{C} & \dfrac{-1}{RC} \end{bmatrix} \begin{bmatrix} V_g & 0 \end{bmatrix} \tag{40}$$

$$\frac{\hat{i}_L}{\hat{d}} = \frac{((s + (1/RC))\dfrac{V_0}{L}) - (\dfrac{i_L}{C}) * (\dfrac{1-D}{C})}{\Delta} \tag{41}$$

$$\frac{\hat{V}_0}{\hat{d}} = \frac{(\dfrac{-V_0}{L})(\dfrac{1-D}{C}) - \dfrac{si_L}{C}}{\Delta} \tag{42}$$

where;

$$i_L = \frac{V_0}{R(1-D)} \tag{43}$$

$$\Delta = s^2 + \frac{s}{RC} + \frac{(1-D)^2}{LC} \tag{44}$$

7.2.3 BUCK BOOST CONVERTER

A Buck Boost converter is a type of SMPS obtained by combining boost and buck converter. A schematic of Buck-Boost converter is shown in Figure 7.7.

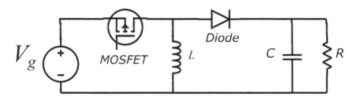

FIGURE 7.7   Schematic of buck boost converter.

When the switch is closed (Figure 7.8).

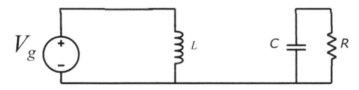

FIGURE 7.8   Switch closed.

$$V_L = V_g - V_0$$

$$i_c = i_L - \frac{V_0}{R}$$

When the switch is opened (Figure 7.9).

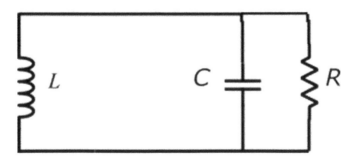

FIGURE 7.9   Switch opened.

$$V_L - V_0$$

$$i_c = i_L - \frac{V_0}{R}$$

Under steady state conditions, $\hat{X} = 0$

$$X = -A^{-1}BU$$

$$A_1 = \begin{bmatrix} 0 & 0 \\ 0 & \dfrac{-1}{RC} \end{bmatrix} \tag{45}$$

$$A_2 = \begin{bmatrix} 0 & \dfrac{1}{L} \\ \dfrac{1}{C} & \dfrac{-1}{RC} \end{bmatrix} \tag{46}$$

$$B_1 = \begin{bmatrix} \dfrac{1}{L} \\ 0 \end{bmatrix} \tag{47}$$

$$B_2 \begin{bmatrix} 0 \\ 0 \end{bmatrix} \tag{48}$$

$$A = \begin{bmatrix} 0 & \dfrac{1-D}{L} \\ \dfrac{1-D}{C} & \dfrac{-1}{RC} \end{bmatrix} \tag{49}$$

$$B = \begin{bmatrix} \dfrac{D}{L} \\ 0 \end{bmatrix} \tag{50}$$

$$C = \begin{bmatrix} 0 & 1 \end{bmatrix} \tag{51}$$

$$\begin{bmatrix} i_L \\ V_0 \end{bmatrix} = \left(\frac{LC}{(1-D)^2} \right) \begin{bmatrix} \dfrac{-DV_g}{RLC} \\ \dfrac{D(1-D)V_g}{L} \end{bmatrix} \tag{52}$$

The small signal can be obtained using Eqn. (23) as:

$$\frac{\hat{V}_0}{\hat{d}} = \frac{\left(\dfrac{V_g - V_0}{L} \right) * \left(\dfrac{D-1}{C} + \dfrac{sV_0}{RD'C} \right)}{\Delta} \tag{53}$$

where;

$$\Delta = s^2 + \frac{s}{RC} + \frac{(1-D)^2}{LC}$$

7.3   SMALL SIGNAL MODELING

In this approach, the steps shown below are followed:

1. Governing equations for inductor and capacitor are determined
2. The quantities are perturbed viz., V_g, V_0, D and i_L
3. The terms are linearized and the product of terms having double hat quantities is eliminated.

7.3.1   BUCK CONVERTER

When the switch is closed,

$$V_L = V_g - V_0$$

$$i_C = i_L - \frac{V_0}{R}$$

When the switch is opened,

$$V_L = -V_0$$

$$i_C = i_L - \frac{V_0}{R}$$

➢   **Step 1:** Volt-sec balance equation for inductor and Amp-second balance for capacitor.

$$V_L = (V_g - V_0)D - (1-D)V_0 \tag{54}$$

$$i_C = (i_L - V_0/R)D + (i_L - V_0/D)(1-D) \tag{55}$$

➢   **Step 2:** Perturbation:

$$V_L + \hat{V}_L = (V_g + \hat{V}_g - V_0 - s\hat{V}_0)(D + \hat{d}) - (V_0 + \hat{V}_0)((1-D) - \hat{d}) \tag{56}$$

Simplifying (*see* Figure 7.10),

$$\hat{V}_L = V_g\,\hat{d} + D\hat{V}_g - \hat{V}_0 \tag{57}$$

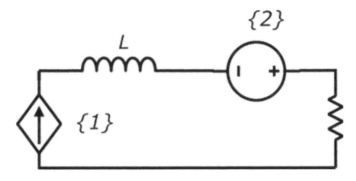

FIGURE 7.10 Equivalent circuit of Eqn. (57).

where; {1} is $\hat{V}_g D$, {2} is $V_g \hat{d}$ and the current in the circuit is $\hat{i}_L$. Similarly (*see* Figure 7.11),

$$I_c + \hat{i}_c = \frac{(I_L + \hat{i}_L) - (V_0 + \hat{V}_0)}{R} \tag{58}$$

$$\hat{i}_c = \hat{i}_L - \frac{\hat{V}_0}{R} \tag{59}$$

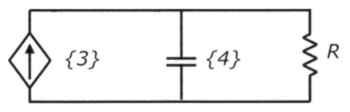

FIGURE 7.11 Equivalent circuit of Eqn. (59).

where; {3} is $\hat{i}_c$ and {4} is $\hat{i}_L$.

➤ **Step 3:** Linearizing:

$$\hat{V}_L = V_g \hat{d} + D\hat{V}_g - \hat{V}_0 \tag{60}$$

$$\hat{i}_c = \hat{i}_L - \frac{\hat{V}_0}{R} \tag{61}$$

The supply current i_g can be expressed as:

$$i_g = Di_L \tag{62}$$

Perturbing the above equation:

$$I_g + \hat{i}_g = (I_L + \hat{i}_L)(D + \hat{d}) \tag{63}$$

Linearizing (*see* Figure 7.12),

$$\hat{i}_g = i_L \hat{d} + \hat{i}_L D \tag{64}$$

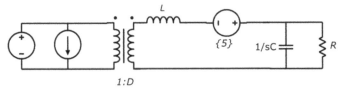

FIGURE 7.12   Equivalent circuit for Eqns. (57), (59), and (64).

where; {1} is $\hat{V}_g$ {3} is $I_L \hat{d}$ {4} $\hat{i}_L d$ {X} is $\hat{V}_g d$ and {b} is $\hat{i}_L$.

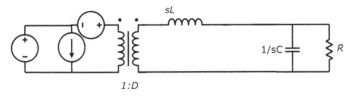

FIGURE 7.13   Combining the three circuits.

where; {5} is $V_g + \hat{d}$.

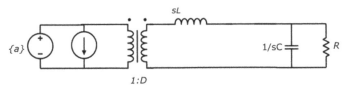

FIGURE 7.14   Canonical model.

FIGURE 7.15   Simplifying the network.

where; {a} is $\hat{V}_g$ (*see* Figures 7.14 and 7.15).

On simplifying the circuit,

$$\frac{\hat{V}_0}{\hat{V}_g} = \frac{DR}{sL*(1+RsC)+R} \tag{65}$$

7.3.2 BOOST CONVERTER

A similar analysis considering the ESR of the inductor is carried out. This ESR tends to decrease V_0 and causes increase in D to meet the desired V_0. A schematic of a boost converter with inductor ESR is shown in Figure 7.16.

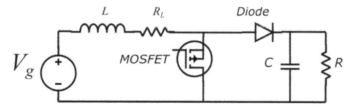

FIGURE 7.16 Equivalent circuit.

From the circuit shown in Figure 7.17, when the switch is closed.

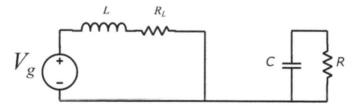

FIGURE 7.17 Switch closed.

Closed condition (*see* Figure 7.18),

$$V_L = V_g - I_L R_L \tag{66}$$

$$i_c = -\frac{V_0}{R} \tag{67}$$

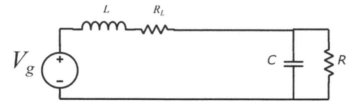

FIGURE 7.18   Switch opened.

Open condition,

$$V_L = V_g - I_L R_L - V_0 \tag{68}$$

$$i_c = i_L - \frac{V_0}{R} \tag{69}$$

Applying volt sec balance and perturbing,

$$V_L = V_g - I_L R_L - V_0 (1 - D) \tag{70}$$

$$V_L + \hat{V}_L = (V_g + \hat{V}_g) - (I_L + \hat{i}_L)R_L - (V_0 + \hat{V}_0)(D' - \hat{d})$$

$$V_L + \hat{V}_L = \hat{V}_g - \hat{i}_L R_L + V_0 \hat{d} - \hat{V}_0 D' \tag{71}$$

Upon linearizing,

$$\hat{V}_L = \hat{V}_g - \hat{i}_L R_L + V_0 \hat{d} - \hat{V}_0 D' \tag{72}$$

The equivalent circuit is shown in Figure 7.19.

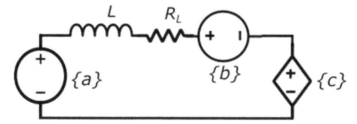

FIGURE 7.19   Equivalent circuit for Eqn. (71).

where; {a} is $\hat{V}_g$ {b} is $V_0 \hat{d}$ and {c} is $\hat{V}_0 D'$.

Applying amp sec balance and perturbing,

$$i_c = \frac{-V_0}{R} D + (1 - D)(i_L - \frac{V_0}{R}) \tag{73}$$

$$i_c = i_L(1-D) - \frac{V_0}{R} \tag{74}$$

Linearizing,

$$\hat{i}_c = -I_L\hat{d} + \hat{i}_L D' - \frac{\hat{V}_0}{R} \tag{75}$$

The equivalent circuit is shown in Figure 7.20.

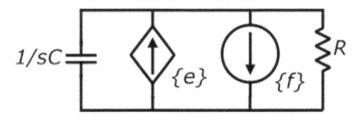

FIGURE 7.20 Equivalent circuit for Eqn. (71).

where; $\{e\} - I_L\hat{d}\{f\}\ i_L D'$, $\{g\}\ \hat{i}_c$ and current in R is $\frac{\hat{V}_0}{R}$

Combining the equations (Figure 7.21).

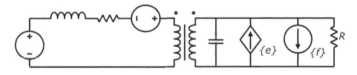

FIGURE 7.21 Combining the circuits shown in Figures 7.19 and 7.20.

Upon simplification,

$$\hat{V}_0 / ((V_0\hat{d}/D')((-sL - R_L)/(RD'^2) + 1) = 1/(1 + (2)(3)) \tag{76}$$

where; (1) RD'^2, (2) $(R + sL)$ and (3) is $(1 + RsC)$

The equation for $\frac{\hat{V}_0}{\hat{V}_g}$ can be derived from the above expression. The design of classical controllers for achieving constant voltage (CV)/current is essential in DC-DC converters. Design of such converters for a Pressure Regulating Valve (PRV) is shown in Ref. [13].

7.4   CIRCUIT AVERAGING (CA) TECHNIQUE

It is one of the simple techniques used to analyze the stability of the converter using the switch voltages currents. This method is a generalized approach and can be applies to any converter having two switches (Diode and MOSFET), operating in CCM or DCM mode. The following steps are used to obtain the switch voltages and currents. It was shown for a Cuk converter that G_{vd} obtained from small signal model and CA matched [11]. The simulation can be easily performed on LTSpice simulation tool which is open-source software.

7.5   CONVERTER SPECIFICATIONS (Tables 7.1–7.3)

TABLE 7.1   Specifications of Buck Converter

Specification	Value
Input voltage (V_g)	18 V
Output voltage (V_0)	5 V
Resistance (R)	5 Ω
Capacitance (C)	2.5 μF
Inductance (L)	1.8 mH
Duty cycle (D)	0.27

TABLE 7.2   Specifications of Boost Converter

Specification	Value
Input voltage (V_g)	15 V
Output voltage (V_0)	25 V
Resistance (R)	100 Ω
Capacitance (C)	10 μF
Inductance (L)	2 mH
Duty cycle (D)	0.4

7.5   RESULTS OF LTSPICE SIMULATION

For the simulation, CCM1 module under average .lib was used. Figure 7.22 shows the simulation model using CCM1 module as the modeling and simulation is performed for CCM operation.

TABLE 7.3 Specifications of Buck-Boost Converter

Specifications	Value
Input voltage (V_g)	5 V
Output voltage (V_0)	−18 V
Resistance (R)	9 Ω
Capacitance (C)	87 μF
Inductance (L)	212.08 μH
Duty cycle (D)	0.78

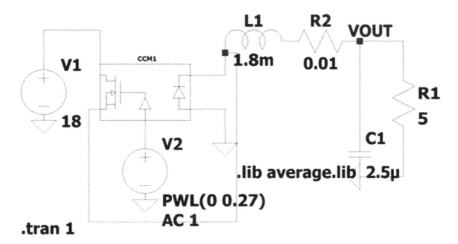

FIGURE 7.22 LTSpice simulation.

Initially, a transient simulation for 1s is analyzed. The PWM is changed from 0.27 to 0.5 from 0 to 3 ms. A plot of output voltage and inductor current is shown in Figure 7.23. Figure 7.24 shows the variation of V_0 under varying D (0.2 to 0.9 in steps of 0.01). Figure 7.25 shows G_{vd} for a buck converter inclusive of Inductor ESR.

Similar analysis for the Boost converter was carried out and the plots are shown in Figure 7.26.

In Figure 7.27, variation of i_L and V_0 for change in the duty cycle is shown. Figure 7.28 shows the variation of i_L and V_0 for change in the duty cycle from 0.2 to 0.9 in steps of 0.01. It is observed that as D increases, i_L and V_0 increase. In Figure 7.29, frequency response of G_{vd} for fixed R is shown. It is to be noted that gain margin (GM) and the phase margin (PM) are negative due to which the system is unstable.

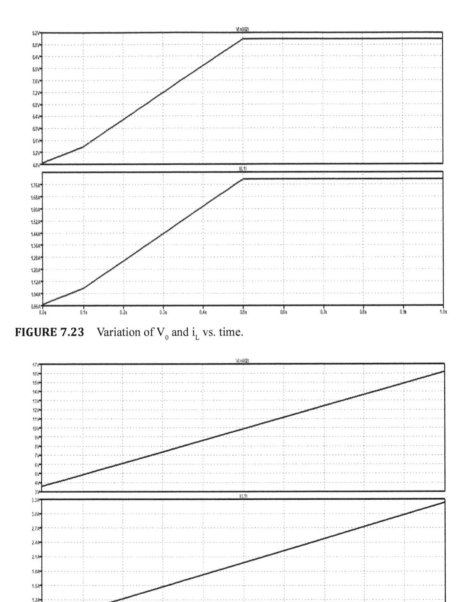

FIGURE 7.23   Variation of V_0 and i_L vs. time.

FIGURE 7.24   Variation of V_0 and i_L vs. time.

In Figure 7.30, bode plot of G_{vd} for varying load is shown. It was observed that as the resistance increases, the value of resonance also increases.

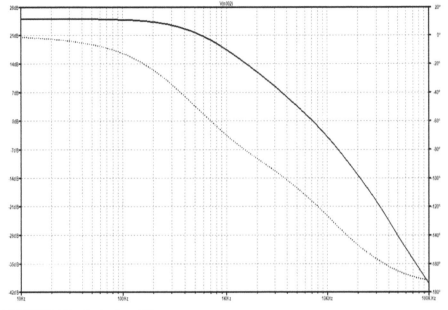

FIGURE 7.25 Bode plot of G_{vd}.

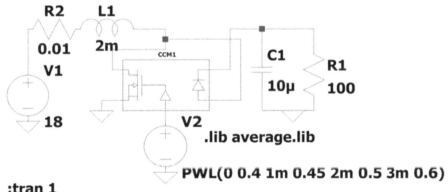

FIGURE 7.26 Circuit averaging for a boost converter.

7.6 RESULTS OF MATLAB SIMULATION

To analyze the open loop response of the converter, MATLAB was used to obtain the frequency response of $\dfrac{\hat{V}_0}{\hat{V}_g}$

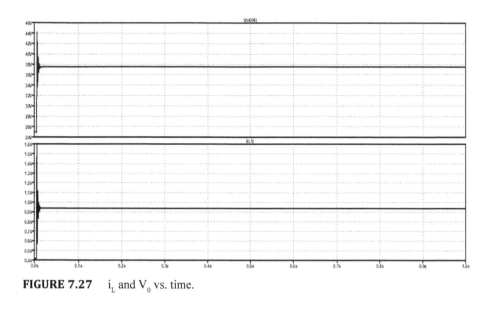

FIGURE 7.27 i_L and V_0 vs. time.

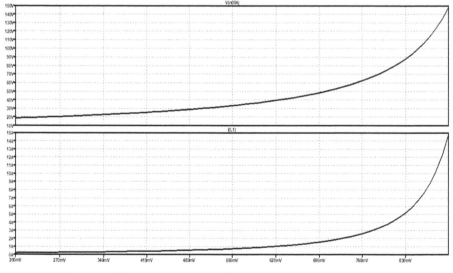

FIGURE 7.28 i_L and V_0 vs. time for varying D.

7.6.1   *BUCK CONVERTER*

$$\frac{5000}{(1.25*10^5\ s^2 + s + 2778)} \tag{77}$$

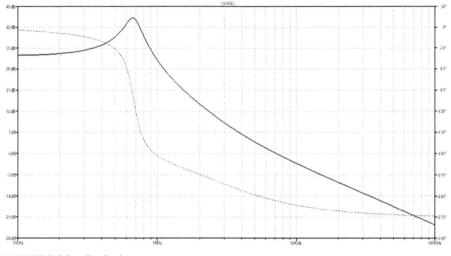

FIGURE 7.29 G_{vd} for boost converter.

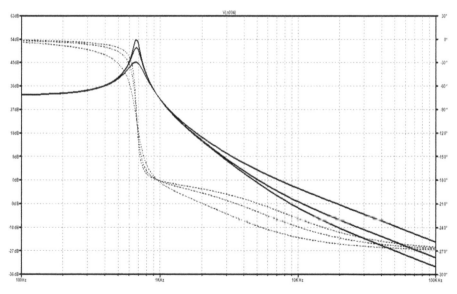

FIGURE 7.30 Bode plot for varying R.

Figures 7.31 and 7.32 show root locus and bode plot for a stable system with two poles.

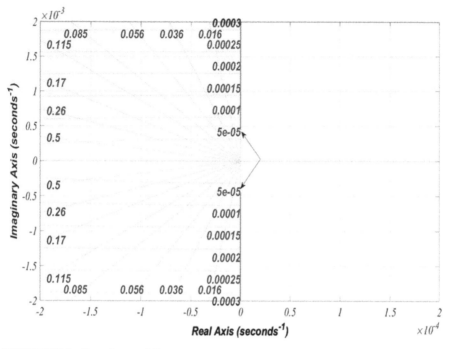

FIGURE 7.31   Root locus of G_{vd}.

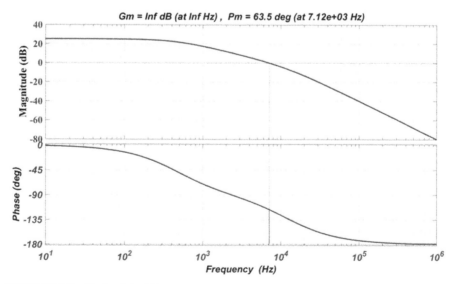

FIGURE 7.32   Bode plot of G_{vd}.

7.6.2 BOOST CONVERTER INCLUSIVE OF INDUCTOR ESR

$$\frac{-41667 * (s - 1.799 * 10^4)}{s^2 + 1005s + 1.8 * 10^7} \tag{78}$$

As observed, the transfer function possesses a Right Half Plane (RHP) zero due to which the system is unstable. The origin of right half plane zero is explained in Ref. [1].

Figures 7.33 and 7.34 show the root locus and bode plot for a two pole and one zero system. The negative GM and PM indicate the instability in the system.

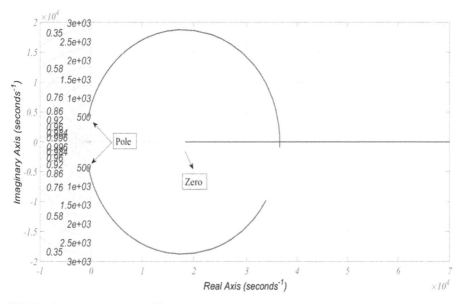

FIGURE 7.33 Root locus of G_{vd}.

7.6.3 BUCK–BOOST CONVERTER INCLUSIVE OF INDUCTOR ESR

$$\frac{-1.0449 * 10^5 (s - 2624)}{(s^2 + 1277s + 2.623 * 10^6)} \tag{79}$$

As observed, the transfer function possesses a RHP zero due to which the system is unstable. Figures 7.35 and 7.36 show the root locus and bode plot for a two pole and one zero system.

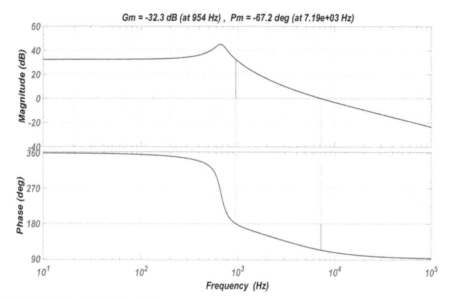

FIGURE 7.34   Bode plot of G_{vd}.

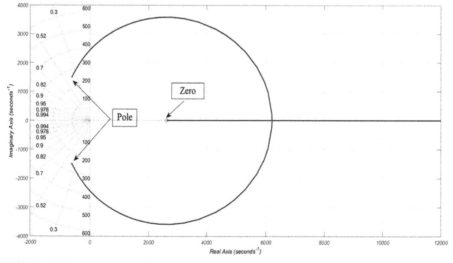

FIGURE 7.35   Root locus of G_{vd}.

7.7   FUTURE SCOPE

Modeling the dynamics of G_{vd} using different control techniques for non-isolated topology is presented in this chapter. Similar analyzes for isolated topologies are recommended. Design of suitable controllers like P, PI, PD and PID to stabilize the open loop system and achieve sufficient Gain and Phase Margins is recommended. Analyzes for average current control, G_{id} (ratio of perturbed duty cycle to inductor current) can be modeled, and its characteristics can be studied for isolated and non-isolated topologies.

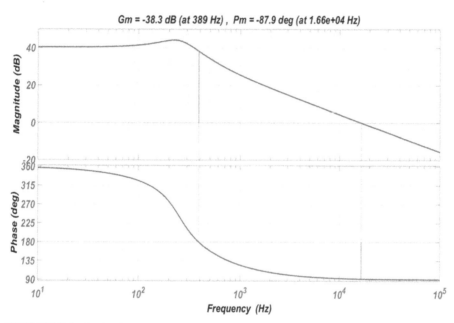

FIGURE 7.36   Bode plot of G_{vd}.

7.8   CONCLUSION

In this chapter, open loop transfer functions for ideal second order converters operating in CCM are derived using different control techniques. Using these techniques, transfer functions like Control to Output (G_{vd}) and output to input voltage (G_{vg}) were derived and analyzed using MATLAB 2018a. A simpler technique like CA provided the frequency response for G_{vd} using built-in libraries in LTSpice software tool. It was observed that the boost and buck-boost converters showed a right-hand side zero, leading to instability in the open loop configuration. However, buck converter showed stability as it had two poles on the LHS of the s-plane. In this work, the open loop transfer functions for the DC-DC converters have been derived using different methods. Amongst various methods, CA is the simplest and provides the frequency response with minimal mathematical analysis. Ideal Buck converter showed high stability whereas the ideal Boost and ideal Buck-Boost converters showed instability due to the presence of RHP zero (Table 7.4).

TABLE 7.4 Nomenclature

SL. No.	Symbol	Parameter
1.	V_g	Input voltage (V)
2.	V_0	Output voltage (V)
3.	i_L	Inductor current (A)
4.	i_c	Capacitor current (A)
5.	R	Resistance (Ω)
6.	L	Inductance (H)
7.	C	Capacitance (F)
8.	D	Duty cycle
9.	R_L	Inductor ESR (Ω)

Note: All *quantities are perturbed quantities.

KEYWORDS

- **buck–boost converter**
- **CCM**
- **circuit averaging**
- **instability**
- **MATLAB/Simulink**
- **small signal modeling**
- **state space modeling**

CONFLICT OF INTEREST

The author declares no conflict of interest. No data or information has been taken from BGSW.

REFERENCES

1. Erickson, R. W., & Dragan, M., (2007). *Fundamentals of Power Electronics*. Springer Science & Business Media.
2. Singer, S., & Robert, W. E., (1992). Canonical modeling of power processing circuits based on the POPI concept. *IEEE Transactions on Power Electronics, 7*(1), 37–43.
3. Canalli, V. M., Cobos, J. A., Oliver, J. A., & Uceda, J., (1996). Behavioral large signal averaged model for DC/DC switching power converters. In: *27ᵗʰ Annual IEEE Power Electronics Specialists Conference* (Vol. 2, No. 5, pp. 1675–1681).
4. Cuk, S., & Middlebrook, R. D., (1979). A general unified approach to modeling switching DC- to-DC converters in discontinuous conduction mode. *Proc. IEEE PESC* (pp. 36–57).
5. Bertoldi, B., et al., (2018). A non-ideal SEPIC DCM modeling for LED lighting applications. In: *2018 IEEE 4ᵗʰ Southern Power Electronics Conference (SPEC)*. IEEE.
6. Eng, V., & Chanin, B., (2009). Modeling of a SEPIC converter operating in discontinuous conduction mode. In: *2009 6ᵗʰ International Conference on Electrical Engineering/ Electronics, Computer, Telecommunications and Information Technology* (Vol. 1). IEEE.
7. Simonetti, D. S. L., Javier, S., & Javier, U., (1997). The discontinuous conduction mode SEPIC and Cuk power factor preregulators: Analysis and design. *IEEE Transactions on Industrial Electronics, 44*(5), 630–637.
8. Chien-Min, L., & Yen-Shin, L., (2007). Averaged switch modeling of dc/dc converters using new switch network. In: *2007 7ᵗʰ International Conference on Power Electronics and Drive Systems*. IEEE.
9. Veerachary, M., (2008). Analysis of fourth-order DC-DC converters: A flow graph approach. *IEEE Transactions on Industrial Electronics* (Vol. 55, No. 1, pp. 133–141).
10. Kathi, L., Ayachit, A., Saini, D. K., Chadha, A., & Kazimierczuk, M. K., (2018). Open-loop small-signal modeling of Cuk DC-DC converter in CCM by circuit-averaging technique. In: *2018 IEEE Texas Power and Energy Conference (TPEC)* (pp. 1–6). IEEE.
11. Surya, S., & Sheldon, W., (2021). Modeling of average current in ideal and non-ideal boost and synchronous boost converters. *Energies, 14*(16), 5158.
12. Middlebrook, R. D., & Cuk, S., (1976). A general unified approach to modeling switching converter stages. *IEEE Power Electronics Specialists Conf.*, 18–34.
13. Surya, S., & Singh, D. B., (2019). Comparative study of P, PI, PD and PID controllers for operation of a pressure regulating valve in a blow-down wind tunnel. In: *2019 IEEE International Conference on Distributed Computing, VLSI, Electrical Circuits and Robotics (DISCOVER)*. IEEE.

CHAPTER 8

Evolution of Hybrid Ultracapacitors in Solar Microgrids

J. PRADEEP KUMAR RAO[1] and H. N. NAGAMANI[2]

[1]Department of Engineering (R&D), Central Power Research Institute, Bangalore, Karnataka, India

[2]Additional Director (Retd.), Central Power Research Institute, Bangalore, Karnataka, India

ABSTRACT

A lead carbon hybrid ultracapacitor (Pb–C HUC) is a hybrid energy storage device comprising a battery-type electrode (PbO2: lead oxide electrode) and an ultracapacitor-type electrode (AC: activated carbon electrode). The chapter discusses the standardization of test protocol for optimizing the performance of 12 V 2500F range Pb–C HUCs for solar microgrid applications. For solar power applications, the charge duration of the energy storage system would be less than 5.5 hours, as the solar power availability is nearly 5.5 hours on a bright sunny day. No standard test protocol is available to date to estimate the charge/discharge protocol of HUCs for solar power applications. A 12 V 2500F HUC is expected to have a charge capacity of 4.167 AH and energy storage capacity of 37.5 Wh. HUCs have been subjected to 12 different protocols of charge/discharge cycles at constant current (CC) (Ich = Idis = 1 A, 2 A, and 3 A) followed by constant voltage (CV) of 13.8 V for tcv charging time durations (tcv = 0 h, 1 h, 2 h, and 3 h) on 150 number of commercially available HUCs. The performance optimization of HUCs is measured in terms of certain key parameters of HUC, namely, Charge input/output, Energy input/output, Charge efficiency, Energy efficiency, and Capacitance offered. It is observed that 96% of the total energy delivered is

The Internet of Energy: A Pragmatic Approach Towards Sustainable Development. Sheila Mahapatra, Mohan Krishna S., B. Chandra Sekhar, & Saurav Raj (Eds.)

© 2024 Apple Academic Press, Inc. Co-published with CRC Press (Taylor & Francis)

due to the capacitance of HUC, and only 4% of energy is lost due to internal resistance. The analysis establishes that 2A-3h charge/discharge protocol is the standard test protocol for charging 12 V/2500F HUCs for solar microgrid applications.

8.1 INTRODUCTION

Grid connected solar photovoltaic power generation is expected to contribute to combating problems associated with global warming [1]. However, direct integration of photovoltaic energy sources (ES) with the main grid has certain practical issues mainly due to intermittent availability of solar power. The development of microgrids with its own energy resources require an energy storage system that can be used to store the electric power generated from solar PV system and then supply it to the main grid as a regulated power using appropriate power controlling units. Typically, the size of the energy storage system required for developing a microgrid ranges between 1 kWh and a few 100 kWh [2–6].

Among the various energy storage systems, electrochemical energy storage systems, like rechargeable batteries, are considered to be the best choice due to their flexibility in size and shape [7, 8]. However, most of the rechargeable batteries have low cycle life, require long periods for charging and slow discharge schedules. For example, lead-acid batteries need constant current (CC) charging at C/10 rate followed by constant voltage (CV) charging for a time duration of 6 h and are discharged at C/5 rate. Lithium batteries can be charged at C/3 to C/5 rate and happen to be expensive and unsafe. These characteristics make them unsuitable for energy storage in microgrids. By contrast, hybrid energy storage devices comprising of a battery type electrode and an ultracapacitor type electrode have higher cycle life and energy density intermediate between batteries and ultracapacitors [9, 10]. Different types of hybrids ultracapacitor systems such as Pb–C, Ni–C, MnO_2–C, etc., are being developed [11–16]. Among them, Pb–C is an attractive system due to its lower cost, abundance, higher recyclability, and simpler manufacturing process.

8.2 LEAD CARBON HYBRID ULTRACAPACITOR

A Pb–C HUC, consists of a lead oxide (PbO_2) cathode and an activated carbon anode with sulfuric acid as an electrolyte. PbO_2 electrode acts as the battery type electrode with the charge/discharge reactions akin to the positive plate in a typical lead-acid battery. The charge and discharge reactions for PbO_2 electrode and its thermodynamic, reversible potential in HUC can be expressed as shown in Eqn. (1).

At cathode:

$$\frac{1}{2}PbSO_4 + H_2O \underset{\text{discharge}}{\overset{\text{charge}}{\rightleftharpoons}} \frac{1}{2}PbO_2 + \frac{1}{2}H_2SO_4 + H^+ + e^- \tag{1}$$

A standard potential of 1.69 V is attained at cathode electrode $\left(E^o_{cathode} = 1.67V\right)$ due to chemical reaction. This potential changes with the concentration of the electrolyte.

Activated carbon anode is an electrical-double-layer-capacitor (EDLC) type electrode. EDLCs are governed by the same physics as parallel-plate electrolytic capacitors. As EDLCs use much thinner dielectric medium and higher surface-area electrodes, EDLCs tend to store relatively larger charge. The total charge stored in the double layer is proportional to the potential of the electrode-electrolyte interface with its capacitance as the proportionality constant, as expressed in Eqn. (2)

$$Q_{anode} \propto C_{anode} V \tag{2}$$

The charge and discharge reactions on the activated carbon anode and its potential can be expressed as in Eqn. (3). The potential at anode is represented by Eqn. (4)

At anode:

$$C_s + H^+ + e^- \underset{\text{discharge}}{\overset{\text{charge}}{\rightleftharpoons}} (C_s^- //H^+) \tag{3}$$

$$E_{anode} - E_{(discharge-anode)} = \frac{Q_{anode}}{C_{anode}} \tag{4}$$

In Eqn. (3), C_v^- represents the carbon atoms at the electrode surface, "//" represents the double layer where charges are accumulated on either side. E_{anode} and $E_{(discharge-anode)}$ refer to the potential of the electrode in its charged and discharged states, respectively. Q_{anode} is the charge on the carbon anode and C_{anode} is its capacitance in sulfuric acid electrolyte. The nature of interaction between C_s and H^+ has not yet been established and remains a subject of further study. However, the double layer behavior of $C_s^-//H^+$ can be inferred from ac-impedance and cyclic voltammetry studies.

The net cell reaction for Pb–C HUC is expressed in Eqn. (5).

$$\frac{1}{2}PbSO_4 + H_2O + C_s \underset{\text{discharge}}{\overset{\text{charge}}{\rightleftharpoons}} \frac{1}{2}PbO_2 + \frac{1}{2}H_2SO_4 + \left(C_s^- //H^+\right) \tag{5}$$

The cell voltage is expressed as Eqn. (6):

$$E_{cell} = E_{cathode} - \frac{Q_{anode}}{C_{anode}}$$ (6)

Accordingly, the cell voltage (E_{cell}) for Pb–C HUC depends on the anode capacitance (C_{anode}) and total charge (Q_{anode}) on the carbon anode in sulfuric acid electrolyte. Typically, the open-circuit voltage (OCV) of the Pb–C cell is about 2 V. When six of Pb–C cells are connected in series, a 12 V Pb–C HUC device is realized.

Pb–C HUCs have peculiar charge and discharge characteristics due to the asymmetric nature of energy storage mechanism associated with the PbO_2 cathode and activated carbon anode. PbO_2 electrode involves faradaic reaction while the carbon electrode involves non-faradaic reactions during their charge and discharge cycles. Hence, the electrochemical characteristics of these two electrodes are different and are governed by different energy storage principles. It is reported that [17] Pb–C hybrid device requires more than one step charging to achieve State of Charge (SoC) of 100%. SoC is estimated from depth of discharge (DoD) of energy storage system as expressed in Eqn. (7)

$$\%SoC = (100 - DoD)\%$$ (7)

Conway et al. reported that the Pb–C asymmetric cell is charged at CC to cut-off voltage of 2.3 V and subsequently the charging current is gradually decreased stepwise to achieve the required SoC [17]. Similarly, Andrew Burke et al. have reported that a PbO_2–C HUC charged in two steps with CC to a cut-off voltage 2.2 V and later at CV at 2.2 V to fully charge the PbO_2–C cell [18]. These studies establish that Pb–C HUCs require more than one step charging.

In Refs. [19, 20], the characteristics of a 12 V/kF range Pb–C HUCs have been reported, which are charged by single step CC to a cut-off voltage of 13.8 V. Although these Pb–C HUCs with substrate-integrated PbO_2 electrodes and activated carbon double layer electrodes can be charged quickly by a single CC step, their capacitance is found to decline during repeated charge and discharge cycles suggesting that these HUCs need two-step charging, first with CC followed by a CV charging.

Performance optimization of 12V 2500F range Pb – C HUCs with standard test procedure is not established for solar microgrid application till date; hence this work is undertaken where HUCs are optimized in terms of charge input/output, Energy input/output, charge efficiency, Energy efficiency and Capacitance offered for various constant current and constant voltage protocols. These parameters are determined from the voltage current (v-i) charge/discharge curves of HUC. Typical v-i characteristics of HUC for one cycle is shown in Figure 8.1.

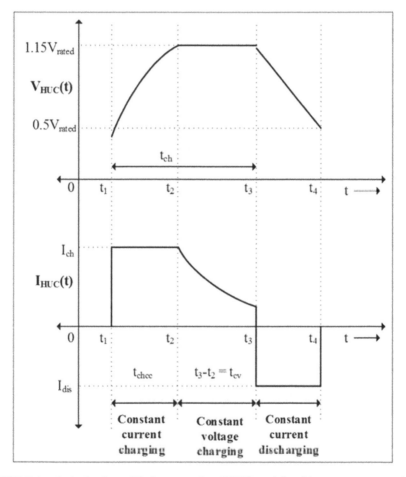

FIGURE 8.1   A single charge/discharge cycle of HUC module with constant current (CC), constant voltage (CV) charging and constant current (C_{dis}) discharging.

From the variables denoted in Figure 8.1, the mathematical representation of key parameters from the v-i curves is explained as follows:

During charging, charge taken by HUC (AH_{input}) in constant current constant voltage (CCCV) mode is mathematically represented by Eqn. (8)

$$AH_{input} = \int_{t=t_1}^{t_3} i_{HUC}(t)dt = I_{ch}*(t_2 - t_1) + \int_{t=t_2}^{t_3} i_{HUC}(t)dt \qquad (8)$$

During discharging, charge delivered by HUC (AH_{output}) in CC mode is mathematically represented by Eqn. (9)

$$AH_{output} = \int_{t=t_3}^{t_4} i_{HUC}(t)dt = I_{dis} * (t_4 - t_3) \tag{9}$$

During charging, input energy (E_{input}) taken by HUC is mathematically expressed by Eqn. (10)

$$E_{input} = \int_{t=t_1}^{t_3} v_{HUC}(t)i_{HUC}(t)dt \tag{10}$$

Similarly, during discharging, energy output (E_{output}) given by HUC is mathematically expressed by Eqn. (11)

$$E_{output} = \int_{t=t_3}^{t_4} v_{HUC}(t)i_{HUC}(t)dt \tag{11}$$

where; i_{HUC} is the HUC current.

HUC current will be constant (I_{ch}) in CC mode and during CV mode it decays exponentially to leakage current.

Energy efficiency and charge efficiencies are calculated from Eqns. (12) and (13).

$$\eta_c = \frac{I_{dis} * (t_4 - t_3)}{I_{ch} * (t_2 - t_1) + \int_{t=t_2}^{t_3} i_{HUC}(t)dt} \tag{12}$$

$$\eta_e = \frac{\int_{t=t_3}^{t_4} v_{HUC}(t)i_{HUC}(t)dt}{\int_{t=t_1}^{t_3} v_{HUC}(t)i_{HUC}(t)dt} \tag{13}$$

Voltage discharge curves are drawn separately in Origin™ software to determine the slope of the linear region of voltage discharge curve using least square linear curve fitting technique. Capacitance C_{HUC} offered by HUC is estimated from the slope of voltage discharge curve of HUC at I_{dis} as calculated by Eqn. (14)

$$C_{HUC} = -\frac{I_{dis}}{(slope)} \tag{14}$$

In order to estimate the capacitance offered by the HUC, the following process is adopted during discharging the HUC. A HUC is generally considered as a healthy/conditioned one, if the slope of the voltage discharge curve is linear and the slope doesn't change with number of charge/discharge cycles. For the present study, the tolerance allowed in slope is 10%.

A fully charged HUC retains an open circuit voltage of 1.15 times the rated voltage and it could be discharged at various load currents. If a healthy HUC is loaded at load current I_{load}, the voltage of HUC linearly drops from V_o (output voltage) at t_1 to V_{uv} (undervoltage) at t_2, as shown in Figure. 8.2.

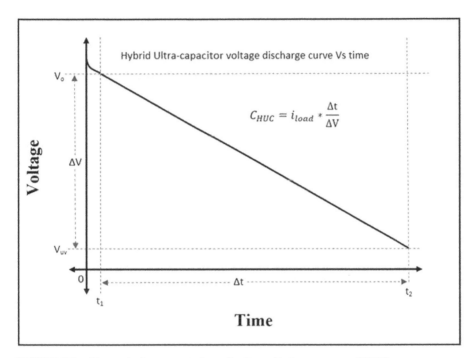

FIGURE 8.2 Theoretical representation of voltage discharge curve of HUC.

Capacitance of HUC is calculated by Eqn. (15)

$$C_{HUC} = I_{load} * \frac{\Delta t}{\Delta v} \tag{15}$$

where; $\Delta v = V_o - V_{uv}$ and $\Delta t = t_2 - t_1$.

Energy output plays an important role in energy storage systems for power applications. In view of this, an analysis has been carried out to estimate the percentage of pure capacitance and internal resistance in HUC in terms of energy output using the voltage discharge curve.

A typical voltage discharge curve of HUC as shown in Figure 8.2 is redrawn as in Figure 8.3. As it is seen from Figure 8.3, the complete voltage discharge curve can be divided into two regions, a nonlinear region from t_0 to t_1 and a linear region from t_1 to t_2.

Applying a CC load onto the HUC, a sudden dip in the voltage is restricted by the internal capacitance of the HUC from its basic physics.

Hence this sudden nonlinear drop of voltage discharge curve from 1.15 times the rated voltage ($1.15*V_{rated}$) to V_2, could be considered as energy consumed by internal resistance of HUC. This part of voltage discharge curve is non-linear represents the ESR of HUCs. Linear voltage drop from V_2 to V_1 represents the energy delivered to CC load from pure capacitance of the HUC.

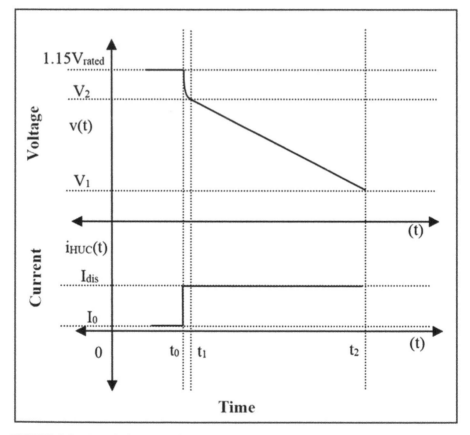

FIGURE 8.3   A typical voltage discharge curve of HUC representing linear and non-linear region.

The total energy output delivered by HUC is represented in Eqn. (11). Applying Eqn. (14) between voltages V_1 and V_2, capacitance of HUC is determined by rewriting Eqn. (15) as Eqn. (16) and the corresponding energy output by Eqn. (17)

$$C_{HUC} = \frac{I_{dis}}{-\left(\dfrac{V_1 - V_2}{t_1 - t_2}\right)} \tag{16}$$

$$E_{linear\_output} = 0.5 * C_{HUC} * \left(V_1^2 - V_2^2\right) \tag{17}$$

The difference between total energy output (under the entire discharge curve) and energy output due to capacitance (under linear region) represents the energy loss due to internal resistance (E_{res}) of HUC, as given by Eqn. (18).

$$E_{res} = E_{output} - E_{linear\_output} \tag{18}$$

Implementing constant current constant voltage charge/discharge profiles on HUC, we can calculate and make a lookup table of performance parameters of HUC as explained in equations (8) – (18). The input parameters of CCCV are chosen such that an optimal performance of charge/ discharging the energy storage unit is seen in 5.5 hours.

The optimal charging protocol of HUCs is estimated through v-i characteristics, for achieving a better tradeoff between capacitance, charge, and energy outputs with better efficiencies.

8.3 EXPERIMENTAL

8.3.1 DESCRIPTION OF COMMERCIAL GRADE 12 V/2500F PB–C HUCS

The development of Pb–C HUCs with substrate-integrated PbO_2 positive electrode and activated-carbon-based double layer capacitor negative electrode is described elsewhere [21, 23]. The present study has been carried out on approximately, 150 numbers of commercially available 12 V/2500F HUCs. These HUC modules consist of six cells connected in series with each cell comprising 9 positive electrodes and 8 negative electrodes. Lead metal sheets of 0.3 mm were employed as substrate for making substrate integrated Pb/PbO_2 electrodes. Flexible graphite sheets of 0.75 mm thickness are coated with activated carbon mixed with Polyvinylidene fluoride (PVDF) binder in dimethylformamide (DMF) solvent.

A 12 V 2500F HUC is expected to have a charge capacity of 4.167 AH and energy storage capacity of 37.5 Wh.

8.3.2   PRECONDITIONING OF PB–C HUCS

The fresh HUC needs to be pre-conditioned before subjecting them for characterization [24, 25]. For preconditioning, HUCs are charged in a CCCV mode of 0.5 A, 13.8 V for 48 h. At the end of 48 h, HUCs having open circuit potential (OCP) less than 12 V were considered as faulty and discarded. HUCs showing OCP as 12 V were considered for further charge/discharge protocols.

8.3.3   CHARGE/DISCHARGE PROTOCOL

HUC charge/discharge protocol is represented as "$I_{ch}/I_{dis}A\_t_{cv}h$" with respect to Figure 8.1. In the representation, I_{ch}/I_{dis} corresponds to charging/discharging currents and t_{cv} is time duration of CV mode (CV mode) charging. For the sake of simplicity, charging current (I_{ch}) and discharging current (I_{dis}) are taken same.

Each charge/discharge cycle of 12 V, 2500F HUC has three modes of operation:

- Constant current charging of I_{ch} till HUC reaches 13.8 V.
- Constant voltage charging of 13.8 V for time duration of "t_{cv};" and
- Constant current discharge of I_{dis} till 6 V.

HUCs have been subjected to 12 different protocols of charge/discharge cycles at CC ($I_{ch} = I_{dis} = 1$ A, 2 A, and 3 A) followed by CV of 13.8 V for t_{cv} charging time durations ($t_{cv} = 0$ h, 1 h, 2 h, and 3 h). The HUCs were discharged till the cut-off voltage of 6 V with rest of 5 minutes between each charge and discharge step. The HUCs were subjected to 5 to 10 cycles of each charge/discharge protocol.

The key parameters namely charge output, energy output, and capacitance are determined from charge/discharge curves [26–28]. An optimal test protocol has been determined from comparing the estimated values of key parameters from different test protocols [29–31].

8.4   RESULTS

Voltage discharge characteristics of two healthy 12 V/2500F HUC samples (HUC25F15L070046) and (HUC25F15L90034) accepted for standardization of test protocol are shown in Figures 8.4 and 8.5, respectively.

From Figure 8.4, it is seen that the voltage discharge curve of HUC sample, HUC25F15L090034 was a steady voltage discharge curve whose capacitance is calculated as 2.7 kF on constant current (CC) load of 5 A. From Figure 8.5, it is seen that the voltage discharge curve of HUC sample, HUC25F15L070046 was a steady voltage discharge curve whose capacitance is calculated as 2.2 kF on CC of 5 A. With a tolerance band of 12%, these HUC samples are accepted. Further these HUC samples are used in standardizing the charge discharge procedure of HUC to deliver optimal performance.

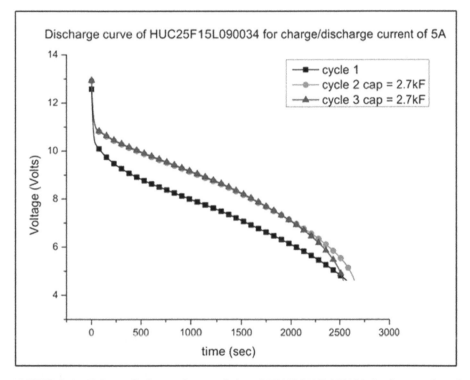

FIGURE 8.4 Voltage discharge characteristics of HUC25F15L090034 for three cycles of 5A constant current discharge.

Figures 8.6 and 8.7 represents the voltage discharge curves of a faulty 12 V/2500F range HUCs. From Figure 8.6, It is seen that the voltage discharge curve of HUC sample, HUC25F15L090038 show a sudden dip of voltage from 13.8 V to 3–5 V on constant load of 5A for seven cycles. This behavior represents the HUC unable to deliver the energy on load. From Figure 8.7, it is seen that the voltage discharge curves of an another HUC sample, HUC25F15L070042 show a droop effect on 2A CC load. This behavior represents the HUC is unable to deliver the steady energy requirements on the CC loads. Hence, the HUC which show inconsistent behavior in voltage discharge characteristics are considered as the faulty capacitors.

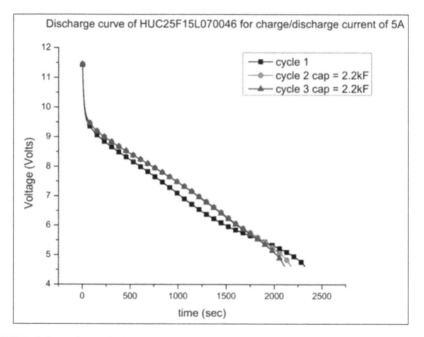

FIGURE 8.5 Voltage discharge characteristics of HUC25F15L070046 for three cycles of 5A constant current discharge.

Out of 150 HUC samples the HUCs whose performance curves are similar to HUC2515FL090038 or HUC25F15L070042 are considered as faulty HUCs and are discarded in development of standard test procedure, i.e., charge/discharge procedure of HUCs. Manufacturing processes and purity of electrodes, electrolyte is some of the reasons for failure of HUCs.

The voltage current (v–i) characteristics of HUC were obtained for charge/discharge currents of 1 A, 2 A, and 3A and CV of 13.8 V for t_{cv} of 0 h, 1 h, 2 h and 3 h. V-I characteristic curves are shown for 0 h, 1 h, 2 h, and 3 h t_{cv} of 1 A, 2 A, and 3A currents in Figures 8.8–8.10, respectively.

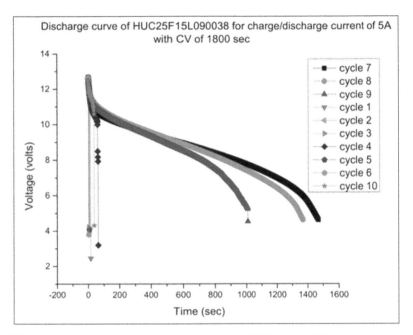

FIGURE 8.6 Voltage discharge characteristics of HUC25F15L090038 with zero charge retention for seven cycles of 5A constant current discharge.

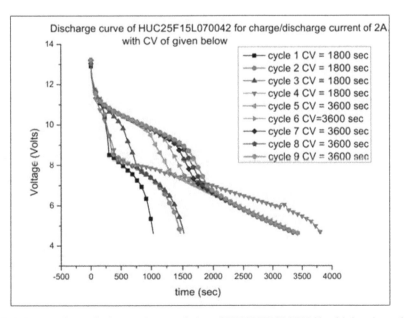

FIGURE 8.7 Voltage discharge characteristics of HUC25F15L070042 with interim voltage dips for nine cycles of 2A constant current discharge.

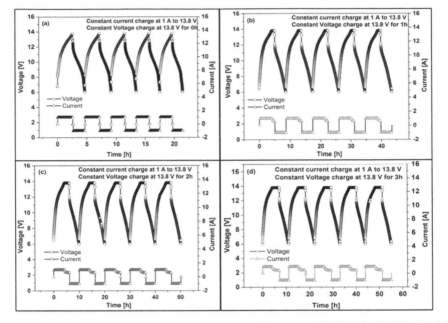

FIGURE 8.8 V-I characteristics of HUC for $1A\_t_{cv}h$ protocol (a) $1A\_0h$; (b) $1A\_1h$; (c) $1A\_2h$; and (d) $1A\_3h$.

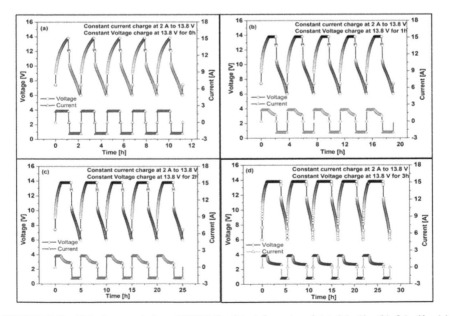

FIGURE 8.9 V-I characteristics of HUC for $2A\_t_{cv}h$ protocol (a) $2A\_0h$; (b) $2A\_1h$; (c) $2A\_2h$; and (d) $2A\_3h$.

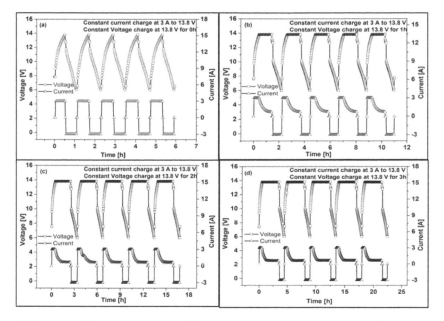

FIGURE 8.10   V-I characteristics of HUC for 3A_t$_{cv}$h protocol (a) 3A_0h; (b) 3A_1h; (c) 3A_2h; and (d) 3A_3h.

The key parameters, charge input/output, energy input/output, charge efficiency, energy efficiency and capacitance offered by HUC are determined through V-I curves and Eqns. (8)–(14). The estimated values of key parameters for 12 charge/discharge protocols are tabulated as shown in Table 8.1.

For the sake of brevity, the results reported are the average of five charge/discharge cycles in each charge/discharge protocol.

The percentage of energy output delivered to CC load and percentage of energy lost in the internal resistance of the HUC are calculated and tabulated in Table 8.2 for the 12 charge/discharge protocols of the 12V/2500F HUCs.

8.5 DISCUSSION

For optimizing HUC charge/discharge protocol, one of the scopes of present work is to find out, whether HUC is a single step or two step charging devices.

Single step charge/discharge cycle will have only CC charging and CC discharging. This process does not have a CV charging followed by CC charging. Hence t$_{cv}$ in single step charging will be zero. Figures 8.8(a)–8.10(a) show single step charge/discharge protocol of HUC at 1 A, 2 A, and 3 A, respectively. Table 8.1 shows that the capacitance delivered is only 50% of the rated capacitance and hence single step charging of HUCs is discarded from optimal protocol.

TABLE 8.1   Key Parameters of HUC Estimated for Different Charge/Discharge Protocols

SL. No.	Protocol	Charge [Ah]		Energy [Wh]		Efficiency [%]		Cap [F]
		Input	Output	Input	Output	Charge	Energy	
1.	1A_0h	2.308	1.832	27.46	17.565	79.382	63.966	1252.78
2.	1A_1h	4.733	3.576	57.356	32.283	75.555	56.285	2223.65
3.	1A_2h	5.888	3.827	72.562	34.506	64.987	47.554	2305.97
4.	1A_3h	6.346	4.039	85.338	34.322	63.646	40.219	2494.3
5.	2A_0h	2.258	1.993	27.52	16.304	88.264	59.244	1739.13
6.	2A_1h	3.715	3.072	47.52	25.62	82.692	53.914	2638.56
7.	2A_2h	4.383	3.248	56.61	30.56	74.104	53.983	2751.72
8.	2A_3h	4.555	3.961	57.84	35.71	86.959	61.739	2873.09
9.	3A_0h	1.813	1.676	21.03	15.12	92.443	71.897	1040.16
10.	3A_1h	2.539	2.021	33.03	16.52	79.598	50.015	1710.88
11.	3A_2h	3.325	2.318	43.58	19.55	69.714	44.86	2079.65
12.	3A_3h	4.141	2.579	54.6	20.93	62.28	38.333	2320.24

TABLE 8.2   Percentage of Energy Output Delivered and Energy Output Lost for 12 12V/2500F HUC Charge Discharge Protocols

SL. No.	Charge/ Discharge Protocol	C_{HUC} [kF]	V_2 [V]	V_1 [V]	t_1 [h]	t_2 [h]	Percentage of Energy Output Due to Capacitance	Percentage of Energy Loss Due to Internal Resistance
1.	1A_0h	1.53	11.91	8.17	0.031	1.590	95.66	4.34
2.	1A_1h	1.84	11.77	7.51	0.031	2.177	95.22	4.78
3.	1A_2h	2.32	11.81	6.17	0.031	3.635	97.94	2.06
4.	1A_3h	2.50	11.30	5.55	0.031	3.993	97.90	2.10
5.	2A_0h	1.74	10.14	6.10	0.031	0.976	97.26	2.74
6.	2A_1h	2.64	10.32	6.15	0.031	1.529	98.29	1.71
7.	2A_2h	2.77	10.18	6.00	0.031	1.608	93.72	6.28
8.	2A_3h	2.99	11.25	7.22	0.031	1.674	97.07	2.93
9.	3A_0h	1.04	11.62	6.07	0.031	0.534	93.95	6.05
10.	3A_1h	1.71	10.13	6.10	0.031	0.638	93.99	6.01
11.	3A_2h	2.08	10.07	6.12	0.031	0.761	94.51	5.49
12.	3A_3h	2.46	10.10	6.16	0.031	0.897	96.55	3.45

Two step charge/discharge cycle will have CCCV charging followed by CC discharging. Figures 8.8(b)–(d)–8.10(b)–(d) show two step charge/discharge protocol with t_{cv} as 1 h, 2 h, and 3 h, respectively. Table 8.1 shows that the capacitance delivered by HUCs is closer to the rated capacitance indicating that, HUCs are two-step charging devices.

Based on the availability of solar power for 5.5 h on a bright sunny day, the following are considered as the deciding criteria for optimizing the standard protocol:

- Total charging time (t_{ch}) should be less than 5.5 h.
- Capacitance deliverance (C) should be as close as possible to the rated capacitance (i.e., 2500F).
- Charge capacity of 4.167 AH and energy storage capacity of 37.5 Wh should be achieved.

From Figures 8.8(b)–(d), it could be observed that charging time (t_{chcc}) is nearly 4 h at 1 A. From Figures 8.9(b)–(d), and 8.10(b)–(d); t_{chcc} is nearly 1.5 h and 0.75 h at 2 A, and 3 A, respectively. The total charging time ($t_{ch} = t_{chcc} + t_{cv}$) of HUC at 1A, 2A, and 3A is summarized in Table 8.3.

TABLE 8.3 Total Charging Time of HUC at Constant Current Charge of 1 A, 2 A, and 3 A and Constant Voltage Charge of 13.8 V for t_{cv} of 1 h, 2 h, and 3 h

SL. No.	I_{ch}	t_{ch}/Capacitance		
		t_{cv} = 1 h	t_{cv} = 2 h	t_{cv} = 3 h
1.	1A	5.0 h/2.2 kF	6.0 h/2.3 kF	7.0 h/2.5 kF
2.	2A	2.5 h/2.6 kF	3.5 h/2.8 kF	4.5 h/2.9 kF
3.	3A	1.75 h/1.7 kF	2.75 h/2.1 kF	3.75 h/2.3 kF

As the charging times of HUC for 1A_2h and 1A_3h protocols are greater than 5.5 h, these protocols are discarded as standard protocols of HUC charge/discharge for solar power applications. Remaining protocols have been further analyzed for determination of standard test protocol.

From Figures 8.8–8.10 and Tables 8.1 and 8.2, following are the important observations:

- Capacitance deliverance of HUC improves with total charging time (t_{ch}).
- For each charging current (I_{ch}), the energy output has increased with t_{ch}.
- Energy output has marginally decreased with increase in I_{ch}.
- Charge efficiency and energy efficiency has no correlation with t_{ch} and I_{ch}. However, for two step charging, charge efficiency and energy efficiency lie in the range of (62% to 92%) and (38% to 72%), respectively.

Charging at 1A is not meeting the requirement for the following reasons:

- Charging time is more than 5.5 h (except for $t_{cv} = 1$ h).
- Capacitance deliverance is less than 2.5 kF.

Following are the observations during charge/discharge at 3A:

- HUCs are getting charged faster with t_{ch} ranging from 1.75 h to 3.75 h.
- Capacitance deliverance is less than 2.5 kF.
- Charge output and energy output are 2.6 AH and 21 Wh which are only 50% of the rated values.

Charging at 3A has also been discarded as the estimated values of key parameters are not meeting the rated values of HUCs, in spite of HUCs getting charged faster.

Regarding 2A charging, following are the observations:

- HUCs charge with t_{ch} ranging from 2.5 h to 4.5 h.
- Capacitance deliverance is in the range of 2.6 kF to 2.9 kF range, which is better than the rated capacitance.
- Charge output and energy output are 3.0 AH, 3.2 AH, 4.0 AH and 26 Wh, 31 Wh, 36 Wh for t_{cv} of 1 h, 2 h, and 3 h, respectively.
- Charge efficiency and energy efficiencies are 54%, 54%, 62% and 83%, 74% and 87%.
- The maximum charge output, energy output and capacitance of 4.0 AH, 36 Wh, and 2.9 kF were delivered in 2A_3h charge/discharge protocol.

As can be seen from Table 8.2, the capacitance deliverance increases with t_{cv}. In order to avoid, overcharging of HUCs, t_{cv} has been limited to 3 h in the present study.

12V/2500F HUCs are 4 AH energy storage systems. To operate HUC safely and ensure battery electrode of HUC is not sulfated, the C-rating of 12V/2500F range HUC is limited to 0.75C, i.e., 3A.

The above analysis establishes that 2A_3h charge/discharge protocol is the standard test protocol for charging 12V/2500F HUCs for solar microgrid applications.

8.5.1   STANDARD PROTOCOL FOR OPTIMAL PERFORMANCE OF HUCS

The experimental work reported, and the analysis made establish that 2A_3h charge/discharge protocol is the standard test protocol for charging 12V/2500F HUCs for solar microgrid applications. The standard procedure for one cycle of charge/discharge of 12V/2500F HUC module with "2A_3h" charge/discharge protocol consists of following sequence:

- ➤ **Step 1:** Rest the HUC module for 5 min.
- ➤ **Step 2:** Charge HUC at 2A till the voltage across HUC reaches 13.8 V
- ➤ **Step 3:** Charge HUC with constant voltage of 13.8 V for 3 h
- ➤ **Step 4:** Rest the HUC module for 5 min
- ➤ **Step 5:** Discharge HUC at 2A till the voltage across HUC reaches 6 V.
- ➤ **Step 6:** Repeat Step 1 to step 5 for 'N' number of cycles.

Number of cycles (N) required to estimate the capacitance deliverance of HUC is of specific interest. As of now, there is no standard protocol for deciding the number of cycles. Based on the experimental study carried out on approximately 150 numbers of 12 V, 2500 kF HUCs, N is chosen between 5 and 10.

8.5.2 ANALYSIS OF ENERGY OUTPUT

Comparing the energy output delivered in the linear and nonlinear regions from Table 8.2, it could be seen that 96% of total energy delivered is due to capacitance of HUC and only 4% of energy is lost due to internal resistance. Therefore, HUC could be considered predominantly as a capacitor having substantially low internal resistance.

8.6 CONCLUSION

The present research work has been carried out on approximately 150 numbers of 12 V Pb–C HUCs, having rated capacitance of 2500F, rated charge capacity of 4.167 AII and rated energy storage capacity of 37.5 Wh. Performance optimization of HUCs has been attempted employing different charge/discharge protocols at appropriate voltage, current and time durations. Key parameters, namely, charge output, energy output and capacitance have been estimated and analyzed to arrive at a standard test protocol for HUCs as energy storage devices for solar power applications. Availability of solar power has been considered as 5 ½ h on a bright sunny day.

Following are the major conclusions of the study conducted on 12V 2500F Pb–C HUC modules:

1. Single step charging is not applicable as it yields only 50% of the rated capacitance. Whereas the capacitance deliverance during two

step charge/discharge cycle is closer to the rated capacitance indicating that HUCs are two-step charging devices.

2. For the nine different two step charge/discharge protocols at CC of 1A, 2A, and 3A followed by CV of 13.8 V, the total charging time is nearly 5 h, 3.5 h, and 3.75 h, respectively.

3. Capacitance deliverance of HUC marginally improves with total charging time. For example, an increase from 2.6 kF @ 2A-1h charging to 2.9 kF @ 2A-3h has been seen. Similarly, energy output improves from 26 Wh to 36 Wh with charging time of 1 h to 3 h.

4. Charge efficiency and energy efficiency has no correlation with charging time and charging current. However, charge efficiency and energy efficiency lie in the range of (62% to 92%) and (38% to 72%), respectively.

5. To operate HUC safely and to ensure that battery electrode of HUC is not sulfated, C-rating is limited to 0.75C, i.e., 3A. In order to avoid, overcharging of HUCs, charging time is limited to 3 h in the present study.

6. HUC can be treated predominantly as a capacitor, as the energy output delivered due to capacitance is 96% of total energy and only 4% of energy is lost due to internal resistance.

7. Charge/discharge protocol of 2A_3h has yielded values of capacitance, charge output and energy output of 2.9 kF, 4.0 AH and 36 Wh, which are close to the rated values of HUC.

8. Also, the total charging time for 2A_3h charge/discharge protocol is 4.5 h. Based on the assumption of availability of 5 ½ h of sunlight in one single day would charge the HUC fully, 2A_3h charge/discharge protocol has therefore been considered as the standard test protocol for charging 12V 2500F HUCs for solar power applications.

ACKNOWLEDGMENT

The authors acknowledge the Central Power Research Institute (CPRI), Bangalore, India for the encouragement and support extended in forming this publication. The authors would also like to thank the officials of the Solid State and Structural Chemistry unit of Indian Institute of Science, Bangalore, Capacitors Division and R&D Management of CPRI for their help in carrying out the research work.

KEYWORDS

- **discharge protocol**
- **hybrid ultracapacitors**
- **lead carbon hybrid ultracapacitor**
- **optimal performance**
- **performance characterization**
- **solar microgrid**
- **standard test protocol**

REFERENCES

1. Chu, S., & Majumdar, A., (2012). Opportunities and challenges for a sustainable energy future. *Nature, 488,* 294–303.
2. Hittinger, E., Wiley, T., Kluza, J., & Whitacre, J., (2015). Evaluating the value of batteries in microgrid electricity systems using an improved energy systems model. *Energy Conversion and Management, 89,* 458–472.
3. Abu-Sharkh, S., Arnold, R. J., Kohler, J., Li, R., Markvart, T., Ross, J. N., Steemers, K., et al., (2006). Can microgrids make a major contribution to UK energy supply? *Renewable and Sustainable Energy Sources, 10*(2), 78–127.
4. Tan, X., Li, Q., & Wang, H., (2013). Advances and trends of energy storage technology in microgrid. *International Journal of Electrical Power and Energy Systems, 44*(1), 179–191.
5. Pradeep, K. R. J., & Nagamani, H. N., (2016). Modeling of hybrid ultra-capacitors for power electronic applications using PSPICE/ORCAD™. *Power Research–A Journal of CPRI, 12, 3,* 547–554.
6. Ravikumar, M. K., Rathod, S., Jaiswal, N., Patil, S., & Shukla, A. K., (2017). The renaissance in redox flow batteries. *J. Solid State Electrochemistry, 21*(9), 2467–2488.
7. Yang, Z., Zhang, J., Meyer, M. C. W. K., Lu, X., Choi, D., Lemmon, J. P., & Liu, J., (2011). Electrochemical energy storage for green grid. *Chem. Rev., 111,* 3577–3613.
8. Dunn, B., Kamath, H., & Tarascon, J. M., (2011). Electrical energy storage for the grid–A battery of choices. *Science, 334,* 928–935.
9. Dubal, D. P., Ayyad, O., Ruiz, V., & Gòmez–Romero, P., (2015). Hybrid energy storage–the merging of battery and supercapacitor chemistries. *Chem. Soc. Rev., 44*(7), 1777–1790.
10. Bélanger, D., Brousse, T., & Long, J. W., (2008). Manganese oxides–battery materials make the leap to electrochemical capacitors. *The Electrochemical Society Interface,* 49–52.
11. Zuo, W., Li, R., Zhou, C., Li, Y., Xia, J., & Jingpin, L., (2017). Battery-supercapacitor hybrid devices: Recent progress and future prospects. *Adv. Sci., 4.* Open Access. https://doi.org/10.1002/advs.201600539.

12. Yu, N., Gao, L., Zhao, S., & Wang, Z., (2009). Electrodeposited PbO_2 thin film as positive electrode in PbO_2/AC hybrid capacitor. *Electrochimica. Acta, 54*(14), 3835–3841.
13. Banerjee, A., Ravikumar, M. K., Jalajakshi, A., Suresh, K. P., Gaffoor, S. A., & Shukla, A. K., (2012). Substrate integrated lead-carbon hybrid ultracapacitor with flooded, absorbent glass mat and silica-gel electrolyte configurations. *Journal of Chemical Sciences, 124*(4), 747–762.
14. Wang, J., Yang, Y., Huang, Z. H., & Kang, F., (2013). A high-performance asymmetric supercapacitor based on carbon and carbon–MnO_2 nanofiber electrodes. *Carbon, 61,* 190–199.
15. Zhao, C., Ren, F., Xue, X., Zheng, W., Wang, X., & Chang, L., (2016). A high-performance asymmetric supercapacitor based on $Co(OH)_2$/graphene and activated carbon electrodes. *Journal of Electroanalytical Chemistry, 782,* 98–102.
16. Lang, J. W., Kong, L. B., Liu, M., Luo, Y., & Kang, L., (2010). Asymmetric supercapacitors based on stabilized α-$Ni(OH)_2$ and activated carbon. *Journal of Solid State Electrochemistry, 14*(8), 1533–1539.
17. Pell, W. G., & Conway, B. E., (2004). Peculiarities and requirements of asymmetric capacitor devices based on combination of capacitor and battery-type electrodes. *Journal of Power Sources, 136*(2), 334–345.
18. Burke, A., & Miller, M., (2010). Testing of electrochemical capacitors: Capacitance, resistance, energy density, and power capability. *Electrochimica. Acta, 55*(25), 7538–7548.
19. Banerjee, A., Ravikumar, M. K., Jalajakshi, A., Gaffoor, S. A., & Shukla, A. K., (2012). A 12 V substrate-integrated PbO_2-activated carbon asymmetric hybrid ultracapacitor with silica-gel-based inorganic-polymer electrolyte. *ECS Transactions, 41*(13), 101–113.
20. Shukla, A. K., Banerjee, A., Jalajakshi, A., & Ravikumar, M. K., (2013). 12 V / kilo-farad range lead-carbon hybrid ultracapacitors and their envisaged applications. *ECS Transactions, 50*(31), 367–376.
21. Shukla, A. K., Ravikumar, M. K., & Gaffoor, S. A., (2015). *Energy Storage Device and Method Thereof.* US Patent US 9,147,529.
22. Shukla, A. K., Banerjee, A., Ravikumar, M. K., & Gaffoor, S. A., (2015). *Energy Storage Device, an Inorganic Gelled Electrolyte and Methods Thereof.* US Patent US 9,036,332.
23. Shukla, A. K., Banerjee, A., & Ravikumar, M. K., (2012). Electrochemical capacitors: Technical challenges and prognosis for future markets. *Electrochimica. Acta, 84,* 165–173.
24. Pradeep, K. R. J., Nagamani, H. N., Vaidhyanathan, V., & Bhavani, S. T., (2017). A study on preliminary screening of 12V/ 2500F range lead carbon hybrid ultracapacitors as energy storage device for solar power applications. *Power Research–A Journal of CPRI, 13*(3), 503–510.
25. Pradeep, K. R. J., Nagamani, H. N., Vaidhyanathan, V., & Bhavani, S. T., (2017). Charge–discharge behavior of lead acid battery and lead carbon hybrid ultracapacitors as an integrated system for solar power applications. *Power Research–A Journal of CPRI, 13*(3), 517–524.
26. Sanjeevi, K. P., Neeraj, P., Mahajan, S. B., Bo Holm–Nielsen, J., & Eklas, H., (2019). A hybrid photovoltaic–fuel cell for grid integration with Jaya based maximum power point tracking: Experimental performance evaluation. *IEEE Access, 7,* 82978–82990.
27. Yong, S., Xu–Wei, G., Ji, X., Xin, W., & Xu, Y., (2019). Large power hybrid soft switching mode PWM full bridge DC–DC converter with minimized turn–ON and turn–OFF switching loss. *IEEE Transactions on Power Electronics, 34*(12), 11629–11644.

28. Ben, S. S., (2020). Intelligent energy management for off–grid renewable hybrid system using multi–agent approach. *IEEE Access, 8*, 8681–8696.

29. Chen, Z., He, Y., & Chengbin, M., (2020). Equivalent series resistance based real–time control of battery–ultracapacitor hybrid energy storage systems. *IEEE Transactions on Industrial Electronics, 67*(3), 1999–2008.

30. Siyuan, C., Quifan, Y., Jianyu, Z., & Xia, C., (2021). A model predictive control method for hybrid energy storage systems. *CSEE Journal of Power and Energy Systems, 7*(2), 329–338.

31. Mince, L., Li, W., Yujie, W., & Zonghai, C., (2021). Sizing optimization and energy management strategy for hybrid energy storage system using multiobjective optimization and random forest. *IEEE Transactions on Power Electronics, 36*(10), 11421–11430.

CHAPTER 9

Speed Sensorless Model Predictive Current Control of Induction Motor-Driven Electric Vehicle

KARUNA KIRAN

Indian Institute of Technology (Indian School of Mines), Dhanbad, Jharkhand, India

ABSTRACT

The chapter deals with an analysis of speed-sensorless model-predictive-current control (MPCC) technique of induction motor (IM) driven electric vehicle (EV). MPCC is a direct control approach in which the errors between the reference stator current and the predicted stator current are converges to minimize the cost function. The cost function is defined for each of the predicted stator current corresponding to every voltage vector of the inverter. The presence of an optimized inner current controller with cost function and absence of pulse width modulator (PWM) are the key features of the proposed MPCC scheme. Moreover, stator current (i_s) based model reference adaptive system (i_s-MRAS) is proposed as speed estimator owing to its accuracy in estimation and simplicity. The absence of PWM as well as the employment of i_s-MRAS aid in making the overall drive system modest and cost effective. The simulation results in MATLAB/Simulink accomplish satisfactory dynamic behavior and uphold smooth steady state performance.

9.1 INTRODUCTION

In recent years, electric vehicles (EVs) are gaining remarkable research efforts due to sharp depletion of petroleum resources as well as climate and

The Internet of Energy: A Pragmatic Approach Towards Sustainable Development. Sheila Mahapatra, Mohan Krishna S., B. Chandra Sekhar, & Saurav Raj (Eds.)

environmental awareness. Electrical machine drive is a crucial component of EV powertrain as it gives the vital power to propel the EV in an efficient manner [1]. From the industrial applications point of view, induction motor (IM) shows strong candidature as EV drive for sensorless speed control in respect of low-cost and high performance [2, 3]. Moreover, it is efficient, trustworthy and offers an extended range of stability at different speeds [1]. Literature survey shows that manifold analytical research and efficacious practical implementations for the speed control of IM have been carried out [3, 4]. Some of the classic control approaches such as direct torque control (DTC) [6] and vector control (VC)/field-oriented control (FOC) [5] have been comprehensively employed during recent few decades. FOC scheme comprises linear proportional integral (PI) controllers as well as pulse-width modulator (PWM) [5]. FOC scheme is widely applicable for low and medium power electrical drives. However, DTC appeared as a plausible solution for the moderate to high-power applications [2]. However, the outcomes of hysteresis torque and flux controllers in DTC vitiate the steady state performance. On contrary, FOC has merit of outstanding performance at steady state [5]. However, inherent slow dynamics of the inner current control loop degrades its dynamic response [4]. Being a linear control strategy, FOC has drawbacks pertaining to limited bandwidth and requirement of extra PWM hardware for pulse generation of voltage source inverter with two-levels (2L-VSI) [7].

To overcome the above-mentioned concerns, the emerging model-predictive-control (MPC) is deputized as encouraging approach to tackle the aforementioned FOC drawbacks [8, 9]. With the advent of high-performance power electronic devices and powerful digital signal processors (DSPs), MPC emerged as an attractive area of research for nonlinear and direct control of electric drives and power electronic devices [10]. It utilizes model of system and a cost/objective function to directly govern the control variables [11]. Optimal solution is determined by meticulous searching of voltage vectors in the specified viable margin, according to the principle of cost function minimization [9]. The scheme can be classified into two mainstream categories: model predictive current and model predictive torque control (MPCC [7], MPTC [12]). Alike classical FOC [4], MPCC makes use of orthogonal current components only in the cost function and hence weighing factor is not mandatory [7, 13]. However, whereas in case of MPTC the cost function contains both the torque and the flux terms and hence a combination of two different units and order of magnitude needs to be made, thus requiring a weighing factor which is challenging to tune [12]. Therefore, MPCC strategy is recommended in this work over MPTC due to its simplicity.

Apart from this, the installation of a physical speed sensor (such as tachometers, optical encoders, etc.) on the rotor shaft of EV motor to get the instantaneous speed information is troublesome [4]. It not only reduces the system reliability in hostile environment but also imposes cost burden and hardware intricacy to the overall drive system [14]. Subsequently, the estimation of speed from electrical measures (i.e., machine terminal phase currents and phase voltages) is favored than speed measurement [15]. A number of sensorless techniques have been suggested for IM drive [14]. Recent literature survey recommends that MRAS-based methods for rotor speed estimation show more promising performance over other model based and signal injection-based speed estimation techniques because of its simplicity and physical exposition. It is, however, difficult to improve the performance of these schemes at very low speeds. Various MRAS configurations have been analyzed on the basis of simplicity, ease in implementation, stability, and high performances in all the four quadrants and immunity towards machine parameter variation. In this context, MRAS speed estimators using rotor flux [16], back EMF [17], electromagnetic torque [18], active power or reactive power [15, 19], stator current as control variables are investigated comprehensively [20]. Rotor flux MRAS is the classical MRAS strategy, first introduced by Schauder [21]. It suffers from pure integrator related issues. Back EMF scheme avoids pure integration process but shows low noise immunity and may become unstable at low or zero supply frequencies. Reactive power MRAS overcomes parameter sensitivity effects and pure integrator related issues like drift or saturation [22] but suffers from instability at few operating points. This instability issue is avoided using active power MRAS.

The present work extends the concept of MPCC as discussed in Ref. [7] for the sensorless operation. In this context, a stator current based MRAS algorithm (i_s-MRAS) in association with MPCC scheme is developed for the first time. The reference model of the estimator is designed using the instantaneous terminal stator current. Moreover, the speed dependent instantaneous predicted current from the MPCC algorithm is unit delayed and employed as adaptive model. The developed i_s-MRAS based MPCC scheme is investigated in MATLAB/Simulink for wide-ranging speed variations.

The present work is divided into six sections. The fundamentals (2L-VSI) and IM drive model formulation are discussed in Section 9.2. In Section 9.3, the overall speed control using MPCC scheme of IM driven EV is described in detail. To make the developed MPCC scheme speed sensorless, a stator current MRAS for speed estimation is integrated in Section 9.4. To investigate the efficacy of the acclaimed scheme, some important outcomes of the simulation in MATLAB/Simulink are discussed in Section 9.5. In the final Section 9.6, the work has been concluded.

9.2 MATHEMATICAL MODELING OF VOLTAGE SOURCE INVERTER AND INDUCTION MOTOR

In this section, simple mathematical models of the three-phase 2L-VSI and IM are presented.

9.2.1 MODELING OF 2L-VSI

The DC-link power is converted to three-phase alternating (AC) power by the power circuit of three-phase 2L-VSI (*see* Figure 9.1(a)). To circumvent short-circuiting of DC source, the complementary switches in each phase operate at a time [8]. Hence, switching states of the six switches (S_x, $\bar{S}_x = 1$, 0 and $x = 1,2,3$) can be represented by switching signals S_a, S_b, S_c as [7]:

$$S_a \Rightarrow \begin{cases} 1 & \text{if } S_1 \rightarrow \text{on, and } \bar{S}_1 \rightarrow \text{off} \\ 0 & \text{if } S_1 \rightarrow \text{off, and } \bar{S}_1 \rightarrow \text{on} \end{cases} \tag{1}$$

$$S_b \Rightarrow \begin{cases} 1 & \text{if } S_2 \rightarrow \text{on, and } \bar{S}_2 \rightarrow \text{off} \\ 0 & \text{if } S_2 \rightarrow \text{off, and } \bar{S}_2 \rightarrow \text{on} \end{cases} \tag{2}$$

$$S_c \Rightarrow \begin{cases} 1 & \text{if } S_3 \rightarrow \text{on, and } \bar{S}_3 \rightarrow \text{off} \\ 0 & \text{if } S_3 \rightarrow \text{off, and } \bar{S}_3 \rightarrow \text{on} \end{cases} \tag{3}$$

The phase to neutral (N) voltages of 2L-VSI are defined in terms of switching signals Eqns. (1)–(3) as:

$$v_{aN} = S_a V_{dc} \tag{4}$$

$$v_{bN} = S_b V_{dc} \tag{5}$$

$$v_{cN} = S_c V_{dc} \tag{6}$$

where; V_{dc}: voltage across the DC-link.

The voltage vector output is the vector sum of three-phase to neutral voltages as:

$$v = \frac{2}{3}\left(v_{aN} + a v_{bN} + a^2 v_{cN}\right). \tag{7}$$

where;

$$a = e^{\frac{j2\pi}{3}}$$ (8)

Therefore, eight switching states are obtained using all feasible arrangements of the gating signals (S_a, S_b and S_c). Consequently, eight different voltage vectors are generated [9]. These voltage vectors are represented in Figure 9.1(b).

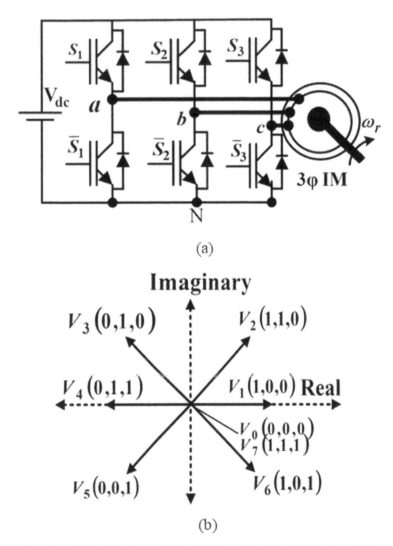

(a)

(b)

SFIGURE 9.1 Three-phase 2L-VSI topology and inverter voltage vectors in complex plane.

9.2.2 MODELING OF IM

In stationary (superscript 's') reference frame, the classical electrical dynamic equations are described below as following set of equations:

$$v_s^s = i_s^s R_s + \frac{d}{dt}\psi_s^s \tag{9}$$

$$0 = i_r^s R_r + \frac{d}{dt}\psi_r^s - j\omega\psi_r^s \tag{10}$$

$$\psi_s^s = L_s i_s^s + L_m i_r^s \tag{11}$$

$$\psi_r^s = L_r i_r^s + L_m i_s^s \tag{12}$$

$$T_e = \frac{3}{2}p\,\mathrm{Im}\{\overline{\psi}_s^r \cdot i_s^s\} \tag{13}$$

where; p,ω and T_e denote number of pole pairs, electrical speed and generated electromagnetic torque of the motor, respectively.

From IM model described in Eqns. (9)–(12), the current in the stator winding can be evaluated as:

$$i_s^s = -\frac{1}{R_\sigma}\left\{\left[L_\sigma\frac{di_s^s}{dt} - k_r\left(\frac{1}{\tau_r} - j\omega\right)\psi_r^s\right] - v_s^s\right\} \tag{14}$$

where;

$$k_r = \frac{L_m}{L_r};\ L_\sigma = \sigma L_s;\ \text{and}\ R_\sigma = R_s + k_r^2 R_r \tag{15}$$

9.3 MODEL PREDICTIVE CURRENT CONTROL

Figure 9.2 represents the comprehensive block diagram of IM driven EV with rotor flux oriented sensorless MPCC scheme. The rotor speed estimates are obtained from the i_s-MRAS block. The direct and quadrature axes reference current vector components are generated using the rotor flux and load torque reference, respectively. The rotor flux estimation is done using the stator current in rotating reference frame to fulfill the rotor flux orientation strategy so that the decoupled torque and flux control could be possible alike VC. However, the remaining control mechanism is executed in the stationary frame. The calculation of unit vector is done to perform conversions from stationary to synchronously rotating frame and contrariwise. For the calculation of unit vector, slip speed is determined using the stator current in stationary reference frame. Hence, the control mechanism becomes very simple. Consequently, the rotating frame reference currents are transferred to the static reference frame as a parameter in the MPCC algorithm. Using the speed estimates from i_s–MRAS, instantaneous stator current and 2L-VSI voltage signals, the stator current for the next sampling interval is predicted corresponding to all feasible inverter voltage vectors. Comparing the predicted currents to that of the stationary frame reference stator current components; a cost function is designed corresponding to all the predicted currents. The optimization of the cost function in the MPCC algorithm is performed using the algorithm depicted in flow diagram – Figure 9.3. The voltage vector that corresponds to the optimal value of cost function is applied for the upcoming control cycle.

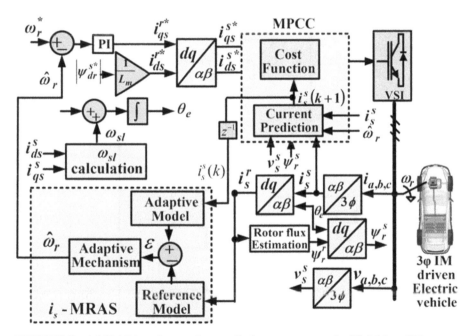

FIGURE 9.2 i_s-MRAS based sensorless predictive current control of IM driven EV.

The entire MPCC scheme can be explained through the algorithm depicted in Figure 9.3. First of all, voltage and current in the stator is measured at the terminal. Then, this information is used for the estimation of rotor flux and speed. These estimated quantities along with the reference values of current (i_{ds}^{r*} and i_{qs}^{r*}) obtained from flux and torque references are utilized to calculate of stator current in stationary frame. Thereafter, prediction of current for the next control cycle is done corresponding to each voltage vectors. On comparison of these predicted currents to their respective references, the cost functions are calculated. The predicted value that minimizes the cost function is selected and its conforming voltage vector is fed to VSI for the next control cycle. The step-by-step process of the scheme as accomplished in Figure 9.3 is discussed as in subsections.

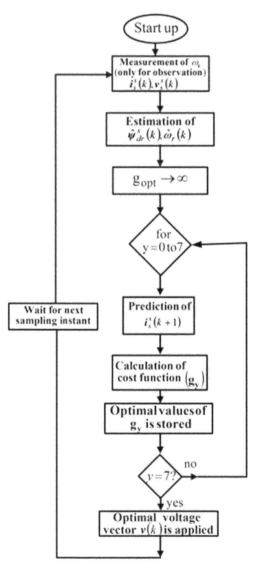

FIGURE 9.3   Model predictive current control algorithm.

9.3.1 GENERATION OF REFERENCE CURRENT

The actual stator current is measured at the terminal. For the execution of the MPCC scheme, the direct and quadrature axes current references need to be generated. The quadrature-axis current reference is obtained using torque reference which is generated as an output of the outer speed loop. However, the direct-axis current reference is obtained by flux reference which is taken as a constant.

$$i_{ds}^{r*} = \left| \psi_{dr}^{s*} \right| / L_m \tag{16}$$

$$i_{qs}^{r*} = \left(K_p + K_i / s \right) T^* \tag{17}$$

where; K_p and K_i are the speed PI regulator's proportional and integral constants.

The stator current reference signals generated from the above direct and quadrature components are further converted to stationary reference frame (i_α^* and i_β^*) for the evaluation of cost function.

9.3.2 PREDICTION OF STATOR CURRENT FOR THE NEXT CONTROL INTERVAL

The stator current in the succeeding control interval $i_s^s(k + 1)$ is predicted for all feasible inverter voltage vectors using system model Eqn. (14) as [7]:

$$i_s^s(k+1) = \left(1 + \frac{T_s}{\tau_\sigma} \right) i_s^s(k) + \frac{T_s}{\tau_\sigma + T_s} \left\{ \frac{1}{R_\sigma} \left[\left(\frac{k_r}{\tau_r} - k_r j\omega \right) \psi_r^s(k) + v_s^s(k) \right] \right\} \tag{18}$$

where; $\tau_\sigma = L_\sigma / R_\sigma$

9.3.3 EVALUATION OF COST FUNCTION

The cost function that is described in terms of the error in predicted currents with respect to the reference current and is evaluated as:

$$g_y = \left| i_\alpha^* - i_\alpha(k+1)_y \right| + \left| i_\beta^* - i_\beta(k+1)_y \right| \Big|_{y=0,1,2..7} \tag{19}$$

where; $i_\alpha(k+1)_y$, $i_\beta(k+1)_y$ are real and imaginary components of predicted current for y^{th} iteration.

The switching state signals corresponding to that voltage vector is selected which minimizes the current error. Finally, the generated output inverter voltage is fed as an input to the stator of the induction motor.

9.4 FORMULATION OF i_s-MRAS-BASED SENSORLESS MPCC SCHEME

The main purpose of a MRAS is to make the state variable adjust itself as per the quantity to be observed in the system. For a i_s-MRAS, the state variable is the instantaneous stator current. The elementary arrangement of i_s-MRAS is shown in Figure 9.4. It comprises of reference model and adaptive model. The reference model incorporates a set of equations that is free from the parameter to be estimated (i.e., speed). The instantaneous current at motor terminal is used as reference model. The adaptive model computes the same state variables using speed (parameter to be estimated) dependent equations. In this case, the predicted current signal at instant k +1 is passed through a unit delay to obtain the instantaneous current for the adaptive model. An adaptive mechanism is employed to do away with the error between these two models to accomplish exact estimation of the desired quantity.

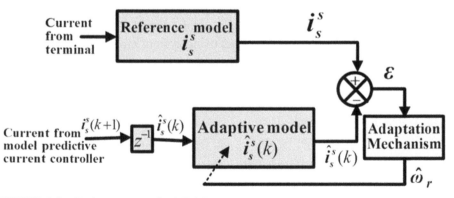

FIGURE 9.4   Basic structure of i_s–MRAS.

9.5 SIMULATION STUDY AND RESULTS

The simulation study in MATLAB/Simulink environment validates developed MPCC based speed control scheme of the motor drive. The motor parameters and PI controller gains for speed and i_s-MRAS are revealed in the Appendix A. The performance of drive for steep step speed variation is represented in this section.

Figure 9.5(a) shows that the speed reference is set in step changing pattern. The starting and throughout speed tracking performance is very smooth. The estimated speed $\hat{\omega}_r$ trails the reference ω_r^* and actual speed ω_r very precisely at all the operating points even during the speed transitions as well. In Figure 9.5(b), absolute speed inaccuracy is always below than 1.5 rad/s at steady state and within the permissible limit during start-up and speed transitions which shows good overall response of the drive system. Figure 9.5(c) shows that electromagnetic torque produced in motor and the applied constant load torque (one-fourth of the rated value). The applied load torque is followed by electromagnetic torque during steady-state and remains in stable range within the transient period as well. The maximum values of absolute torque error at steady-state and transient states are always within the reasonably satisfactory range as shown in Figure 9.5(d). In Figure 9.5(e), the amplitude of one of the stator phase currents remains constant at constant load torque despite of change in speed. Figure 9.5(f) shows the rotating frame rotor flux which validates the rotor flux orientation.

9.6 CONCLUSION

The present research work is dedicated to the speed sensorless MPCC scheme of IM drive which is suitable for electric vehicle applications. To mitigate the defects of degraded reliability and increased cost, stator current MRAS observer-based speed estimation algorithm is integrated with the MPCC technique. In i_s-MRAS, the reference model is the instantaneous terminal stator current, however the predicted stator current for the next control cycle of model predictive current controller is unit delayed in order to acquire the estimated current for the adaptive model. The adaptive mechanism constructed by employing a PI controller provides the rotor speed estimates. Moreover, the independent control of electromagnetic torque and flux (rotor) is also attained using FOC method. This inventive incorporation comprises the merits of both the intuitive sensorless method and MPCC scheme. Consequently, the system is robust, cost-effective, and easy to implement. It shows good speed tracking and dynamic performances for a wide speed range. Due to the presence of model predictive controller, the scheme works well even at closely zero speeds. The simulation results validate the soundness of the proposed scheme.

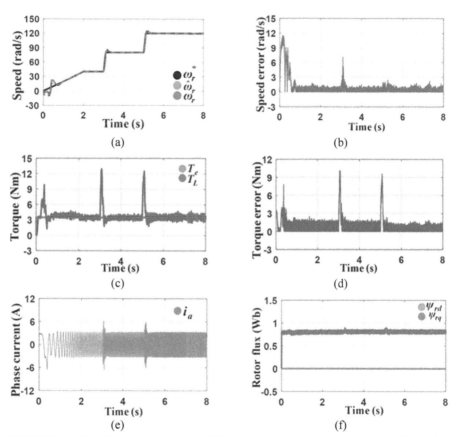

FIGURE 9.5 Simulation results for variable speed constant torque operation of IM drive: (a) Reference, estimated and actual speed; (b) absolute error between estimated and actual speed; (c) electromagnetic and load torque; (d) torque error; (e) stator phase current; and (f) rotating frame rotor flux.

KEYWORDS

- electric vehicle
- induction motor
- model predictive current control
- MRAS current estimator
- sensorless
- simulation study
- voltage source inverter

REFERENCES

1. Hadraoui, H., Zegrari, M., Chebak, A., Laayati, O., & Guennouni, N., (2022). A multi-criteria analysis and trends of electric motors for electric vehicles. *World Electric Vehicle Journal, 13*(65), 1–28.
2. Rodríguez, J., et al., (2011). High-performance control strategies for electrical drives: An electrical assessment. *IEEE Trans. Ind. Electron,* 812–820.
3. Wallace, A., Spee, R., & Alexander, G., (1993). Adjustable speed drive and variable speed generation systems with reduced power converter requirements. In: *IEEE International Symposium on Industrial Electronics, Conference Proceedings, (ISIE'93)*. Budapest, Hungary.
4. Bose, B. K., (2002). *Modern Power Electronics and AC Drives*. Prentice–Hall.
5. Karuna Kiran (2018) An improved rotor flux space vector-based MRAS for field oriented control of induction motor drives. *IEEE Transactions on Power Electronics, 33*(6), 5131–5141.
6. Das, S., Pal, A., & Manohar, M., (2017). Adaptive quadratic interpolation for loss minimization of direct torque controlled induction motor driven electric vehicle. In: *IEEE 15th International Conference on Industrial Informatics (INDIN)* (pp. 641–646). Emden, Germany.
7. Wang, F., Mei, X., Tao, P., Kennel, R., & Rodriguez, J., (2017). Predictive field-oriented control for electric drives. *Chinese Journal of Electrical Engineering, 3*(1).
8. Rodriguez, J., et al., (2007). Predictive current control of a voltage source inverter. *IEEE Transactions on Industrial Electronics, 54*(1), 495–503.
9. Rodriguez, J., & Cortes, P., (2012). *Predictive Control of Power Converters and Electrical Drives*. Valparasio, Chile: A John Wiley & Sons Ltd.
10. Cortes, P., Kazmierkowski, M. P., Kennel, R. M., Quevedo, D. E., & Rodriguez, J., (2008). Predictive control in power electronics and drives. *IEEE Trans. Ind. Electron., 55*(12), 4312–4324.
11. Kennel, R., Rodríguez, J., & Espinoza, J., (2010). High performance speed control methods for electrical machines: An assessment. In: *IEEE International Conference on Industrial Technology (ICIT)*. Vina del Mar, Chile.
12. Wang, F., et al., (2015). Model-based predictive direct control strategies for electrical drives: An experimental evaluation of PTC and PCC methods. *IEEE Trans. Ind. Informat., 11*(3), 671–681.
13. Fuentes, E., Rodrigues, J., Silva, C., Diaz, S., & Quevedo, D., (2009). Speed control of a permanent magnet synchronous motor using predictive current control. In: *IEEE 6th International Power Electronics and Motion Control Conference (IPEMC)*, 390–395.
14. Holtz, J., (2006). Sensorless control of induction machines-with or without signal injection? *IEEE Trans. Ind. Electron, 53*(1), 7–30.
15. Ravi, T. A. V., Verma, V., & Chakraborty, C., (2015). A new formulation of reactive-power-based model reference adaptive system for sensorless induction motor drive. *IEEE Trans. Ind. Electron., 62*(11), 6797–6808.
16. Rehman, H. U., Derdiyok, A., Guven, M. K., & Xu, L., (2001). An MRAS scheme for on-line rotor resistance adaptation of an induction machine. In: *IEEE 32nd Annual Power Electronics Specialists Conference, 2001; PESC: 2001*. Vancouver, BC, Canada.
17. Rashed, M., & Stronach, A. F., (2004). A stable back-EMF MRAS-based sensorless low-speed induction motor drive insensitive to stator resistance variation. *IEE Proceedings–Electric Power Applications, 151*(6), 685–693.

18. Kojabadi, H. M., Abarzadeh, M., & Chang, L., (2015). A comparative study of various methods of IM's rotor resistance estimation. In: *IEEE Conference on Energy Conversion Congress and Exposition (ECCE), 2015*. Montreal, QC, Canada.

19. Ta, C. M., Uchida, T., & Hori, Y., (2001). MRAS-based speed sensorless control for induction motor drives using instantaneous reactive power. In: *The 27th Annual Conference of the IEEE Industrial Electronics Society, 2001: IECON '01*. Denver, CO, USA.

20. Kumar, R., Das, S., & Chattopadhyay, A. K., (2016). Comparative assessment of two different model reference adaptive system schemes for speed-sensorless control of induction motor drives. *IET Electric Power Applications, 10*(2), 141–154.

21. Schauder, C., (1992). Adaptive speed identification for vector control of induction motors without rotational transducers. *IEEE Transactions on Industry Applications, 28*(5), 1054–1061.

22. Maiti, S., Chakraborty, C., Hori, Y., & Ta, M. C., (2008). Model reference adaptive controller-based rotor resistance and speed estimation techniques for vector controlled induction motor drive utilizing reactive power. *IEEE Trans. Ind. Electron., 55*(2), 594–601.

APPENDIX A

Specifications of IM

Shaft power (rated)	2.2 kilo Watt
Current (A) rating	2/2 Amperes
Voltage (V) rating	415/415 Volts
Frequency of stator supply (Hz)	50 Hertz
Synchronous speed (revolution per minute)	1,441 rpm
p	2
R_s, R_r'	4.125 Ω, 2.486 Ω
L_m, L_{ls}, R_{lr}'	0.2848 H, 0.01557H, 0.01557H
J	0.0182 kg-m^2
X_m	89.477
Speed PI controller gains (K_p, K_i, S_{lim})	0.3, 3, 6
i_s-MRAS adjustable mechanism PI controller gains (mK_p, mK_i, mS_{lim})	0.6, 10, 600
Sampling frequency (discrete) for simulation study	20 kHz

Performance Analysis of Hybrid Off-Grid Two-Axis Photovoltaic Tracking System/ Fuel Cell Energy System Incorporating High-Efficiency Solar Cell

SHUBHASHISH BHAKTA,[1] MESFIN MEGRA,[1] PIKASO PAL,[2] and ASHEBIR BERHANU[1]

[1]*Department of Electrical Power and Control Engineering, Adama Science and Technology University, Adama City, Ethiopia*

[2]*Department of Electrical Engineering, Indian Institute of Technology (Indian School of Mines), Dhanbad, Jharkhand, India*

ABSTRACT

The present chapter deals with a hybrid off-grid two-axis photovoltaic (PV) tracking system with a hydrogen-based fuel cell (FC) system for Adama City, Ethiopia. The hybrid system is designed and simulated using HOMER software. The high efficiency (20%) monocrystalline silicon PV panel is considered to analyze the performance of the PV-FC energy system. From the simulation results, it is found that with the assumed typical load profile of 11.7 kW peak and average annual solar radiation of 6.06 kWh/m²/d, the proposed hybrid system is capable of generating total annual power of 107.39 kW and 19.9498 kW from PV array and FC, respectively, with no unmet electricity. Moreover, the renewable fraction obtained for the hybrid energy system obtained is 100%.

The Internet of Energy: A Pragmatic Approach Towards Sustainable Development. Sheila Mahapatra, Mohan Krishna S., B. Chandra Sekhar, & Saurav Raj (Eds.)
© 2024 Apple Academic Press, Inc. Co-published with CRC Press (Taylor & Francis)

10.1   INTRODUCTION

Nowadays, electricity generation using fossil fuel-based energy sources (ES) is still considered the major energy resource and holds most of the energy sectors worldwide. These conventional ES are finite and depleting rapidly. Due to the profound impact on environmental issues as the production of harmful gases and nuclear waste products and rapidly growing electricity demand by end-users, renewable energy system-based electricity generation from renewable energy resources (RESs) has become an essential and urgent need across the globe [1]. The RES examples are wave, solar, geothermal, wind, and biomass energy. These RES can be properly utilized for the generation of electricity to mitigate future electricity generation issues for the end-users. Among all the RESs, electricity generation using solar photovoltaic (PV) technologies is considered more popular and adopted by developing and developed countries. However, PV system output is considerably impacted by the variation of solar radiation during the daytime, environmental factors such as cloudiness, and ambient temperature. Hence, to generate a steady supply of electricity, the PV system connected to electrical loads needs to incorporate suitable energy storage systems that can act as backup units during the nighttime, rainy season, and even in the cloudy days. The backup units in PV systems are optimal-sized battery banks, diesel generators, and fuel cells run from hydrogen tanks. In the case of the battery, despite the high efficiency of the most common batteries used in a PV system, such as a lead-acid battery which has the disadvantage of a short life period and degradation in capacity [2]. On the other hand, diesel fuel-operated diesel generators produce harmful gases, which is a concern for environmental issues. Therefore, PV systems incorporating hydrogen-based FC systems as backup systems called hybrid energy systems may be the viable option to generate clean electricity for end-user load demand applications, especially in developing African countries where the population is expected to increase to 2.4 billion by 2050 [3]. The electricity consumption in Africa per person is relatively low; however, in 2019, the net electricity generated in Africa is only about 804 TWh compared to the United States, which has total net electricity generated only 20% in the same year. Considering Ethiopia, where 3% of urban electricity have access to off-grid electricity and in rural areas have off-grid electricity access up to 49% only [4], yet as of 2018, approximately 44.98% of the populace have access to electricity [5].

Numerous studies being carried in the research field of RESs technologies by researchers. Castañeda et al. [6] proposed control strategies and sizing technique in a standalone PV/hydrogen-FC/battery energy system in order to well utilization of ES to satisfy the load demand, maintain hydrogen energy reserve and state of charge of the battery bank based on Simulink Design Optimization. Khalid et al. [7] analyzed PV/wind/hydrogen-FC/ battery system for the application of power supply to the residential building to equalize the disparity between supply and demand utilizing the excess electricity generation by the considered energy system using Hybrid Optimization of Multiple Energy Resources (HOMER) software. Tzamalis et al. [8] examined an autonomous power supply of PV/hydrogen-FC/battery and PV/diesel generator/battery system for the rural and remote building with the aid of HOMER simulation software. The simulation results indicated that the increasing nominal PV capacity from 5.5 to 19 kW may eradicate the diesel generator in end-user power systems in both optimal-sized PV and hydrogen technologies. To supply electricity to an inaccessible population in the Amazon region (Tocantins, Brazil), Silva et al. [9] examined an optimized PV/FC/battery system based on technical and economic issues. Ghenai et al. [10] optimized an off-grid PV/FC system to a required residential electrical load consisting of 150 homes located in the Emirate of Sharjah (United Arab Emirates). The renewable fraction acquired from the considered system was 40.2% only.

In the present study, the investigation geographical location considered is Adama City, Ethiopia. The PV system incorporating a two-axis tracker connected to a hydrogen-based FC system and typical electrical load profiles is considered to carry out the performance analysis of the hybrid PV/ hydrogen FC system through HOMER simulation software.

10.2   SYSTEM ARCHITECTURE

The hybrid PV/hydrogen-based FC system architecture is depicted in Figure 10.1. The hybrid off-grid energy system comprises PV modules, direct current (DC) electrolyzer, hydrogen tank and DC FC system. The DC busbar is interlinked with PV modules, FC, electrolyzer and electrical load. The PV modules generate electricity when a sufficient amount of solar irra-diation falls on the top of solar panels. In contrast, FC generates electricity by ingesting hydrogen gas as fuel from a hydrogen tank through a chemical reaction. The hydrogen fuel is produced by electrolyzer through electrolysis of water and the produced hydrogen gas is accumulated in the hydrogen tank. The mathematical equations related to hybrid energy components of the considered system are described in the literature [11].

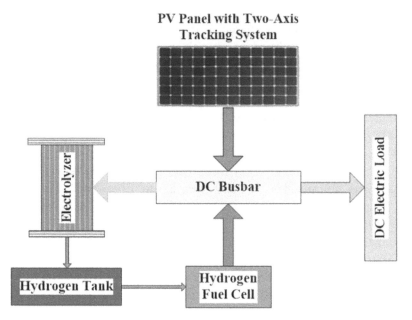

FIGURE 10.1   Hybrid off-grid PV/hydrogen FC energy system.

The PV technology considered is monocrystalline silicon having average efficiency equal to 20%. The nominal operating cell temperature, PV panel lifetime, temperature coefficient and derating factor of power are assumed as 45°C, 25 yrs., –0.35%/°C and 90%, respectively. The ground reflectance value considered is 20%. The PV tracking system incorporated in the present study is a two-axis PV tracking system. The FC lifetime operation hours and minimum load ratio considered are 40,000 hrs. and 0%, respectively. The considered values of electrolyzer lifetime, efficiency and minimum load ratio are 15 yrs., 85% and 0%, respectively. The initial intake level of hydrogen tank and lifetime assumed is 10% and 25 yrs., respectively. The estimated electrical load connected to the hybrid energy system having scaled annual average value obtained is 99.5 kWh/d with average power, peak power and load factor estimated to be 4.14 kW, 11.7 kW and 0.355, respectively. The daily load profile is shown in Figure 10.2.

10.3   SOLAR RADIATION AND TEMPERATURE DATA

The geographical site under investigation is Adama city, Ethiopia. From the National Aeronautics and Space Administration (NASA) Langley Research

Center (LaRC) POWER website [12], the monthly average global horizontal solar radiation and temperature data incurred using geographical coordinates of considered location, i.e., latitude: 8.56331°N and longitude: 39.2884°E as shown in Figure 10.3. It can be seen from Figure 10.3 that the estimated range of monthly average daily solar radiation is found to be 5.23 (July)–6.57 (February) kWh/m²/d, while the average monthly temperature range is estimated to be 16.77°C (December)–20.98°C (June). The average annual solar radiation and temperature are calculated as 6.06 kWh/m²/d and 18.94°C, respectively.

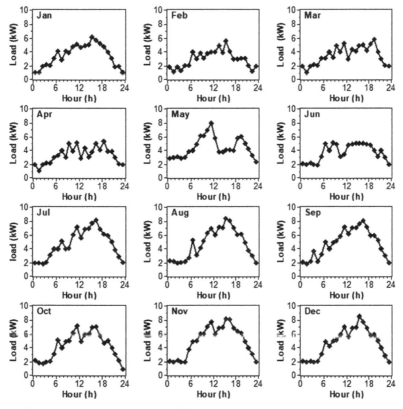

FIGURE 10.2   Typical electric load profile.

10.4   RESULTS AND DISCUSSIONS

In the present study, the hybrid off-grid two-axis PV/FC system is designed, and simulation is carried out using HOMER software. The input parameters such as the size of the PV system considered to be 0, 10, 20, 25, 30 kW. The size considered for FC is 5 kW and 10 kW, hydrogen tank capacity is 0, 20, 25, 30, 35, 40, 45 kg and electrolyzer is 5, 10 kW. The random variability such as day-to-day and time-step-to-time-step are considered to be 2.02% and 12.2%, respectively, for the considered load profile. Moreover, in the simulation, the effect of monthly averaged ambient temperature on PV panels is considered. HOMER accounts for the maximum power point tracker efficiency, which is included in the derating factor. Concerning the connected electrical load, the optimal sizing of the considered energy system is 10 kW FC, 45 kg hydrogen tank, 30 kWp rated PV array and 20 kW electrolyzer.

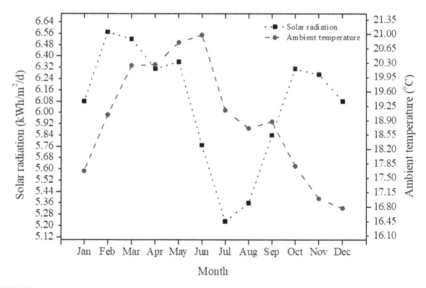

FIGURE 10.3   Monthly average solar radiation and ambient temperature of Adama City, Ethiopia.

The net annual electrical generated from rated PV array found to be 78,336 kWh/yr. (84%) and 14,597 kWh/yr. (16%) from the FC, while the total yearly electrical production from the energy system estimated to be 98,933 kWh/yr. The monthly average electric generation of an optimally sized energy system is presented in Table 10.1. From the rated PV array, the maximum and minimum monthly mean electric power production obtained is 10.204 (December) kW and 7.121 (July) kW, respectively, while the corresponding values for FC were found to be 20.359 (September) kW and

1.3762 (January) kW, in order. The net power generated by rated PV array obtained is 107.39 kW and 19.9498 kW from the FC. The DC primary load and the electrolyzer annual electric energy consumption were found to be 36,317 (46%) kWh/yr. and 42,447 (54%) kWh/yr., respectively. The total annual electric utilization by corresponding electric load estimated as 78,764 kWh/yr. Besides, the energy system is capable of producing excess annual electricity, which is found to be 14,169 (15.2%) kWh/yr. The excess electric production profile is presented in Table 10.2. Also, the energy system found to be zero unmet electric loads and zero capacity shortage throughout the year, while the renewable fraction obtained is 100%.

TABLE 10.1 Monthly Average Electric Production from the PV/Hydrogen FC Energy System

Month	Two-Axis PV Array (kW)	Fuel Cell (kW)
Jan	9.887	1.3762
Feb	9.998	1.1599
Mar	9.276	1.5219
Apr	8.711	1.4152
May	9.044	1.9311
Jun	8.071	1.4872
Jul	7.121	1.9372
Aug	7.349	1.8995
Sep	8.226	2.0359
Oct	9.372	1.4883
Nov	10.131	1.8235
Dec	10.204	1.8739
Total	**107.39**	**19.9498**

With PV penetration of 216%, the estimated daily average energy output and the average power output from the rated PV array obtained is 2,015 kWh/d and 8.9 kW over the year, respectively, while the PV capacity factor obtained is 29.8%. The total PV hour of operation over the year is seen to be 4,421 hr./yr., with the maximum PV power output value found to be 32.7 kW over the year. The hourly power output from the PV array over the year is shown in Figure 10.4.

In the case of the FC system, the net run time of FC generator over the year obtained is 5,249 hr./yr. and the number of times FC generator started annually is estimated to be 428 start/yr., while the operational life incurred by the FC before replacement found to be 7.62 yr. The net annual hydrogen

TABLE 10.2 Daily Profile of Excess Electric Production for Considered PV/Hydrogen FC Energy System

Hour	Jan	Feb	Mar	Apr	May	Jun	Jul	Aug	Sep	Oct	Nov	Dec
0	0	0	0	0	0	0	0	0	0	0	0	0
1	0	0	0	0	0	0	0	0	0	0	0	0
2	0	0	0	0	0	0	0	0	0	0	0	0
3	0	0	0	0	0	0	0	0	0	0	0	0
4	0	0	0	0	0	0	0	0	0	0	0	0
5	0	0	0	0	0	0	0	0	0	0	0	0
6	0	0	0	0	0	0	0	0	0	0	0	0
7	0.56	0.47	0.58	0.18	0.26	0.11	0.11	0.05	0.07	0.32	0.72	0.65
8	1.73	0.97	1.84	0.65	0.10	0.18	0.53	0.23	0.51	0.53	0.99	1.51
9	2.64	2.61	1.68	0.60	0.27	0.44	0.75	0.21	0.76	1.22	1.65	2.00
10	3.06	3.02	1.69	1.14	0.15	1.54	0.38	0.19	0.42	0.42	1.12	2.00
11	2.79	12.67	1.37	1.78	0.01	3.29	0.07	0.11	0.34	0.33	0.86	1.44
12	4.87	18.49	9.82	9.03	0.62	4.99	0.32	0.36	0.69	1.27	2.91	2.72
13	5.11	18.23	13.12	11.56	2.85	6.15	0.05	0.27	0.30	1.07	6.23	6.96
14	5.31	19.50	14.00	13.94	5.47	7.11	0.75	0.21	0.76	0.88	8.60	10.51
15	3.52	16.49	12.84	12.84	7.84	7.08	0.28	0.21	0.26	0.74	7.89	10.33
16	3.21	14.86	10.16	9.82	6.06	5.56	0.21	0.03	0.21	0.23	7.46	9.79
17	2.29	11.33	7.56	7.48	4.46	4.29	0.02	0.01	0.01	0.10	2.28	3.29
18	0	0.04	0	0	0	0	0.19	0	0	0	0	0
19	0	0	0	0	0	0	0	0	0	0	0	0
20	0	0	0	0	0	0	0	0	0	0	0	0
21	0	0	0	0	0	0	0	0	0	0	0	0
22	0	0	0	0	0	0	0	0	0	0	0	0
23	0	0	0	0	0	0	0	0	0	0	0	0

consumption by FC is found to be 876 kg/yr., which is equivalent to the total annual fuel energy input of 29,194 kWh/yr. fed to the FC generator. The capacity factor, average electrical efficiency maximum electrical output and mean electrical output of the FC found to be 16.7%, 50%, 8.4 kW and 2.78 kW, respectively. The hourly FC power output over the operational year is shown in Figure 10.5.

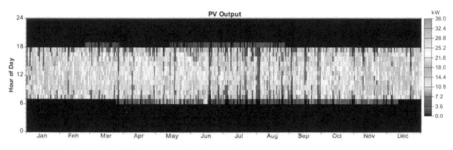

FIGURE 10.4 Hourly PV power output over the year.

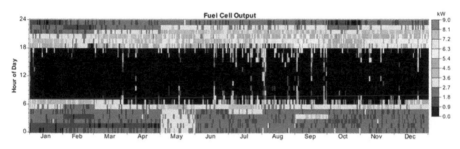

FIGURE 10.5 Hourly FC power output over the year.

The overall hydrogen production by the rated electrolyzer (20 kW) over the year is found to be 915 kg/yr, with a capacity factor value equal to 0.242. The autonomy of the hydrogen tank obtained is 362 hours. Considering the electrolyzer, the monthly average hydrogen production is presented in Table 10.3. The monthly average maximum hydrogen production by electrolyzer obtained is 3.2556 (January) kg/d and minimum hydrogen production is 1.6804 (February) kg/d. The annual average value of hydrogen production by electrolyzer is found to be 2.4983 kg/d. The hourly hydrogen tank storage level is shown in Figure 10.6, which shows from February–June and November-December month, the hydrogen level is above 40 kg.

TABLE 10.3   Hydrogen Production by the Electrolyzer

Month	Monthly Average Hydrogen Production (kg/d)
Jan	3.2556
Feb	1.6804
Mar	2.1694
Apr	2.0441
May	2.7722
Jun	2.1559
Jul	2.1806
Aug	2.2964
Sep	2.7241
Oct	3.2809
Nov	2.7233
Dec	2.6973
Average	2.4983

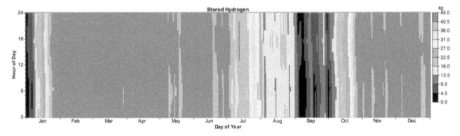

FIGURE 10.6   Hourly hydrogen tank level over the year.

10.5   CONCLUSION

In this chapter, a hybrid off-grid two-axis PV tracking system with a hydrogen-based FC system is designed for Adama city, Ethiopia. A typical load profile is assumed to have a peak demand of 12 kW. To obtain and analyze the system component's performance, the energy system is simulated in HOMER software. From the simulation results, it is found that the energy system is capable of generating access electricity. The hydrogen storage tank is capable of storing hydrogen up to 40 kg from February–June and November–December month.

Furthermore, from the two-axis PV array tracking system and FC system, the total annual power obtained is 107.39 kW and 19.9498 kW, respectively.

In addition, during the energy system operation it is found that no unmet electric load and capacity shortage over the year. Hence, implementing such an energy system may mitigate energy security issues in the nearer future for the end-user electric load demand applications.

KEYWORDS

- **hybrid energy system**
- **hydrogen fuel cell**
- **photovoltaic**
- **photovoltaic tracking system**
- **renewable fraction**
- **solar cell**
- **solar radiation**
- **temperature data**

REFERENCES

1. Salameh, T., Abdelkareem, M. A., Olabi, A. G., Sayed, E. T., Al-Chaderchi, M., & Rezk, H., 2021. Integrated standalone hybrid solar PV, fuel cell and diesel generator power system for battery or supercapacitor storage systems in Khorfakkan, United Arab Emirates. *International Journal of Hydrogen Energy, 46*(8), 6014–6027.

2. Zhang, Y., & Wei, W., (2020). Decentralized coordination control of PV generators, storage battery, hydrogen production unit and fuel cell in islanded DC microgrid. *International Journal of Hydrogen Energy, 45*(15), 8243–8256.

3. *Section Issue Africa.* Available from: https://www.eia.gov/outlooks/ieo/section_issue_Africa.php (accessed on 12 June 2023).

4. *Multi-Tier Framework for Energy,* (2018). Available from: https://trackingsdg7.esmap.org/data/files/download-documents/2019-Tracking%20SDG7-Full%20Report.pdf (accessed on 12 June 2023).

5. *Access to electricity-Ethiopia.* Available from: https://data.worldbank.org/indicator/EG.ELC.ACCS.ZS?end=2018&locations=ET&start=2018 (accessed on 12 June 2023).

6. Castañeda, M., Antonio, C., Francisco, J., Higinio, S., & Luis, M. F., (2013). Sizing optimization, dynamic modeling and energy management strategies of a stand-alone PV/hydrogen/battery-based hybrid system. *International Journal of Hydrogen Energy, 38*(10), 3830–3845.

7. Khalid, F., Ibrahim, D., & Marc, A. R., (2016). Analysis and assessment of an integrated hydrogen energy system. In: *International Journal of Hydrogen Energy.*

8. Tzamalis, G., Zoulias, E. I., Stamatakis, E., Varkaraki, E., Lois, E., & Zannikos, F., (2011). Techno-economic analysis of an autonomous power system integrating hydrogen technology as energy storage medium. *Renewable Energy, 36*(1), 118–124.

9. Silva, S. B., Severino, M. M., & De Oliveira, M. A. G., (2013). A stand-alone hybrid photovoltaic, fuel cell and battery system: A case study of Tocantins, Brazil. *Renewable Energy, 57*, 384–389.

10. Ghenai, C., Tareq, S., & Adel, M., (2020). Technico-economic analysis of off grid solar PV/fuel cell energy system for residential community in desert region. *International Journal of Hydrogen Energy*.

11. Singh, A., Prashant, B., & Bhupendra, G., (2017). Techno-economic feasibility analysis of hydrogen fuel cell and solar photovoltaic hybrid renewable energy system for academic research building. *Energy Conversion and Management*.

12. *Worldwide Renewable Resources Data Access Viewer*. Available from: https://power. larc.nasa.gov/data-access-viewer/ (accessed on 12 June 2023).

CHAPTER 11

Efficient Integration of Distributed Generation in Radial Distribution Network for Voltage Profile Improvement and Power Loss Minimization via Particle Swarm Optimization

PRAGYA GURU,[1] NITIN MALIK,[1] and SHEILA MAHAPATRA[2]

[1]*School of Engineering and Technology, The NorthCap University, Gurgaon, Haryana, India*

[2]*School of Engineering and Technology, Alliance College of Engineering and Design, Bangalore, Karnataka, India*

ABSTRACT

The distribution system is a last and the final link between the generating stations and the consumers end. The integration of distributed generation (DG) resources in distribution system is crucially important due to ever-growing increased energy demand. The ideal allocation (sizing and siting) of DG could lead to accomplish the various benefits like improved voltage profile and reduced losses. The load is modeled as constant complex power. The power flow solution for the type-1 DG integrated test system is computed using direct load flow approach under normal loading conditions for a balanced distribution system. The problem is formulated as single-objective constrained optimization problem. PSO algorithm is used for optimal allocation of DG in radial distribution system. The performance is tested and validated on four IEEE standard bus systems viz. 12-bus, 33-bus, 34-bus and 69 bus systems. The percentage of real power loss reduction

The Internet of Energy: A Pragmatic Approach Towards Sustainable Development. Sheila Mahapatra, Mohan Krishna S., B. Chandra Sekhar, & Saurav Raj (Eds.)

using PSO approach is by 48.30%, 47.36%, 57.73% and 63.05% for IEEE 12-bus, 33-bus, 34-bus, and 69-bus system respectively. The appreciable rise in the minimum voltage using proposed methodology is in the range of 3.8%–6.5% for all the four test systems respectively. The annual energy loss savings for type I DG also shows the remarkable benefits compared to other published results. The proposed methodology is further compared and tabulated with analytical approach as well as other methods to exemplify the superiority of the proposed work.

11.1 INTRODUCTION

11.1.1 *MOTIVATION*

The incorporation of distributed generation (DG) resources in a distribution system are important due to ever-growing increased energy demand, complexity and economic concern with traditional power generation sources, gradual depletion of conventional resources, and environmental concern [1]. The term DG at distribution level represents small, dispersed generation units close to load center to fulfill the local energy demand. If the operation of the DG system is technically efficient, it helps to cleaner power production, i.e., reduction of greenhouse gas emission, because it uses non-conventional resources like small hydro, fuel cells, photovoltaic systems, wind turbines, etc.

With regards to the ideal allocation (location and capacity) of DG, we can accomplish the various merits, including improving the power quality, increasing the voltage profile while reducing the contamination of the environment. But inappropriate allocation may lead to adverse effects on the system stability and some other parameters like excessive temperature and power losses [2, 3]. The impacts of DG may manifest them either positively or negatively according to the DG characteristics or system operating conditions. For the operation of distribution power utilities, future planning, and power transfer between utilities, load flow (LF) calculations are required.

11.1.2 *LITERATURE REVIEW*

The increased load demand and the allocation schemes of the DG are categorized as either analytical method or a meta-heuristic method. Analytical methods use mathematical expressions to identify locations. A new expression is proposed in Refs. [4, 5] to calculate the DG capacity but without cost-benefit consideration. Backtracking algorithm was used in Ref. [6] to find out the most effective capacity and place for DG. In Ref. [7], authors' presents mixed-integer non-linear programming for the most appropriate DG allocation. The impact of ZIP load model has been analyzed in Ref. [8] but without cost analysis. In Ref. [9], author represents a new voltage stability index for most advantageous DG penetration for different types of DGs. The authors presented an index-based multi-objective approach [10] for finding the ideal size and place of DG but only type I DG was considered. A methodology [11] established from the real and reactive branch current components was evolved to minimize the loss in RDN by locating DG. The sensitivity analysis-based approach [12] to power loss minimization is applied for allocation.

A variety of meta-heuristic approaches have been utilized for DG allocation problems, including non-dominated sorting genetic algorithm (GA) [13] and ant lion optimization algorithm [14]. GA [15] is employed for DG incorporation at various loading conditions. The allocation problem can be resolved with bacterial foraging algorithm [16]. To determine the optimal allocation under varying loading conditions, other approaches like bat algorithm (BA) [17], artificial bee colony (ABC) algorithm [18] and hybrid GA-particle swarm optimization (PSO) algorithm [19] are employed.

11.1.3 PAPER TARGET

A massive motivation of the presented work is developing an efficient methodology to discover the optimal place and capacity of DG for reduced network losses and achieving a better voltage profile in the RDN using population based PSO [20] method. The recommended approach has been examined on various standard IEEE test networks. The final evaluation of the approach is compared to the existing techniques and is found out to be superior in terms of minimum losses and refinement in voltage profile.

11.1.4 ORGANIZATION

The systematized paper arrangement is as follows: Section 11.2 explains the mathematical modeling of RDN. Explanation of the objective function formation is in Section 11.3. Section 11.4 is dealing with the direct load flow (DLF) solution. A short explanation of the PSO algorithm is depicted in Section 11.5. In Sections 11.6 and 11.7, solution methodology and results of the test network simulations are shown, respectively. Finally, Section 11.8 concluded the overall work carried out.

11.2 MATHEMATICAL MODELING OF RDN

11.2.1 LINE MODELING

Figure 11.1 demonstrates a sample RDN taken for a case study and Figure 11.2 shows an equivalent circuit diagram of single branch section connecting sending node i and receiving node j, respectively. From Figure 11.2, we find,

$$S_i = P_i + jQ_i \tag{1}$$

$$S_i = V_i \times I_i^* \tag{2}$$

where; V_i, Q_i, and P_i shows the receiving end voltage, reactive power load, and the real power load at bus i, respectively.

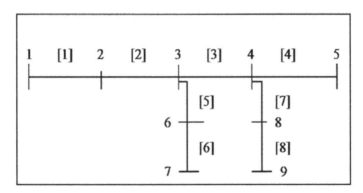

FIGURE 11.1   Sample RDN.

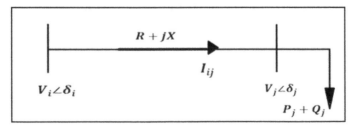

FIGURE 11.2   Equivalent circuit network of single branch.

The corresponding injected branch current at the ith bus (I_i) is given as:

$$I_i = \left(\frac{S_i}{V_i}\right)^* = \frac{P_i - jQ_i}{V_i^*} \tag{3}$$

where; sending end voltage is V_i. The branch real power loss (P_{loss}) between two nodes is given by Eqn. (4)

$$P_{loss}(i, j) = \left(\frac{P_i^2 + Q_i^2}{V_i^2}\right) R \tag{4}$$

11.2.2   LOAD MODELING

The load actions can be modeled by analyzing the variation in their reactive and active power demands due to changes in network voltage. The load model is assumed as constant power load type.

$$P_i = P_0 V^0 \tag{5}$$

$$Q_i = Q_0 V^0 \tag{6}$$

where; Q_0 and P_0 are the nominal reactive and active power corresponding to the initial operating voltage, respectively.

11.2.3   DISTRIBUTED GENERATION (DG)

With small size, DG supplies are generally served as constant negative PQ load (constant power mode) with current injecting at a node. If we choose node i for DG penetration, then Eqn. (7) and (8) can be used to calculate the equivalent real as well as reactive load at bus i, respectively.

$$P_{eq,i} = P_i - P_{DG,i} \tag{7}$$

$$Q_{eq,i} = Q_i - Q_{DG,i} \tag{8}$$

where; $P_{DG,i}$ is generated real and $Q_{DG,i}$ is the reactive power of DG at i^{th} node.

11.2.4   COST ANALYSIS FOR ANNUAL ENERGY LOSSES AND DG COST

The total cost of energy losses per year (CL) is given by Eqn. (9).

$$CL = ((Total\,real\,power\,loss) * (E_C * T)\,\$ \tag{9}$$

where; E_C and T are the energy rate and annual time duration, respectively with values of 0.06 \$/kWh and 8,760 hours.

The cost characteristics of DG for real power are based on information available from [21] and expressed in terms of system's real power generation.

$$C(P_{DG}) = a * P_{DG}^2 + b * P_{DG} + c\,\$/h \tag{10}$$

Based on Ref. [21], cost coefficients are: $a = 0$, $b = 20$ and $c = 0.25$

11.3 PROBLEM FORMULATION

The key aim for DG placement is to minimize power losses subject to constraints [22]. The objective function is given by Eqn. (11)

$$F = minimise\,(P_{loss}(i, j)) \tag{11}$$

Subjected to the following constraints:

$$PG_i - PD_i - PL_i = 0 \text{ Real power balance} \tag{12}$$

$$QG_i - QD_i - QL_i = 0 \text{ Reactive power balance} \tag{13}$$

$$0.95\,p.u \le V_i \le 1.05\,p.u \tag{14}$$

$$i_i \le i_i^{Max} \text{ line current at node } I \tag{15}$$

$$P_{Gi}^{Min} \le P_{Gi} \le P_{Gi}^{Max} \text{ Real power limits} \tag{16}$$

$$Q_{Gi}^{Min} \le Q_{Gi} \le Q_{Gi}^{Max} \text{ Reactive power limits} \tag{17}$$

$$0 \le P_{DG,i} \le P_{DG,i}^{Max} \tag{18}$$

$$0 \le Q_{DG,i} \le Q_{DG,i}^{Max} \tag{19}$$

Q_{Gi} and P_{Gi} are the generated reactive and real power at i^{th} bus, respectively. QL_i and PL_i are the reactive and real power loads at i^{th} bus, respectively. P_{Gi}^{Max}, P_{Gi}^{Min}, Q_{Gi}^{Min} and Q_{Gi}^{Max} are the max real, the min real, min reactive and the max reactive power generation at i^{th} bus, respectively. $P_{DG,i}^{Max}$ is the maximum real power and $Q_{DG,i}^{Max}$ is the reactive power propagated by DG at the ith node.

11.4 LOAD FLOW ANALYSIS

Conventional LF approaches, for example, Newton-Raphson and fast decoupled methods are less effective in distribution networks because of its low X/R ratio. A number of specially designed LF approaches have been proposed in the literature [23–29] for distribution systems. In this technique a DLF approach [16] has been used, developed on the basis of topological structure and widely implemented for distribution system LF analysis.

This LF approach is established from the analog's current injection. With the help of the Kirchhoff's current law, the correlation among the bus current injections (from Eqn. (3)) and branch currents is accomplished.

$$B1 = I1 + I2 + I3 + I4 + I5 + I6 + I7 + I8 \tag{20}$$

$$B3 = I3 + I4 + I5 + I6 + I7 + I8 \tag{21}$$

$$B4 = I4 + I5 + I6 + I7 + I8 \tag{22}$$

$$B6 = I6 \tag{23}$$

$$B7 = I7 + I8 \tag{24}$$

$$B8 = I8 \tag{25}$$

The matrix formation of above equations can be given by,

$$
\begin{bmatrix} B1 \\ B2 \\ B3 \\ B4 \\ B5 \\ B6 \\ B7 \\ B8 \end{bmatrix} =
\begin{bmatrix}
1 & 1 & 1 & 1 & 1 & 1 & 1 & 1 \\
0 & 1 & 1 & 1 & 1 & 1 & 1 & 1 \\
0 & 0 & 1 & 1 & 1 & 1 & 1 & 1 \\
0 & 0 & 0 & 1 & 1 & 1 & 0 & 0 \\
0 & 0 & 0 & 0 & 1 & 1 & 0 & 0 \\
0 & 0 & 0 & 0 & 0 & 1 & 0 & 0 \\
0 & 0 & 0 & 0 & 0 & 0 & 1 & 1 \\
0 & 0 & 0 & 0 & 0 & 0 & 0 & 1
\end{bmatrix}
\begin{bmatrix} I1 \\ I2 \\ I3 \\ I4 \\ I5 \\ I6 \\ I7 \\ I8 \end{bmatrix}
\tag{26}
$$

By rewriting Eqn. (26),

$$[B] = [BIBC][I] \tag{27}$$

where; bus-injections to branch-currents matrix are shown by BIBC. The voltages at bus 3, bus 4, and bus 5 can be given as:

$$V3 = V2 - (B3 * Z23) \tag{28a}$$

$$V4 = V3 - (B4 * Z34) \tag{28b}$$

$$V5 = V4 - (B5 * Z45) \tag{28c}$$

From the above equations,

$$V5 = V2 - (B3 * Z23) - (B4 * Z34) - (B5 * Z45) \tag{29}$$

Similarly, the voltage drops of a sample RDN given in Figure 11.1 can be obtained by:

$$
\begin{bmatrix} V1 \\ V1 \\ V1 \\ V1 \\ V1 \\ V1 \\ V1 \\ V1 \end{bmatrix}\begin{bmatrix} V2 \\ V3 \\ V4 \\ V5 \\ V6 \\ V7 \\ V8 \\ V9 \end{bmatrix} = \begin{bmatrix} Z12 & 0 & 0 & 0 & 0 & 0 & 0 & 0 \\ Z12 & Z23 & 0 & 0 & 0 & 0 & 0 & 0 \\ Z12 & Z23 & Z34 & 0 & 0 & 0 & 0 & 0 \\ Z12 & Z23 & Z34 & Z45 & 0 & 0 & 0 & 0 \\ Z12 & Z23 & Z34 & Z45 & Z56 & 0 & 0 & 0 \\ Z12 & Z23 & Z34 & Z45 & Z56 & Z67 & 0 & 0 \\ Z12 & Z23 & Z34 & 0 & 0 & 0 & Z78 & 0 \\ Z12 & Z23 & Z34 & 0 & 0 & 0 & Z78 & Z89 \end{bmatrix}\begin{bmatrix} B1 \\ B2 \\ B3 \\ B4 \\ B5 \\ B6 \\ B7 \\ B8 \end{bmatrix} \tag{30}
$$

Eqn. (30) can be written as,

$$[\Delta V] = [BCBV][B] \tag{31}$$

Where branch currents and buses voltages are correlated by BCBV matrix. From the Eqns. (27) and (31),

$$[\Delta V] = [BCBV][BIBC][B] \tag{32}$$

$$[\Delta V] = [DLF][I] \tag{33}$$

Solutions of the Eqns. (34)–(36) gives LF results.

$$I_i^k = \left(\frac{S_i}{V_i^k} \right)^* \tag{34}$$

$$[\Delta V^k] = [DLF][I^k] \tag{35}$$

$$[V^{k+1}] = [V_0][\Delta V^{k+1}] \tag{36}$$

where; initial voltage V_0 and the iteration count is 'k.'

11.5 PARTICLE SWARM OPTIMIZATION

PSO is a probabilistic computation inspired by a simplified societal system like fish schooling, bird flocking, etc. [31]. PSO is a successful optimization approach having balanced global and local search throughout the run and has been tested in many application areas. PSO gets faster and better results with minimum computational burden when compared to the other methods. In this approach, the current state of each particle changes in a search space. If $S_{id} = (S_{i1}, S_{i2}, \ldots, S_{in_d})$ and $v_{id} = (v_{i1}, v_{i2}, \ldots, v_{in_d})$ represent the position and velocity of particle p, respectively; $d = 1, 2, \ldots n_d$ and $i = 1, 2, \ldots N$. Swarm size represented by N and dimension represented by n_d. The velocity (v) and position (S) of particles in d-dimensional hyperspace are updated by using Eqns. (37) and (38), respectively.

$$v_{id}^{(q+1)} = \omega_i v_{id}^{(q)} + C_1 rand1 * \left(Pbest_{id} - S_{id}^{(q)} \right) + C_2 rand2 * \left(Gbest_{id} - S_{id}^{(q)} \right) \quad (37)$$

$$S_{id}^{(q+1)} = S_{id}^{(q)} + v_{id}^{(q+1)} \quad (38)$$

The inertia weight given to each velocity is decreased linearly with iteration number as in Eqn. (39).

$$\omega_i = \omega_{max} - \frac{[(\omega_{max} - \omega_{min}) * currentiterationnumber]}{Maximumiteration} \quad (39)$$

where; $S_{id}^{(q)}$ is position and $v_{id}^{(q)}$ is particle (p) current velocity at q^{th} iteration. The random values rand1 and rand2 are generated in the span of [0,1]. Cognitive learning factor is C_1 and social coefficient is C_2. The term Gbest and Pbest represent the global best value and the personal best value of the particle, depending on its personal experience and experience of the overall swarm, respectively.

11.6 SOLUTION METHODOLOGY

The algorithmic steps in optimal DG allocation are as follows:

➤ **Step 1:** Initialize line data, bus data and bus voltage.
➤ **Step 2:** Run the direct power flow algorithm for the base case as explained in Section 11.4 to determine system power losses, branch current, phase angle and bus voltage magnitude at all the nodes.
➤ **Step 3:** Choose the algorithm parameters swarm size, inertia weights and acceleration coefficients for objective function minimization. The selected parameters are given in Table 11.1.
➤ **Step 4:** Initialize the iteration count as zero k=0.
➤ **Step 5:** For each particle, compute the objective function value as given in Eqn. (11) and compare it with the individual Pbest.
➤ **Step 6:** If there is no violation in operational constraints, calculate the total power loss (TRPL) for each particle. Else particle is rejected, regarded as infeasible solution.
➤ **Step 7:** Make a comparison of the objective function value of every particle over the individual Pbest value. A lower function value than Pbest is set to become the new Pbest.

➢ **Step 8:** Among all the Pbest values, lowest Pbest is assigned as the Gbest.

➢ **Step 9:** Utilizing Eqns. (31)–(33) for updating velocity, position and weight, respectively for each particle.

➢ **Step 10:** Increase the iteration count by 1 and repeat steps 5 through 9 until convergence criterion determined by maximum number of iterations is achieved.

➢ **Step 11:** Print out the optimal solution for the best possible place and size for DG. The corresponding fitness values indicate minimal system loss.

The flowchart for DG placement is shown in Figure 11.3.

11.7   SIMULATION RESULTS

The suggested technique is simulated for the TRPL, bus voltages, CL and cost of DG power. The obtained outcomes are analyzed with the previously published outcomes for the identical load models and base values. The simulations are running on Intel Pentium dual-core processor 2 GHz, 3 GB RAM and carried out in commercial software MATLAB R2018a with Windows 10 operating system. The 12-bus RDN has (0.4350+0.4050i) MVA power of demand. There are three laterals in 33-bus system [32] consuming (3.715+2.3i) MVA power demand. The 34-bus system has three laterals and (4.6365+2.8735i) MVA power demand whereas the large scale 69-bus system [33] has seven laterals and 48 load points with (3.803+j.22693) MVA load. The base values for all four systems are chosen as 12.66 kV and 100 MVA. The four test systems are examined for two individual cases.

➢ **Case 1:** In the absence of DG. Base case results are shown in Table 11.2 for all four test systems.

➢ **Case 2:** With DG. Tables 11.3–11.6 shows the test results intended for standard IEEE bus systems.

11.7.1   RESULTS FOR 12 BUS SYSTEM

The TRPL for 12 bus system and minimum voltage value (at bus 12) has been calculated as 0.0207 MW and 0.9434 p.u, respectively. The CL before placing the DG is 10,887 $. With DG allocation, the power loss is decreased to 0.0107 MW. The favorable size of DG is 0.2370 MW at bus 9. The minimum voltage at bus 12 has been increased to 0.9835 p.u. The CL after DG allocation is decreased to 5,662 \$. The cost of PDG is 5.006 \$/h (Table 11.4).

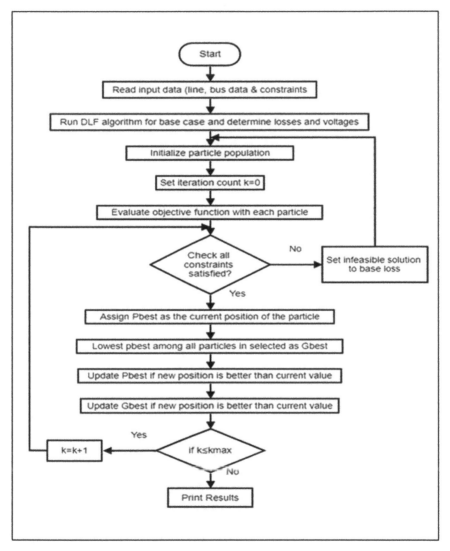

FIGURE 11.3 Flow chart for DG allocation in RDN using PSO algorithm.

TABLE 11.1 Implementation Parameters for PSO

Population Size	Max Iteration	C1, C2	$\omega_{max}, \omega_{min}$
10	100	0.7, 0.7	0.9, 0.1

TABLE 11.2   Base Case Results

Bus Network	TRPL (MW)	Voltage (p.u.)	Cost of Energy Loss ($)
12	0.0207	0.9434	10,887
33	0.2109	0.9038	1,10,890
34	0.2217	0.9417	1,16,550
69	0.2249	0.9417	1,18,160

TABLE 11.3   12-Bus Results After DG Placement

Test Network	With DG
Bus no. for DG allocation	9
Size of DG (MW)	0.2370
TRPL (MW)	0.0107
Percentage reduction in loss	48.30
Voltage (p.u.)	0.9835
CL ($)	5,662
Cost of PDG ($/h)	5.006

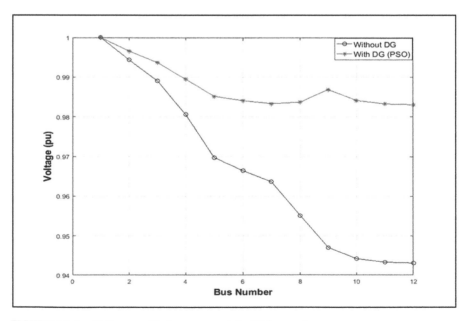

FIGURE 11.4   Voltage profile improvement for 12-bus system.

11.7.2   RESULTS FOR 33 BUS SYSTEM

The TRPL for the 33-bus network and minimum voltage value (at 18 bus) have been calculated as 0.2109 MW and 0.9038 p.u. The CL before placing the DG is 110890 $. The system real power losses after DG allocation are 0.1110 MW. The favorable DG size is 2.5902 MW and placed at 6th bus. The voltage at 18th bus is enhanced to 0.9424 pu. The CL after DG allocation is minimized to 58356 $. The cost of PDG is 52.04 $/h.

TABLE 11.4   Results for the 33-Bus Network with DG

Test Network	With DG
Bus no. for DG allocation	6
Size of DG (MW)	2.5902
TRPL (MW)	0.1110
% Reduction in loss	47.36
Voltage (p.u.)	0.9424
CL ($)	58,356
Cost of PDG ($/h)	52.04

11.7.3   RESULTS FOR 34 BUS SYSTEM

The TRPL for 34-bus network and minimum voltage value (at 27 bus) has been calculated as 0.2217 MW and 0.9417 p.u. The CL before the DG installation is 116550 $. The system real power losses with DG allocation are 0.0937 MW. The most favorable size of DG is 2.9666 MW and placed at 21st bus. The voltage at the bus 27 is increased to 0.9835 p.u. The CL after DG allocation is decreased to 49,272 $. The cost of PDG is 59.58 $/h (Table 11.5).

TABLE 11.5   Results for 34-Bus System After DG Placement

Test System	With DG
Bus no. for DG allocation	21
Size of DG (MW)	2.9666
Real power loss (MW)	0.0937
% Reduction in loss	57.73
Voltage (p.u.)	0.9835
CL ($)	49,272
Cost of PDG ($/h)	59.58

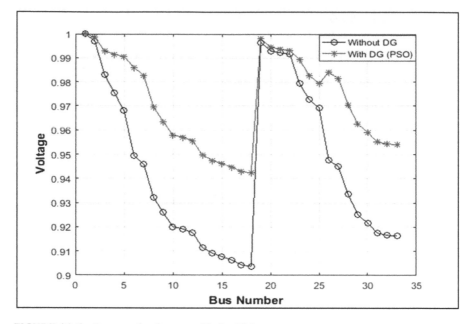

FIGURE 11.5   Improved voltage profile for 33-bus system.

11.7.4   RESULTS FOR 69 BUS SYSTEM

The TRPL for 69 bus network and minimum voltage value (at 65 bus) has been calculated as 0.2249 MW and 0.9417 p.u. The CL before the DG installation is 1,18,160 $. The TRPL with DG is 0.0831 MW. The most favorable DG size is 1.8717 MW and placed at bus 61. The voltage at 65 bus is increased to 0.9789 p.u. The CL after DG allocation is reduced to 43,710 $. The cost of PDG is 36.40 $/h (Tables 11.6 and 11.7).

TABLE 11.6   Results for 69-Bus Network After DG Placement

Test Network	With DG
Bus no. for DG allocation	61
Size of DG (MW)	1.8717
TRPL (MW)	0.0831
% Reduction in loss	63.05
Voltage (p.u.)	0.9789
CL ($)	43,710
Cost of PDG ($/h)	36.40

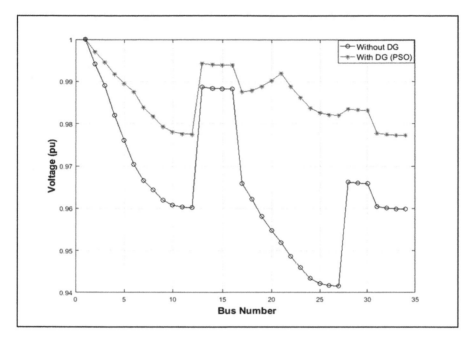

FIGURE 11.6 Improved voltage profile for 34-bus network.

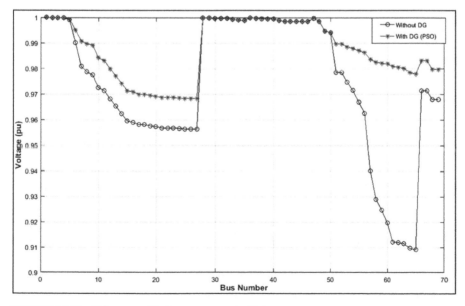

FIGURE 11.7 Voltage profile improvement in 69-bus system.

TABLE 11.7 Comparative Study of Outcomes for Different Test Networks

Bus Network	Total Load (MVA)	Method	Optimum DG Allocation	Optimum DG Size (MW)	∑ Ploss with DG (MW)	Ratio (Reduction in Power Loss/DG Size)
12	0.4350 + 0.4050i	Proposed	9	0.2378	0.0107	0.042
		Acharya [3]	9	0.2272	0.0108	0.043
		Gozel [4]	9	0.2272	0.0113	0.041
33	3.8021 + 2.6945i	Proposed	6	2.5902	0.1110	0.038
		Kansal [34]	6	3.150	0.1150	0.030
		Lalitha [30]	6	2.4948	0.1114	0.038
34	4.6365 + 2.8735i	Proposed	21	2.9666	0.0937	0.043
		Acharya [3]	21	2.8848	0.0938	0.044
		Gozel [4]	21	2.8848	0.0990	0.042
69	3.8021 + 2.6945i	Proposed	61	1.8717	0.0831	0.075
		Acharya [3]	61	1.8078	0.0833	0.078
		Gozel [4]	61	1.8078	0.0920	0.073

11.8 CONCLUSIONS

In this chapter, PSO algorithm is used for optimum positioning of DG in RDN. This analysis is mapped out as a constrained single-objective optimization problem. The power flow solution incorporated with DG is computed using DLF approach under normal loading conditions. The CL and expenditure on DG are also taken into account. The upgraded outcomes are observed for Type I DG placement. The appreciable rise in minimum voltage, significant diminution in network power loss, reduced CL, cost of DG and the ALS is also analyzed. The percentage diminution in real power loss by employing PSO is 48.30%, 47.36%, 57.73% and 63.05% for IEEE 12-bus, 33-bus, 34-bus, and 69-bus system, respectively. The appreciable rise in the minimum voltage by using proposed methodology is in the range of 3.8%–6.5% for all the four test systems, respectively. The ALS for Type I DG also showed significant advantages over the published results. The proposed methodology is further compared and tabulated in Table 11.7 with analytical approach as well as other methods to exemplify the supremacy of the planned work.

KEYWORDS

- bus system
- cost analysis
- distributed generation
- line modeling
- load flow analysis
- optimal positioning
- particle swarm optimization
- radial distribution network

REFERENCES

1. Dey, B., Raj, S., Mahapatra, S., & Fausto, P. G. M. (2022). Optimal scheduling of distributed energy resources in microgrid systems based on electricity market pricing strategies by a novel hybrid optimization technique. *International Journal of Electrical Power & Energy Systems, 134*, 107419.
2. Borges, C. L. T., (2012). An overview of reliability models and methods for distribution systems with renewable energy distributed generation. *Renewable and Sustainable Energy Reviews, 16*, 4008–4015.
3. Escalera, A., Hayes, B., & Prodanović, M., (2018). A survey of reliability assessment techniques for modern distribution networks. *Renewable and Sustainable Energy Reviews, 91*, 344–357.
4. Acharya, N., Mahat, P., & Mithulananthan, N., (2006). An analytical approach for DG allocation in primary distribution network. *International Journal of Electrical Power & Energy Systems, 28*, 669–678.
5. Gözel, T., & Hocaoglu, M. H., (2009). An analytical method for the sizing and siting of distributed generators in radial systems. *Electric Power Systems Research, 79*(6), 912–918.
6. El-Fergany, A., (2015). Optimal allocation of multi-type distributed generators using backtracking search optimization algorithm. *International Journal of Electrical Power & Energy Systems, 64*, 1197–1205.
7. Rueda-Medina, A. C., Franco, J. F., Rider, M. J., Padilha-Feltrin, A., & Romero, R., (2013). A mixed-integer linear programming approach for optimal type, size and allocation of distributed generation in radial distribution systems. *Electric Power Systems Research, 97*, 133–143.
8. Parihar, S. S., & Malik, N., (2021). Optimal allocation of multiple DG in RDS using PSO and its impact on system reliability. *Facta Universitatis, Series: Electronics and Energetics, 34*, 219–237.

9. Parihar, S. S., & Malik, N., (2020). Optimal integration of multi-type DG in RDS based on novel voltage stability index with future load growth. *Evolving Systems*, 1–15.
10. Aman, M. M., Jasmon, G. B., Mokhlis, H., & Bakar, A. H. A., (2012). Optimal placement and sizing of a DG based on a new power stability index and line losses. *International Journal of Electrical Power & Energy Systems*, *43*, 1296–1304.
11. Viral, R., & Khatod, D. K., (2015). An analytical approach for sizing and siting of DGs in balanced radial distribution networks for loss minimization. *International Journal of Electrical Power & Energy Systems*, *67*, 191–201.
12. Kashem, M. A., Le, A. D., Negnevitsky, M., & Ledwich, G., (2006). Distributed generation for minimization of power losses in distribution systems. In: *2006 IEEE Power Engineering Society General Meeting* (p. 8).
13. Singh, D., Singh, D., & Verma, K. S., (2008). GA based energy loss minimization approach for optimal sizing & placement of distributed generation. *International Journal of Knowledge-based and Intelligent Engineering Systems*, *12*, 147–156.
14. Ali, E. S., Abd, E. S. M., & Abdelaziz, A. Y., (2017). Ant lion optimization algorithm for optimal location and sizing of renewable distributed generations. *Renewable Energy*, *101*, 1311–1324.
15. Shukla, T. N., Singh, S. P., Srinivasarao, V., & Naik, K. B., (2010). Optimal sizing of distributed generation placed on radial distribution systems. *Electric Power Components and Systems*, *38*, 260–274.
16. Devi, S., & Geethanjali, M., (2014). Application of modified bacterial foraging optimization algorithm for optimal placement and sizing of distributed generation. *Expert Systems with Applications*, *41*, 2772–2781.
17. Candelo-Becerra, J. E., & Hernández-Riaño, H. E., (2015). Distributed generation placement in radial distribution networks using a bat-inspired algorithm. *Dyna*, *82*, 60–67.
18. Abu-Mouti, F. S., & El-Hawary, M. E., (2011). Optimal distributed generation allocation and sizing in distribution systems via artificial bee colony algorithm. *IEEE Transactions on Power Delivery*, *26*, 2090–2101.
19. Moradi, M. H., & Abedini, M., (2012). A combination of genetic algorithm and particle swarm optimization for optimal DG location and sizing in distribution systems. *International Journal of Electrical Power & Energy Systems*, *34*, 66–74.
20. Mahapatra, S., Badi, M., & Raj, S., (2019). Implementation of PSO, it's variants and hybrid GWO-PSO for improving reactive power planning. In: *2019 Global Conference for Advancement in Technology (GCAT)* (pp. 1–6). IEEE, Bangalore.
21. Murthy, V. V. S. N., & Kumar, A., (2013). Comparison of optimal DG allocation methods in radial distribution systems based on sensitivity approaches. *International Journal of Electrical Power & Energy Systems*, *53*, 450–467.
22. Mahapatra, S., Malik, N., & Jha, A. N., (2020). Cuckoo search algorithm and ant lion optimizer for optimal allocation of TCSC and voltage stability constrained optimal power flow. In: *International Conference on Intelligent Computing and Smart Communication* (Vol. 2019, pp. 889–905). Springer, Singapore.
23. Zimmerman, R. D., & Chiang, H. D., (1995). Fast decoupled power flow for unbalanced radial distribution systems. *IEEE Transactions on Power Systems*, *10*, 2045–2052.
24. Zhang, F., & Cheng, C. S., (1997). A modified Newton method for radial distribution system power flow analysis. *IEEE Transactions on Power Systems*, *12*, 389–397.

25. Expósito, A. G., & Ramos, E. R., (1999). Reliable load flow technique for radial distribution networks. *IEEE Transactions on Power Systems, 14*, 1063–1069.
26. Chang, G. W., Chu, S. Y., & Wang, H. L., (2007). An improved backward/forward sweep load flow algorithm for radial distribution systems. *IEEE Transactions on Power Systems, 22*, 882–884.
27. Kresting, W. H., & Mendive, D. L., (1976). An application of ladder network theory to the solution of three phase radial load flow problem. In: *IEEE PES Winter Meeting*, 76044–76048.
28. Jabr, R. A., (2006). Radial distribution load flow using conic programming. *IEEE Transactions on Power Systems, 21*, 1458–1459.
29. Wang, Z., Chen, F., & Li, J., (2004). Implementing transformer nodal admittance matrices into backward/forward sweep-based power flow analysis for unbalanced radial distribution systems. *IEEE Transactions on Power Systems, 19*, 1831–1836.
30. Lalitha, M. P., Reddy, V. V., Reddy, N. S., & Reddy, V. U., (2011). DG source allocation by fuzzy and clonal selection algorithm for minimum loss in distribution system. *Distributed Generation & Alternative Energy Journal, 26*, 17–35.
31. Guru, P., Malik, N., & Mahapatra, S., (2019). Optimal allocation of distributed generation for power loss minimization using PSO algorithm. In: *2019 3rd International Conference on Recent Developments in Control, Automation & Power Engineering 2019* (pp. 22–26). IEEE, Noida.
32. Baran, M. E., & Wu, F. F., (1989). Network reconfiguration in distribution systems for loss reduction and load balancing. *IEEE Power Engineering Review, 9*, 101–102.
33. Ranjan, R., Venkatesh, B., & Das, D., (2003). Load-flow algorithm of radial distribution networks incorporating composite load model. *International Journal of Power & Energy Systems, 23*, 71–76.
34. Kansal, S., Kumar, V., & Tyagi, B., (2013). Optimal placement of different type of DG sources in distribution networks. *International Journal of Electrical Power & Energy Systems, 53*, 752–760.

CHAPTER 12

Classical and Predictive Control of Interior Permanent Magnet Synchronous Motor for Railway Application

MANNAN HASSAN,[1] MUHAMMAD SUHAIL SHAIKH,[2]
MUHAMMAD SHAHID MASTOI,[1] RAO ATIF,[1] MUHAMMAD FARHAN,[3]
MUHAMMAD AMJAD,[4] MUHAMMAD BILAL SHAHID,[1,4] and
ABDUL LATIF SHAH[5]

[1]*Ministry of Education Key Laboratory of Magnetic Suspension Technology and Maglev Vehicle, School of Electrical Engineering, Southwest Jiaotong University, Chengdu, China*

[2]*School of Physics and Electronic Engineering, Hanshan Normal University, Guangdong, 52104, PR China*

[3]*Department of Electrical Engineering and Technology, Government College University Faisalabad, Pakistan*

[4]*Department of Electronic Engineering, Faculty of Engineering, The Islamia University of Bahawalpur, Pakistan*

[5]*Newports Institute of Communications and Economics, Karachi, Pakistan*

ABSTRACT

The classical and predictive controllers for interior permanent-magnet synchronous machines (IPMSM) railway traction drives were covered in this chapter. Introduction to traction drive systems, mathematical modeling of ipmsm in three phases and dq, predictive control, direct torque control, field-oriented controllers are described in the sections.

The Internet of Energy: A Pragmatic Approach Towards Sustainable Development. Sheila Mahapatra, Mohan Krishna S., B. Chandra Sekhar, & Saurav Raj (Eds.)

12.1   INTRODUCTION

According to DIN 57115 (VDE 0115) part 1, "track-bound and non-track-bound transport systems for passenger and freight transportation" [1] is the most complete definition of a "railway." A strength converter, an everlasting magnet ac machine (PMAM), sensors, and a managed set of rules are defined because of the 4 principal additives of an ac force in modern-day transportation systems. Permanent magnet synchronous automobiles are presently broadly hired withinside the railway traction business (trains, trams, locomotives, and so on). Due to their advantages, higher efficiency drive systems, smaller size, low electrical losses, power density, low torque ripples, flux weakening performance, and torque-speed cures of PMSM smooth and favorable for traction drives as compared to other drives [3, 4].

The basic traction drive system is described in Figure 12.1. Traction drive system owing to their performance capacity, robustness, cost, etc., in a simplified way, that it is composed of:

- An electric motor, which generates the rotational movement;
- A power electronic converter, which supplies the electric motor taking the energy from a specific source of energy, enabling the controlled rotational movement of an electric motor;
- A control algorithm, which is in charge of controlling the power electronic converter to obtain the desired performance of the electric motor; and
- An energy source, which in some cases is part of the electric drive and in other cases is considered an external element.

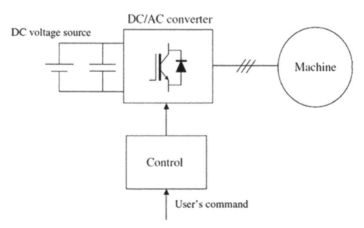

FIGURE 12.1   Basic electric drive configuration.

12.2 MATHEMATICAL MODELING OF INTERIOR PERMANENT MAGNET SYNCHRONOUS MOTOR

12.2.1 INTRODUCTION OF PMSM

Synchronous AC machines are those in which the rotor speed in electrical units is equal to the frequency of the stator current in steady-state circumstances. In general, synchronous machines outperform induction motors in terms of performance, but at a greater cost and with less resilience, due to a more complex rotor design. Synchronous machines have been utilized for decades in high-power applications, and they continue to play an important role. Synchronous machines (rotating or linear and three-phase or multiphase) are different types. Wound field (WFSM), Synchronous reluctance (SyRM), Permanent magnet (PM) with different structures such as Axial, Transversal, Radial, Interior, Surface, Trapezoidal (brushless DC, BLDC), and Sinusoidal.

The design of electrical machines was presented in materials to hold the magnetic field in the 1950s. Permanent magnets are used in the rotor of the PMSM design to set it apart from conventional motors. A permanent magnet for a high-speed application is mounted on an inside magnet, and magnets are surface mounted for maximum power density. The internal and surface mount motor rotor designs are shown in Figure 12.2. The salient pole is the mathematical modeling of the motor with three-phase stator winding, sinusoidal back emf, and rotor construction. For the following differential equations, eddy current, iron saturation, and hysteresis losses are considered to be zero.

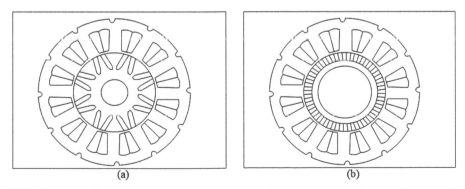

(a) (b)

FIGURE 12.2 Structure of permanent magnet synchronous machine (a) IPMSM; and (b) SPMSM.

Magnets embedded in the rotor generate the rotor flux. The magnets are buried in the rotor in Figure 12.2(a), resulting in an internal permanent magnet synchronous machine. As a result of the magnet, the d axis has a huge air gap. The rotor on the q axis, on the other hand, is largely made of iron. As a result, the permeability along the d axis is much lower than the permeability along the q axis (i.e., Lds< Lqs). In Figure 12.2(b), magnets are placed on the rotor surface, resulting in a surface permanent magnet synchronous machine (SPMSM).

12.2.2   *DYNAMIC MODELS OF SYNCHRONOUS MACHINES USING THREE-PHASE VARIABLES*

Major properties of the stator of three-phase synchronous machines are similar to those of the induction machine covered in the previous chapter. A three-phase synchronous machine's stator voltage equations are thus similar to those of an induction machine. For your convenience, we've replicated them here:

$$
\begin{bmatrix} v_{as} \\ v_{bs} \\ v_{cs} \end{bmatrix} = R_s \begin{bmatrix} i_{as} \\ i_{bs} \\ i_{cs} \end{bmatrix} + \frac{d}{dt} \begin{bmatrix} \psi_{as} \\ \psi_{bs} \\ \psi_{cs} \end{bmatrix}
\tag{1}
$$

The fluxes in the stator are as follows:

$$
\begin{bmatrix} \psi_{as} \\ \psi_{bs} \\ \psi_{cs} \end{bmatrix} = \begin{bmatrix} L_{aa} & L_{ab} & L_{ac} \\ L_{ba} & L_{bb} & L_{bc} \\ L_{ca} & L_{ca} & L_{cc} \end{bmatrix} \begin{bmatrix} i_{as} \\ i_{bs} \\ i_{cs} \end{bmatrix} + \begin{bmatrix} \psi_{am} \\ \psi_{bm} \\ \psi_{cm} \end{bmatrix}
\tag{2}
$$

The basic wave of the fluxes created in the rotor by the magnets or the rotor circuit may be described as follows:

$$
\begin{bmatrix} \psi_{apm} \\ \psi_{bpm} \\ \psi_{cpm} \end{bmatrix} = \begin{bmatrix} \psi_{pm} \cdot \cos \theta_m \\ \psi_{pm} \cdot \cos \left(\theta_m - \dfrac{2\pi}{3} \right) \\ \psi_{pm} \cdot \cos \left(\theta_m + \dfrac{2\pi}{3} \right) \end{bmatrix}
\tag{3}
$$

Asymmetric synchronous machine's stator-voltage equation:

$$
\begin{bmatrix} v_{as} \\ v_{bs} \\ v_{cs} \end{bmatrix} = R_s \cdot \begin{bmatrix} i_{as} \\ i_{bs} \\ i_{cs} \end{bmatrix} + \begin{bmatrix} L_s & 0 & 0 \\ 0 & L_s & 0 \\ 0 & 0 & L_s \end{bmatrix} \cdot \frac{d}{dt} \begin{bmatrix} i_{as} \\ i_{bs} \\ i_{cs} \end{bmatrix} - \omega_m \cdot \psi_{pm} \begin{bmatrix} \sin\theta_m \\ \sin\left(\theta_m - \dfrac{2\pi}{3}\right) \\ \sin\left(\theta_m + \dfrac{2\pi}{3}\right) \end{bmatrix} \tag{4}
$$

12.2.3 DYNAMIC MODELS IN A SYNCHRONOUS REFERENCE FRAME

A revolving d–q reference frame with the d axis aligned with the rotor flux can be used to change the dynamic model of synchronous machines. It should be noted that the rotor flux angle for synchronous machines is the same as the rotor angle. This transformation has significant advantages for both analysis and control.

$$
V_{ds} = R_s i_{ds} - \omega_m \psi_{qs} + \frac{d\psi_{ds}}{dt} \tag{5}
$$

$$
V_{qs} = R_s i_{qs} + \omega_m \psi_{ds} + \frac{d\psi_{qs}}{dt} \tag{6}
$$

$$
\frac{di_{ds}}{dt} = -\frac{R_s}{L_d} i_{ds} + \frac{L_q}{L_d} \omega_e i_{qs} + \frac{V_{ds}}{L_d} \tag{7}
$$

$$
\frac{di_{qs}}{dt} = -\frac{L_{ds}}{L_q} \omega_e i_{ds} - \frac{R_s}{L_q} i_{qs} + \frac{V_{ds}}{L_q} + \frac{\psi_m}{L_q} \omega_e \tag{8}
$$

The following are the motor's electromagnetic torque and speed equations:

$$
T_e = 1.5p\left(\psi_m i_{qs} + \left(L_d - L_q\right) i_{ds} i_{qs}\right) \tag{9}
$$

$$
\frac{d\omega_m}{dt} - \frac{1}{J}\left(1.5p\left(\psi_m i_{qs} + \left(L_d - L_q\right) i_{ds} i_{qs}\right) - D\omega_m - T_l\right) \tag{10}
$$

The d–q axis stator currents and voltages are represented by i_q, i_d, V_q and V_d, respectively. The permanent magnet's flux linkage is m, the stator resistance is Rs, the mechanical rotor angular speed is m, and the d–q axis inductances are L_q and L_d.

The flow equation for the Stator is:

$$
\begin{aligned}
\psi_{qs} &= i_{qs} L_{qs} \\
\psi_{ds} &= i_{ds} L_{ds} + \lambda_m \\
\psi_s &= \sqrt{\psi_{ds}^2 + \psi_{qs}^2}
\end{aligned} \tag{11}
$$

As shown in Figure 12.3, a variety of control strategies are employed to regulate the variables of power converters. The most well-known and easiest to implement of these are hysteresis and linear controls with PWM [5–7]. Highly configurable digital signal processors and microcontrollers are being developed. However, the implementation of current and more intelligent control methods is possible.

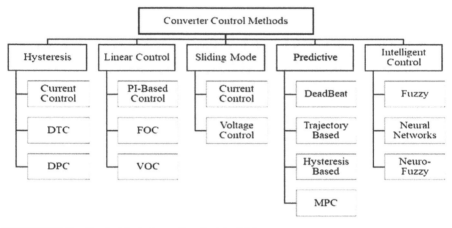

FIGURE 12.3   Control techniques for electrical drives.

12.2.4   REGION OF OPERATION

Figure 12.4 shows a typical IPMSM current trajectory on the id iq plane, which may consist of three or four lines or regions. The maximum torque per ampere (MTPA) methdology delivers desired torque with the least amount of phase current in Region 1 shown in Figure 12.4 with the curve OX. Along the torque curve that represent as hyperbola, Region 2 (hyperbola XY) pulls the existing trajectory away from the MTPA curve. Optional Region 3 (arc YB) initiates the flux-weakening process by moving the current trajectory along the current restriction circle. In Region 4 (curve BE), the maximum torque per voltage (MTPV) operation provides the greatest torque possible under the inverter voltage limitation [26].

12.3   FIELD ORIENTED CONTROL

Blachke [8] proposed the first work on FOC in 1971 for induction motors, and the approach has since evolved to its current state. The use of a rotating coordinate system enables decoupled control of electrical torque and rotor flux, which is one of FOC's main advantages. Eqn. (9) describes the torque and flux caused by PMSM currents Eqn. (11) There are three steps in the FOC process: The three-phase currents are first converted to dq currents. The park inverse transformation is then used to calculate voltage signals. Finally, the SVPWM modulation may be used to compute the inverter switch signals. Figure 12.5 depicts the fundamental diagram of PMSM's FOC.

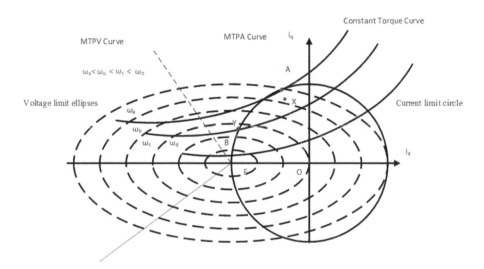

FIGURE 12.4   Region of operation of IPMSM drives.

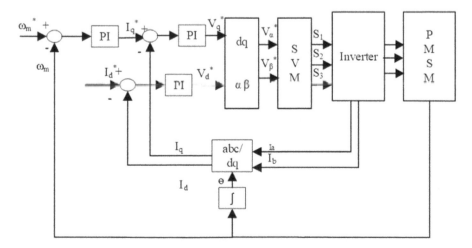

FIGURE 12.5   Control block diagram of field-oriented control.

The sensor measured i_a, i_b, i_c three-phase currents and motor speed, then converts ia and ib into identity notification and iq using the Park and Clarke transformation. The outside velocity loop provides the velocity sign mistakes, whereas the inside modern-day loop provides the modern-day sign blunders. The PI controller receives the velocity sign blunders, which creates the iq reference modern-day, and the PI controller receives the modern-day sign blunders, which generates the ud and uq reference voltages. Inverse park transformation is applied for voltage signals u_a and u_b generated from u_d and u_q. To manage the inverter detailed below, the PWM responsibility cycle is computed, and the SVPWM modulator delivers pulsating alerts.

12.3.1   PULSE WIDTH MODULATION SPACE VECTOR THEORY

SVM (space vector PWM) is a sophisticated and computationally demanding PWM algorithm. It might be the finest PWM approach among other inverter PWM methods. SVM (space vector modulation) is a pulse width modulation control method (PWM). It is most typically used to drive three-phase AC motors at varied DC speeds utilizing several Class D amplifiers to create alternating current (AC) signals. The decrease of total harmonic distortion (THD) produced by the frequent changes in these algorithms is an important topic of research.

Ts=1/fs samples the reference signal V_{ref} at a frequency of fs. With the help of the transform, the reference signal may be created from three independent phase references. After that, the reference vector is created by combining the two neighboring energetic switching vectors and one or each of the zero vectors. The order of the vectors and the zero vector(s) to utilize can be determined using a variety of ways. The harmonic content material and switching losses will be affected by the strategy used.

Switching sequence of voltage vector is shown in Figure 12.6. A 60° angle occurs between two non-zero vectors that are contiguous. The two zero vectors (V_0 and V_7) are the origins and deliver zero voltage to the load at the same time. V_0, V_1, V_2, V_3, V_4, V_5, V_6, and V_7 are the eight vectors that make up the spatial basis vectors. In the d-q plane, the required output voltage may be translated in the same way to provide the required reference voltage vector V_{ref}. Using eight switching modes, the PWM space vector method approximates the reference voltage vector V_{ref}. When an investor trades for a short period of time, a simple approximation is to acquire an average result, in which case T equals V_{ref} for the same time period.

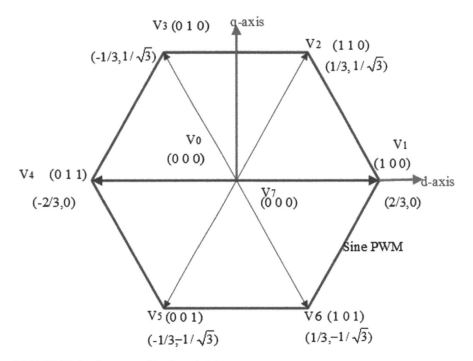

FIGURE 12.6 Sectors and basic switching vectors.

As a result, space vector PWM can be applied using following methods below:

> **Step 1:** Determine V_d, V_q, V_{ref}, and angle (α) in step one.

$$V_d = V_{an} - V_{bn}\cos(60) - V_{cn}\cos(30) = V_{an} - \frac{V_{bn}}{2} - \frac{V_{cn}}{2} \qquad (12)$$

$$V_q = 0 \quad V_{bn}\cos(30) - V_{cn}\cos(30) - \sqrt{3}\frac{V_{bn}}{2} - \sqrt{3}\frac{V_{cn}}{2} \qquad (13)$$

$$V_{ref} = \sqrt{V_d^2 + V_q^2} \qquad (14)$$

$$\alpha = arctan\left(\frac{V_q}{V_d}\right) = \omega t = 2\pi f t \qquad (15)$$

> **Step 2:** T_1, T_2, and T_0 are the time intervals to be determined (Figure 12.7). The following formula may be used to calculate the change over time duration:

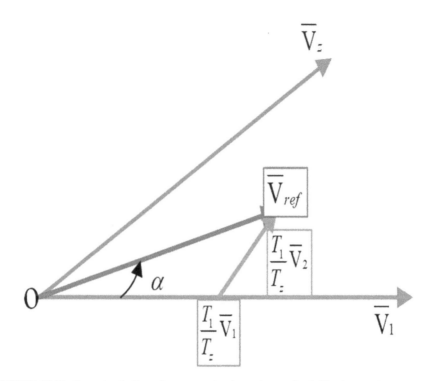

FIGURE 12.7   In sector 1, the reference vector is made up of neighboring vectors.

- Sector 1 switching time period:

$$\int_0^{Tz} \overline{V}_{ref}\ dt = \int_0^{T_1} \overline{V}_1\ dt + \int_{T_1}^{T_1+T_2} \overline{V}_2\ dt + \int_{T_1+T}^{Tz} \overline{V}_0\ dt \tag{16}$$

$$\therefore Tz \cdot \overline{V}_{ref} = \left(T_1 \cdot \overline{V}_1 + T_2 \cdot \overline{V}_2 \right)$$

$$\Rightarrow Tz \cdot |\,\overline{V}ref\,| \cdot \begin{bmatrix} cos(\alpha) \\ sin(\alpha) \end{bmatrix} = T_1 \cdot \frac{2}{3} \cdot V_{dc} \cdot \begin{bmatrix} 1 \\ 0 \end{bmatrix} + T_2 \cdot \frac{2}{3} \cdot V_{dc} \cdot \begin{bmatrix} cos(60) \\ sin(60) \end{bmatrix} \tag{17}$$

where; $(0 \leq \alpha \leq 60)$

$$\therefore T1 = Tz.a.\frac{sin(\frac{\pi}{3} - \alpha)}{sin(\frac{\pi}{3})} \quad \text{and } T_0 = Tz - (T_1 + T_2), \text{ where } Tz = \frac{1}{fz} \text{ and}$$

$$\alpha = \frac{|\overline{V}_{ref}|}{\frac{2}{3}V_{dc}}$$

- Switching time in the sector:

$$T_1 = \frac{\sqrt{3}.T_z.|V_{ref}|}{V_{dc}}\left(sin\left(\frac{\pi}{3} - \alpha + \frac{n-1}{3}\pi\right)\right) = \frac{\sqrt{3}.T_z.|V_{ref}|}{V_{dc}}\left(sin\left(n\frac{\pi}{3} - \alpha\right)\right)$$

$$= \frac{\sqrt{3}.T_z.|V_{ref}|}{V_{dc}} sin(n\frac{\pi}{3})cos(\alpha) - cos(n\frac{\pi}{3})sin(\alpha) \tag{18}$$

$$\therefore T_2 = \frac{\sqrt{3}.T_z.|V_{ref}|}{V_{dc}}\left(sin\left(\alpha - \frac{n-1}{3}\pi\right)\right)$$

$$= \frac{\sqrt{3}.T_z.|V_{ref}|}{V_{dc}} sin(\alpha)cos(\frac{n-1}{3}\pi) - cos(\alpha)sin(\frac{n-1}{3}\pi) \tag{19}$$

where; n = 1, 2, 3, 4, 5, 6 (i.e., sector 1 to 6).

➢ **Step 3:** Sector determination: Each PWM cycle samples the reference vector at a specified input sampling frequency (fs). The sector is selected at this time, and the modulation vector is assigned to two neighboring sectors. The non-zero vector looks like this:

$$\bar{v}_m = \frac{2}{3}v_{dc}e^{j(m-1)\frac{\pi}{3}} \tag{20}$$

Therefore the non-zero vector for $\bar{v}_m$ and $\bar{v}_{m+1}$ become

$$\bar{v}_m = \frac{2}{3}v_{dc}\left[cos(m-1)\frac{\pi}{3} + j\,sin(m-1)\frac{\pi}{3}\right] \tag{21}$$

$$\bar{v}_{m+1} = \frac{2}{3}v_{dc}\left[cos(m\frac{\pi}{3}) + j\,sin(m\frac{\pi}{3})\right] \tag{22}$$

The reference voltage vector $\bar{v}_{ref}$ can be expressed as a function of $\bar{v}_m$ and $\bar{v}_{m+1}$ as follows:

$$\bar{v}_{ref}\frac{T_s}{2} = \bar{v}_m T_a + \bar{v}_{m+1}T_b$$

$$\bar{v}_{ref} = \bar{v}_d + j\bar{v}_q \tag{23}$$

The active state vector and the time required for each sample period are represented by Ta and Tb, respectively, while k is the sector number indicating the reference point. Calculate Ta and Tb, then apply them to the switch to construct the PWM vector space's sector switching pattern (Table 12.1).

TABLE 12.1 Duty Cycle Calculation

Sector	Degrees	Ta	Tb
1	$0 \leq \theta \leq 60°$	$\dfrac{3v_\alpha}{4} - \sqrt{3}\,\dfrac{v_\beta}{4}$	$\sqrt{3}\,\dfrac{v_\beta}{2}$
2	$60° \leq \theta \leq 120°$	$\dfrac{3v_\alpha}{4} + \sqrt{3}\,\dfrac{v_\beta}{4}$	$\dfrac{-3v_\alpha}{4} + \sqrt{3}\,\dfrac{v_\beta}{4}$
3	$120° \leq \theta \leq 180°$	$\sqrt{3}\,\dfrac{v_\beta}{2}$	$\dfrac{-3v_\alpha}{4} - \sqrt{3}\,\dfrac{v_\beta}{4}$
4	$180° \leq \theta \leq 240°$	$\dfrac{-3v_\alpha}{4} + \sqrt{3}\,\dfrac{v_\beta}{4}$	$-\sqrt{3}\,\dfrac{v_\beta}{2}$
5	$240° \leq \theta \leq 300°$	$\dfrac{-3v_ä}{4} - \sqrt{3}\,\dfrac{v_ä}{4}$	$\dfrac{3v_\alpha}{4} - \sqrt{3}\,\dfrac{v_\beta}{4}$
6	$300° \leq \theta \leq 360°$	$-\sqrt{3}\,\dfrac{v_\beta}{2}$	$\dfrac{3v_\alpha}{4} + \sqrt{3}\,\dfrac{v_\beta}{4}$

> ➢ **Step 4:** Control the switching time of each switch (S1 to S6). The switching sequence should be configured to minimize the switching frequency of each inverter branch. You can run SVPWM using different switching modes. Only two adjacent active vectors and two zero vectors are used to avoid switching losses. To meet this optimal condition, each switching cycle begins with a zero vector and ends with a different zero vector.

12.4 DIRECT TORQUE CONTROL

The inverter's optimal voltage switching vectors directly regulate the stator flux linkage and electromagnetic torque, whereas the voltage source inverter that supplies the PM synchronous motor is controlled by direct torque. The major aim is to determine the voltage conversion vectors that will result in the quickest electromagnetic torque response. Both flux error and torque error are inputs to the flux linkage and torque delay comparator, and VSI's DTC PM synchronous motor control diagram has all of the necessary information. Flux error and torque error are used as inputs for flux linkage and torque-delay comparison.

The electromagnetic torque comparator is a two-level comparator, whereas the magnetic flux coupling comparator is a three-level comparator. The hysteresis comparator's discretized output is utilized as an entry for top-of-the-line voltage switching. The stator's location, on the other hand, is undetermined. Within the lookup table, the flux linkage area vector (area number) is also provided. Six voltage switching vectors and six voltage switching vectors are shown in Figure 12.8 Angles u (1).... u Eqn. (6).

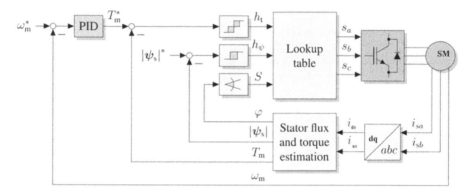

FIGURE 12.8   Block diagram of DTC.

Using the converter's active space vectors, the stator flux may be influenced in several ways throughout the time period Ts. Regardless of flux location, three of these voltage vectors enhance flux magnitude while others reduce it. In addition, three of the space vectors induce an anticlockwise rotation, which raises the relative angle, while the others create a clockwise rotation, which reduces it.

To separately regulate stator flux and torque, it must be assured that the amount of flux is exclusively regulated by the application of suitable voltage vectors, while torque may be raised or lowered by modifying the relative angle (–) within the interval [–/2; /2]. The direct component of the voltage vector-only affects the magnitude of the stator flux vector, but the indirect component affects torque by changing the angle of the stator flux vector.

Furthermore, both have indirect components that cause torque to rise, whereas and may be utilized to decrease torque. As a consequence, both the amplitude of the flux and the angle of the torque may be increased [9, 10]. The voltage vector can also be employed to lower stator flux and torque. The link between voltage space vectors and their effects on flux and torque within each field is summarized in a look-up table (Table 12.2). The control error sampled at the time instant $t_0 = nTs$ determines the flux controller's output value. As a result, a similar relationship exists between the torque controller's controller error eT and controller output hT. The torque controller's output can be one of three values: -1, 0, or 1. Only the numbers -1 and 1 are accepted by the flux controller. The corresponding control variable must be lowered if the controller output is -1, whereas the control variable must be increased if the controller output is 1. The torque controller, on the other hand, must keep the control variable constant if the output is zero.

TABLE 12.2   Lookup Table for DTC

h_ψ	h_T	S_1	S_2	S_3	S_4	S_5	S_6
1	1	V_2	V_3	V_4	V_5	V_6	V_1
1	0	V_0	V_7	V_0	V_7	V_0	V_7
1	-1	V_6	V_1	V_2	V_3	V_4	V_5
-1	1	V_3	V_4	V_5	V_6	V_1	V_2
-1	0	V_7	V_0	V_7	V_0	V_7	V_0
-1	-1	V_5	V_6	V_1	V_2	V_3	V_4

12.5   MODEL PREDICTIVE CONTROL

Predictive control has emerged as a viable alternative to converter control among all of these current control strategies. It includes a variety of controllers, including MPC, deadbeat, hysteresis-based, and trajectory-based control. All predictive control is based on obtaining a mathematical model of the system in order to anticipate the future response of the quantities to be controlled. As a result, based on the established optimization condition, select the appropriate actuation. The goal of the hysteresis-based predictive control approach is to keep the controlled variable inside the boundaries of an area, whereas the variables in the trajectory-based control technique must follow a predetermined course.

Figure 12.9 provides a basic concept of the prediction horizon of MPC. Furthermore, MPC's fundamental principle is to employ the system's model to predict a few steps forward in time (Prediction horizon) based on present conditions, which may be observed or approximated. A cost function is used to decrease the differences between the reference and predicted variables of interest. This technique results in an optimal control law of a certain dimension (Control horizon). In the next sample, just the first entry of this dimension will be used, and the operation will be repeated (Receding Horizon).

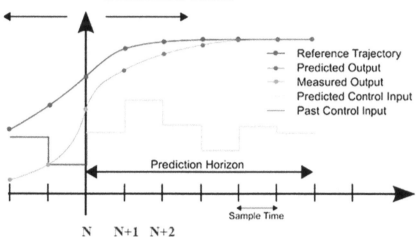

FIGURE 12.9   The basic concept of MPC.

The optimization criteria in the deadbeat control approach are to pick actuation that causes the error to be near to zero at future sampling instants. Cost function The distinction between these two control methods is that continuous control-set MPC and deadbeat control both need the use of a modulator to provide the appropriate voltage. As a result, the switching frequency will remain constant. The other predictive controllers generate switching signals directly to operate the converter without the need for a modulator. As a result, the fluctuating switching frequency will result [11]. Predictive control offers a number of advantages that make it a good choice for power converter control: minimization or optimization, which is employed in MPC, is a more flexible optimization criterion.

- Nonlinearities may be accounted for in modeling without the necessity for model linearization;
- Concepts are natural and easy to grasp;
- Technique is applicable to a wide range of constructions;
- Nonlinearities and limitations may be dealt with easily;
- The controller is simple to set up;
- Avoiding cascaded schemes is simple;
- It is possible to produce fast transient behavior.

The books [12–14] and survey studies [15–18] cited above provide applications and theoretical conclusions. MPC has the virtue of being able to govern limited nonlinear MIMO systems collectively and transparently. According to the control set of the power converter, there are two basic types of MPC.

- MPC continuous control set (CCS-MPC);
- MPC finite control set (FCS-MPC).

Model predictive control (MPC) is a sophisticated process control technology that is used to govern a process while meeting a set of criteria. The main benefit of MPC is that it allows you to optimize the current timeslot while also considering future timeslots. This is accomplished by optimizing a finite time horizon while only implementing the current timeslot and then optimizing again and again, as opposed to a linear–quadratic regulator (LQR). In addition, MPC can predict future occurrences and execute appropriate control actions.

12.5.1   MODEL PREDICTIVE CURRENT CONTROL

A shown in Figure 12.10. To simplify the FCS-MPC optimization process, the modulation stage can be removed by adopting a discrete model of the power converter. Because there are a finite number of switching states, the optimization process can be shortened by selecting the optimum state from among all those that correspond to the lowest cost function value. Furthermore, if the horizon length is set to N=1, the computation overhead is minimized and implementation becomes simple.

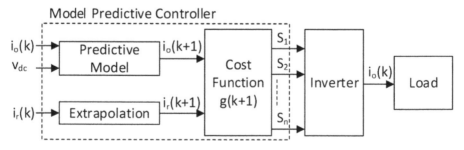

FIGURE 12.10   Block diagram of FCS-MPC.

Voltage and current sensors are employed in this stage to monitor control variables including phase current, capacitor voltages, and dc-link at the nth instant. The continuous-time model in Eqns. (7) and (8) is converted to discrete-time using the Euler approximation approach to forecast d–q axis stator currents.

$$i_{ds}(N+1) = \frac{L_d - T_s R_s}{L_d} i_{ds}(N) + \frac{T_s L_q}{L_q} \omega_e(N) i_{qs}(N) + \frac{T_s V_{ds}(N)}{L_d} \tag{24}$$

$$i_{qs}(N+1) = -\frac{L_d - T_s R_s}{L_q} i_{qs}(N) - \frac{T_s L_d}{L_q} \omega_e(N) i_{ds}(N) + \frac{T_s V_{ds}(N)}{L_q} + \frac{T_s \psi_m}{L_q} \omega_e(N) \tag{25}$$

Block diagram of model predictive current control is shown in Figure 12.11. In the discrete-time domain, the traditional MPC algorithm is developed. The MPC control technique necessitates several calculations, resulting in action delay. If the delay problem is not addressed, the controller control process will be delayed and will continue, potentially deteriorating control performance. As a result, time-delay compensation must be considered. Time delay compensation is a simple and effective method of computing time, and prediction control will anticipate the amount of delay compensation in a single value. A PI controller is used to reference the torque-producing current, which is then used to regulate the speed. This current is tracked using a predictive current controller. The stator current components for the seven various voltage vectors generated by the inverter are predicted using the machine's discrete-time model in the predictive approach. For the whole sampling interval, the voltage vector that minimizes a cost function is chosen and used [19].

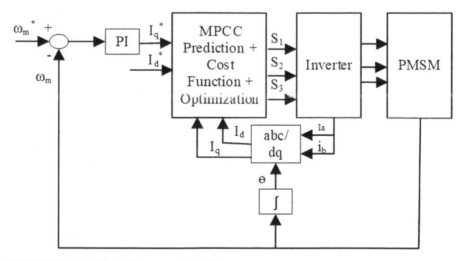

FIGURE 12.11 Block diagram of MPCC.

The goals of the predictive current control scheme are as follows:

- Current magnitude limiting;
- Torque current reference tracking;
- Torque by ampere optimization.

Tracking changes in reference and decreasing tracking errors are key components of model predictive control. As a result, cost functions must play an important role in control systems depending on the cost functions selected.

$$g(N) = (i_{ds}^*(N) - i_{ds}(N+1))^2 + (i_{qs}^*(N) - i_{qs}(N+1))^2 + \hat{L} \qquad (26)$$

A non-linear function is inserted in the cost function for the present restriction, which may be stated as:

$$\hat{L}i_{ds}(N+1), i_{qs}(N+1) = \begin{cases} \infty \, if \, |i_{ds}(N+1)| \geq i_{max} \, or \, |i_{qs}(N+1)| \geq i_{max} \\ 0 \, if \, |i_{ds}(N+1)| \leq i_{max} \, or \, |i_{qs}(N+1)| \leq i_{max} \end{cases} \qquad (27)$$

12.5.2 MODEL PREDICTIVE TORQUE CONTROL

The following steps are included in model predictive torque control measurement and estimate, prediction, and cost function reduction. The suggested method measures the stator current, voltage, and speed directly from the inverter terminals and the motor shaft. The PI-speed controller provides the torque reference. To develop a current prediction equation and forecast the current under the operation of eight voltage vectors, first construct a model predictive torque control equation. Second, the following control cycle's flux and torque are projected using the expected current, which is scrolled and compared. Finally, in the following control cycle, the voltage vector with the effect closest to the goal flux and torque is chosen and applied. Calculate the difference between the anticipated flux and the torque using the provided flux and torque Eqn. (28). Compare and choose the voltage vector that results in the minimum flux and torque differential, which is then used in the next control cycle. Because the variables had various units and orders of magnitude in value, the control technique was implemented by including the control variables in the cost function with weighting factors. The initial weighting factor λ in the cost function modifies the trade-off between the torque and flux terms. It's worth noting that changing the weighting factor between two primary variables like torque and flux has a direct impact on how well they compare.

$$g(k) = \left(\frac{1}{T_{en}} \times \left| T_e^*(k) - T_e(k+1) \right| + \lambda \times \frac{1}{\psi_n} \times \left| \psi_s^* - \psi_s(k+1) \right| \right. \tag{28}$$

Because MPTC combines the system model directly with the finite switching states, it is more effective and accurate in voltage vector selection than direct torque control. The FCS-MPTC has no inner current controller loop. Therefore, it has a fast dynamic response. The speed of the motor is controlled by an outer speed-loop with PI controller. According to the research, in comparison with industry practiced control strategies such as FOC and DTC, FCS-MPTC's structure is simple and it can achieve similar and in some cases, better performance. According to the literature, a detailed comparison between DTC, FOC, and FCS-MPTC is listed in Table 12.3. Nevertheless, FCS-MPTC has some limitations such as high computational time, variable switching frequency, weighting factor tuning, availability of the finite number of voltage vectors and higher current distortion which are the new research focuses of the MPC-researchers [20–27].

TABLE 12.3 Comparison of MPTC. MPCC, DTC, and FOC Methods

Index	MPTC	MPCC	FOC	DTC
Algorithm	Simple	Simple	Complex	Simple
Axis transformation	Not required	Not required	Required	Not required
Parameter sensitivity	Low	Low	High	Low
Position sensor	Not required	Not required	Required	Not required
Switching frequency	Variable	Variable	Constant	Variable
Dynamic response	Fast	Fast	Slow	Fast
Switching table	Can be used	Not required	Not required	Required
System's nonlinearity	Can be included	Can be included	Hard to include	Hard to include
Modulation block	No	No	Yes	No
Weighting factor	Yes	No	No	No
Computational time	High	High	Low	Low

12.6 SIMULATION RESULTS

Variations in load and speed are used to test the performance of the proposed design and other controllers. IPMSM motor specifications used in the railway traction drive system [27–31] is used in this chapter. For the PI speed controller, the anti-windup technique clamping is employed. The given design has been tested for high-speed operation. In the suggested design, the Maximum Torque per Ampere (MTPA) principle is applied for reference current.

12.6.1   STEADY-STATE RESPONSE

The speed response findings show significant disparities between the predictive-based flux-weakening algorithm and the conventional controller, even though both predictive control methods can track the orders well. Analysis of steady-state response predictive controls and classical controls is performed at rated speed and rated loaded torque of the system. Time 0-sec reference speed of 189 rad/sec is applied and load torque is 800 Nm. Speed analysis of describe controller is shown in Figure 12.12.

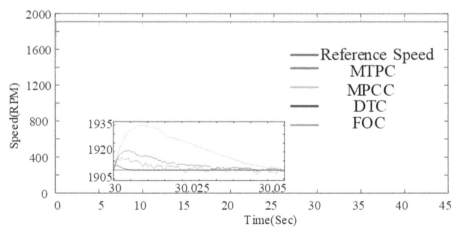

FIGURE 12.12   Steady-state response of speed of MPTC, MPCC, DTC, and FOC.

12.6.2   DYNAMIC RESPONSE

At step input of reference speed of 50 rad/sec up to 2 sec and 189 rad/sec up to 5 sec, a constant load of 800 Nm is utilized to validate the dynamic performance of the IPMSM. Figure 12.13 depicts the speed performance of all methods. In Figure 12.14, the simulation results of MPTC, MPCC, FOC, and MMPC proposed methods are shown.

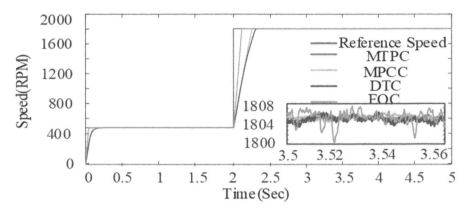

FIGURE 12.13 Dynamic performance of speed at constant load and step reference speed.

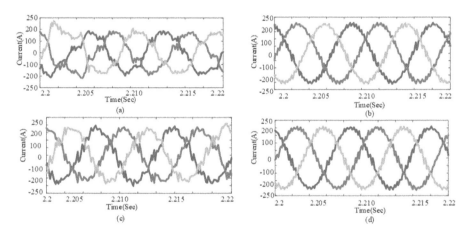

FIGURE 12.14 Comparison of current dynamic performance. (a) The FOC method; (b) method of the MPCC; (c) the DTC method; and (d) method of MTPC.

12.6.3 DYNAMIC PERFORMANCE OF CONSTANT LOAD AND RAMP SPEED

The train goes through acceleration, uniform motion, and a slowdown in the railway traction system. These sorts of procedures will now have to be certified for dynamic analysis. Table 12.4 shows the operating status in detail. Figure 12.15 depicts the speed performance of a constant load and a ramp. The conclusion is that the suggested architecture performs better than alternative controllers. As a result, during acceleration and deceleration, the suggested mixed model predictive control is efficient and resilient.

TABLE 12.4   Reference Ramp Speed Details

Application Time (Sec)	Load (Nm)	Speed (rad/sec)
0–5	800	100
5–10		20*(t–5) + 100
10–15		200
15–20		–20*(t–5) + 200
20–25		100

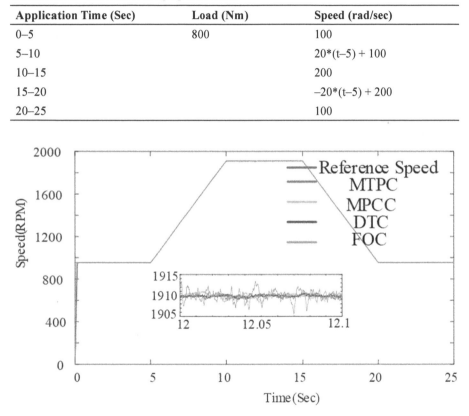

FIGURE 12.15   Dynamic performance of speed at constant load and ramp reference speed of MPTC, MPCC, DTC, and FOC.

12.6.4   DISCUSSION

In this chapter, four control methods MPCC, MPTC FOC, and DTC are applied to the traction drive system of PMSM. The comparison is carried out to examine several aspects such as torque performance in stable and dynamic states, torque, and flux ripple. To guarantee smooth reference tracking under varying operating circumstances, a PI compensator is employed. It is critical to consider objectives and restrictions, for example, for an application, before deciding on a control algorithm.

KEYWORDS

- **dynamic models**
- **dynamic performance**
- **dynamic response**
- **field oriented control**
- **modulation space vector theory**
- **ramp speed**
- **steady-state response**

REFERENCES

1. Steimel, A., (2008). *Electric Traction-Motive Power and Energy Supply: Basics and Practical Experience*. Oldenbourg Industrieverlag.
2. Krause, P. C., Oleg, W., Scott, D. S., & Steven, D. P., (2013). *Analysis of Electric Machinery and Drive Systems* (Vol. 75). John Wiley & Sons.
3. Türker, T., Umit, B., & Faruk, B. A., (2016). A robust predictive current controller for PMSM drives. *IEEE Transactions on Industrial Electronics, 63*(6), 3906–3914.
4. Jöckel, A., & Knaak, H. J., (2002). INTRA ICE-A novel direct drive system for future high speed trains. In: *Proceedings ICEM*.
5. Kazmierkowski, M. P., Ramu, K., & Frede, B., (2002). *Control in Power Electronics* (Vol. 17). San Diego: Academic press.
6. Mohan, N., Tore, M. U., & William, P. R., (2003). *Power Electronics: Converters, Applications, and Design*. John Wiley & Sons.
7. Linder, A., (2006). *Modellbasierte Praediktivregelung in der Antriebstechnik*. Logos-Verlag.
8. Blaschke, F., (1972). The principle of field orientation as applied to the new transvektor closed-loop control system for rotating field machines. *Siemens Review, 34*(1).
9. Takahashi, I., & Youichi, O., (1989). High-performance direct torque control of an induction motor. *IEEE Transactions on Industry Applications, 25*(2), 257–264.
10. Sorchini, Z., & Philip, T. K., (2006). Formal derivation of direct torque control for induction machines. *IEEE Transactions on Power Electronics, 21*(5), 1428–1436.
11. Cortés, P., Marian, P. K., Ralph, M. K., Daniel, E. Q., & José, R., (2008). Predictive control in power electronics and drives. *IEEE Transactions on Industrial Electronics, 55*(12), 4312–4324.
12. Camacho, E. F., & Carlos, B., (1999). *Model Predictive Control (Advanced Textbooks in Control and Signal Processing)*.
13. Maciejowski, J. M., (2002). *Predictive Control: With Constraints*. Pearson education.
14. Goodwin, G., María, M. S., & José, A. De. D., (2006). *Constrained Control and Estimation: An Optimisation Approach*. Springer Science & Business Media.
15. Qin, S. J., & Thomas, A. B., (2003). A survey of industrial model predictive control technology. *Control Engineering Practice, 11*(7), 733–764.

16. Mayne, D. Q., James, B. R., Christopher, V. R., & Pierre, O. M. S., (2000). Constrained model predictive control: Stability and optimality. *Automatica, 36*(6), 789–814.
17. Garcia, C. E., David, M. P., & Manfred, M., (1989). Model predictive control: Theory and practice—A survey. *Automatica, 25*(3), 335–348.
18. Cortés, P., José, R., Daniel, E. Q., & Cesar, S., (2008). Predictive current control strategy with imposed load current spectrum. *IEEE Transactions on Power Electronics, 23*(2), 612–618.
19. Rodriguez, J., & Patricio, C., (2012). *Predictive Control of Power Converters and Electrical Drives.* John Wiley & Sons.
20. Habibullah, M., Dah-Chuan Lu, D., Dan, X., & Muhammed, F. R., (2016). Finite-state predictive torque control of induction motor supplied from a three-level NPC voltage source inverter. *IEEE Transactions on Power Electronics, 32*(1), 479–489.
21. Kennel, R., Rodriguez, J., Espinoza, J., & Trincado, M., (2010). High performance speed control methods for electrical machines: An assessment. In: *2010 IEEE International Conference on Industrial Technology* (pp. 1793–1799). IEEE.
22. Rodríguez, J., Ralph, M. K., José, R. E., Mauricio, T., César, A. S., & Christian, A. R., (2011). High-performance control strategies for electrical drives: An experimental assessment. *IEEE Transactions on Industrial Electronics, 59*(2), 812–820.
23. Geyer, T., (2011). Computationally efficient model predictive direct torque control. *IEEE Transactions on Power Electronics, 26*(10), 2804–2816.
24. Ma, Z., Saeid, S., & Ralph, K., (2014). FPGA implementation of model predictive control with constant switching frequency for PMSM drives. *IEEE Transactions on Industrial Informatics 10*(4), 2055–2063.
25. Rojas, C. A., Jose, R., Felipe, V., José, R. E., César, A. S., & Mauricio, T., (2012). Predictive torque and flux control without weighting factors. *IEEE Transactions on Industrial Electronics, 60*(2), 681–690.
26. Ming-Shyan, W., Min-Fu, H., & Hsin-Yu, L., (2018). Operational improvement of interior permanent magnet synchronous motor using fuzzy field-weakening control. *Electronics, 7*(12), 452.
27. Zhang, Z., Xinglai, G., Zisi, T., Xiaohua, Z., Qidi, T., & Xiaoyun, F., (2017). A PWM for minimum current harmonic distortion in metro traction PMSM with saliency ratio and load angle constrains. *IEEE Transactions on Power Electronics, 33*(5), 4498–4511.
28. Woldegiorgis, A. T., Xinglai, G., Songtao, L., & Mannan, H., (2019). Extended sliding mode disturbance observer-based sensorless control of IPMSM for medium and high-speed range considering railway application. *IEEE Access, 7*, 175302–175312.
29. Woldegiorgis, A. T., Xinglai, G., Huimin, W., & Mannan, H., (2020). A new frequency adaptive second-order disturbance observer for sensorless vector control of interior permanent magnet synchronous motor. *IEEE Transactions on Industrial Electronics, 68*(12), 11847–11857.
30. Woldegiorgis, A. T., Xinglai, G., Songtao, L., & Yun, Z., (2021). An improved sensorless control of IPMSM based on pulsating high-frequency signal injection with less filtering for rail transit applications. *IEEE Transactions on Vehicular Technology, 70*(6), 5605–5617.
31. Woldegiorgis, A. T., Xinglai, G., Huimin, W., & Yun, Z., (2022). An active flux estimation in the estimated reference frame for sensorless control of IPMSM. *IEEE Transactions on Power Electronics.*

CHAPTER 13

Electric Mobility

ASHWINI KUMAR SHARMA,[1] SRIMANTI ROYCHOUDHURY,[2] and
SUNAM SAHA[3]

[1]*Graphic Era Deemed to be University, Dehradun, India*

[2]*School of Engineering and Technology, Adamas University, Kolkata,
West Bengal, India*

ABSTRACT

Climate is considered as one of the pillars for sustainable development. In order to curb the challenges posed by the air pollution on the environment now a days, companies have started rolling out the different vehicle with technologies including electric vehicles in the market. In present chapter the authors discussed about the scenario of electric vehicles in India their types and challenges in adopting the new technologies. It has been found that the three-wheeler market has adopted the change rapidly followed by four wheelers and two wheelers. The government regulation has also encouraged the acceptability of electric vehicles in different segments by defining new policies. Overall, a fundamental change is happening with the automotive industry for sustainable environment.

13.1 INTRODUCTION

With electric vehicles (EV) the first thought that comes is, why EVs are becoming popular around the world? There are many strong reasons behind this fact as in the present scenario many countries are facing huge pollution problems because of the use of conventional vehicles, especially

The Internet of Energy: A Pragmatic Approach Towards Sustainable Development. Sheila Mahapatra, Mohan Krishna S., B. Chandra Sekhar, & Saurav Raj (Eds.)

in the more populated cities. The increasing cost of fuel and scarcity of the same has compelled the engineers and industries around the world to think about a better alternative that can mitigate the issue. Increasing customer consciousness about the comfort and luxury experience in the vehicle has led the engineers to think seriously about the option of electric mobility as a promising alternative.

As shown in Figure 13.1 there is a continuous rise in the electric vehicle demand since 2019 and China is in the lead with 1.149 million units followed by Europe. This indicates a new era of development in the mobility segment with possibilities for the future.

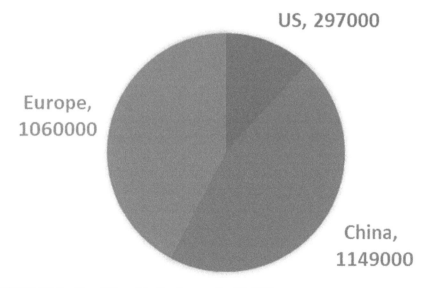

FIGURE 13.1   New EV registration by area for H1 2021.
Source: EV Volumes.

Though the present-day roads are populated by conventional vehicles, in near future, say within three to four years, markets are going to be captured by EVs only. For the past five years, Governments in different countries especially in Southeast Asia have undertaken many initiatives to promote and force the use of electric vehicles in the market. Based on this many automotive companies are shifting toward EVs and lined up several models in the coming years. We can understand these electric vehicles have some add-on advantages which pass over initial resistance towards accelerated adaptation of electric vehicles.

In many countries, electric mobility has already been embraced positively. For example, this technology has proved its worth in terms of public transport as e-rickshaw are being used by people everywhere, especially in India, China, Europe, USA. Now the industries are venturing into the other domains of transport such as luxury segment as well as heavy-duty transport with this alternative.

13.2 HISTORY BEHIND ELECTRIC MOBILITY

Electric vehicle, also known as EVs operates with the help of an electric motor. The internal combustion engine of the current generation automobile, which operates by burning fuel and gases has been replaced by the electric motor. The electric motor in an EV is operated with electric power, which is clean energy. As, a result, electric vehicles are seen as a replacement for the current automobiles. The outline of the historical timeline of electric vehicles is shown in Figure 13.2.

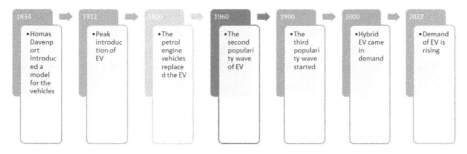

FIGURE 13.2 Electric vehicle history chart.

Sibrandus Stratingh traced that the first Electrical vehicle was used in the Netherlands run by Non-rechargeable batteries. Vehicles running with Non-rechargeable primary batteries were found to be used till 1859. Using rechargeable batteries in place of primary batteries was not a viable option as the cost of rechargeable batteries that were required in huge quantities was very high, which increased the cost of the e-vehicle. This drawback was addressed with the introduction of the first lead-acid battery by Gaston Plante. The introduction of this battery made a revolution in the entire world and many countries started to produce electric cars. And the U.S.A. was the first to invent the six-seater car with 22 kilometers per hour speed.

Today we refer the Electric vehicles as the transportation of the future due to their efficiency and minimal carbon emission. This was the same situation with electric vehicles in comparison with engine-based automobiles at the commencement of the 20th century. The first motor for running the electric vehicles was prototyped in 1828 by Anyos Jedlik while a model for the vehicles was conceptualized by Homes, Davenport in 1834 [1]. In that era, three alternatives had the same probabilities for development and use in the future. Out of three, steam engines were fast and cheap, but they required a lot of time to prepare before driving, also it is needed to stop to get water. The gasoline-powered vehicle was fast with a satisfactory covering distance but their problem was noise, complication, and too many parts. Electric-powered vehicles were not able to match the speed of combustion engines but were environment friendly, comparatively silent, and handling was also easy. It has been traced and surprising that New York, Boston, and Chicago had one-third of vehicles electric powered in 1904 [2]. But in the 1920s petrol engine vehicles replaced the e-vehicles because the cost of the electric vehicle was very high, the travel speed of the e-vehicle was only around 20–30 miles per hour, which was very less, and with the development of the roads, fast vehicles were the requirement [3]. Also, the development of an electric self-starter to start the engine and the abundance and low prices of gasoline led to gain the mass market of gasoline-powered vehicles.

The second popularity period of the electric vehicle was when environmental issues gained interest around the year the 1960s and were shadowed by the oil emergency in the year 1973. During this time most electric vehicles were developed for government institutions [4].

The third admiration wave for electric vehicles started around the 1990s due to environmental awareness, pollution after-effects, and development in e-vehicle technology which is continued till today. Across the globe, several governments regulation has been passed on reducing the use of fossil fuels owing to its environmental impact and scarcity of fossil fuels. The electric vehicle seems to have become a trend, showing its interest in environmental problems. Every company is developing and in process of launching more and more electric vehicles in different segments such as 2 wheeler, 3 wheeler, and 4-wheelers [3]. As the concern of environmental issues and oil depletion remains as severe as ever, governments are proactively considering different other options to old-fashioned transport technologies.

Battery electric vehicles (BEVs) are now being considered an encouraging option that has the potential for the vehicle independence from oil. The principle of operation is simple. Here the internal combustion engine and the tank of a conventional vehicle are replaced by the electric motor powered by a battery. When not in use the battery of the vehicle is charged by plugging into a charging spot [5–7]. But the challenges in BEVs are the technology readiness of the battery and the energy storage systems. Some of the popular battery technologies have been discussed below.

Lead-acid batteries are well-known batteries and their improvement potential is low. It has low specific energy (20–40 Wh/kg) compared to Lithium-ion batteries. Typically to achieve a particular range for example, 200 km the weight of led acid batteries is three times more than the Lithium-ion batteries. This has rendered the lead-acid batteries not being used for the electric vehicles, particularly for the long-range [8, 9].

The other popular battery is Nickel Metal Hydride Battery. This is now being adopted for Hybrid vehicles by Toyota in their Prius model. This is less costly compared to a lithium-ion battery [10]. Nickel Metal Hydride Battery is considered an established technology and has achieved its greatest prospective in reduced cost reduction and operation characteristics. Its energy density is between 60 and 80 Wh/kg and is insufficient for BEVs [11].

To date, in the ever-changing battery technologies, Lithium-ion batteries are being adopted as the supreme encouraging option for future use in BEVs. Therefore, Lithium-ion batteries are researched and studied in detail since it has the highest electrochemical potential resulting in high energy density and low equivalent mass [12, 13]. It has high effectiveness and a longer lifecycle, and its potential to further improve has led it to be used virtually for all the vehicles [14, 15]. There are certain disadvantages of the lithium-ion battery as well like it is expensive and it presents safety issues. Overcharging may lead to fire and destruction. The energy density is still not sufficient as per the requirement [16–18].

Another popular option for battery in research is the Sodium Nickel Chloride battery. This battery has many advantages. It is considered to have a longer life span and a less costly and safer option and also can be drained entirely without degrading its life prospect. Though the specific energy it has (less than 150 W/Kg) is satisfying, because of the lower specific power, it is not a good option to power BEVs. But efforts are being made to associate it with other sources like supercapacitors which can make it usable to power electric vehicles [19].

Improvement in battery technology can also come from a battery management system (BMS) that manages the use of batteries in different situations also known as BMS. The BMS has two major roles to play. Firstly, it optimizes the charging and discharging behavior of the battery, and secondly, it controls the battery with safety and efficiency [20, 21]. Studies are being done especially in the area of improving battery efficiency through BMS.

13.3 PRESENT SCENARIO

As a formal definition, we know these EVs are run by rechargeable batteries which can easily be charged by external sources of electricity. For EVs, we need separate supply equipment, which is formally known as EV Supply Equipment or EVSE. This entire EVSE system consists of a separate power source for an EV for providing an easy charging facility, network connectivity, smart metering, etc. The charging of EVs is simple in terms of the process however it requires thorough planning and procedure for installation to be used by the public as there should not be any hazard due to mishandling of charging equipment. The charging stations may be supplied by conventional electricity or suitable renewable energy such as wind, and solar. With the decresing cost of photovoltaic cells day by day the cost of solar-based charging is going to reduce shortly and viable which will make the charging independent of fossil fuels. This is the promising option being studied and implemented by many countries and the lead taken by the Netherlands.

Nowadays the electric vehicle industry is becoming a center of the new manufacturing hub, with added advanced technologies in BMS, and its control mechanism. All over the world Government is giving different subsidies and technological support for advancements in this new era of industry. India with the added advantage of being the world's largest producer of electric vehicles has proactively taken an initiative to produce the batteries of global standard which will be the other growth drivers for the Indian EV industry. Recently, in September 2021, the cabinet approved the production linked incentive scheme which will aid in supporting the automotive industry in manufacturing electric vehicles and hydrogen fuel cell vehicles. The world-wide conventional automotive industry is experiencing an exemplar shift to switch to alternative smart vehicles with higher efficiency, which is nothing other than electric vehicles. In the list of all those countries, India is also serving a prominent role in the electric mobility shift.

Figure 13.3 indicates the continuously rising demand for electric cars owing to the different reasons, possibly the prominent one is rising pollution due to conventionally powered vehicles. In the last few decades, the burden of rising pollution, greenhouse effects, and global climate changes has perpetually become major factors that motivated the rapid transition from traditional automotive vehicles to electric mobility. This has created a global awareness of a sustainable solution that can take care of these factors. A combination of Electric-powered vehicles with renewable energy-based charging infrastructure is one of the most sought-after alternatives being researched by researchers across the globe. There are countries especially in Europe using this method thereby trying to reduce the carbon emission to a net-zero level.

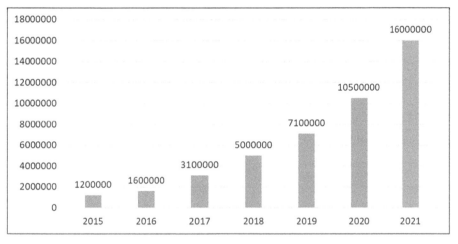

FIGURE 13.3   Global electric car fleet.

Sources: EV Volumes and Global EV Outlook 2021.

As per Table 13.1, we can see the promises being offered by electric vehicles over the conventional ones.

13.4   TYPES OF EVS

Broadly if we divide the EVs based on their construction, they can be of three types:

1. **Battery Electric Vehicles (BEVs):** These are powered only by a battery. This battery is recharged by an external source of electricity that may include plugging into conventional electricity or through renewable energy-based charging stations adding a further advantage to vehicular systems. Most electric vehicles including two-wheelers are BEVs.

2. **Hybrid Electric Vehicles (HEVs):** Hybrid cars have a battery and a combustion engine. In this case, both the IC engine and electric propulsion system work together to add efficiency to the existing system. The regenerative braking process used in EVs enhances the efficiency of the entire vehicular system. So, that energy will be stored in the battery for future use.

3. **Plugin Hybrid Electric Vehicles (PHEVs):** Here, both the internal combustion engine and electric battery are present. The battery plugin cables with the vehicle can be attached to an outside electric power source to recharge the battery. Similar to other electric vehicles, PHEV reduces greenhouse gas emissions. Plugin Hybrid vehicles are making it possible to reduce emissions to a great extent by tapping on the advantages of both conventional and electric vehicles.

TABLE 13.1 Comparison of Conventional Vehicle and Electric Vehicle

SL. No.	Conventional Vehicle	Electric Vehicle
1.	Power is generated by conventional IC engines.	In Electric vehicles, electric motors are used in place of IC engines.
2.	The efficiency of the IC engine vehicle is approximately 30%, which is very low and makes its running cost high.	With the use of electric motors, the efficiency of EVs has reached up to 85% resulting in lower running costs.
3.	It emits greenhouse gases, which cause global warming, and nowadays it is a severe issue.	By using electric energy, there is no harmful emission.
4.	One of the main reasons behind lower efficiency is the loss of braking energy, which cannot be restored for future use.	Using a regenerative braking mechanism, the braking energy can be restored and reused.
5.	IC engine vehicle uses complex gear systems.	Electric Vehicle employs electric motor.
6.	High maintenance cost.	Low maintenance cost as it uses fewer components.
7.	Reuse after its intended life is costly.	The reuse of batteries makes it a viable option and increases the total lifetime of the vehicle.

13.5 CONSTRUCTIONAL DETAILS OF ELECTRIC VEHICLES

The electric vehicle consists of an electric traction motor, charging port, onboard charger, DC/DC Converter, Thermal system, Traction battery pack, transmission, BMS, etc.

If we think about, electric traction motors, we use motors of high efficiency like BLDC, Induction motors, etc., where the efficiency goes up to 88%. These motors use power from the traction battery pack, which stores electrical energy. Electric vehicles have an added advantage of a charging port, with the help of which we can charge the battery using an external power supply. The onboard charger used in the electric vehicle helps in charging the traction battery. DC/DC converter helps in power modulation and here, it takes the power from the traction battery pack and converts it into the required amount for the accessories of the electric vehicle. A thermal system helps in maintaining the proper temperature of the electric motor, battery pack, power electronics devices, and other components of an electric vehicle. A transmission transfers mechanical power from the motor to the driving wheels. The block diagram of a typical basic electric vehicle with different components is shown in Figure 13.4.

FIGURE 13.4   Block diagram of electric car.

As the electric vehicle operates at high voltage there lies a risk of what may be a fail-safe condition? Therefore, a special management system is required to control the operation of the battery arrangement in the electric vehicle which is referred to as a BMS. The primary purpose of the BMS is to continue the workflow with the battery and also to enhance the safety limit of the entire system. Another important role of the BMS is to provide optimization over battery operation, which in turn enhances the life of a battery. BMS also monitors each battery cell to reduce voltage fluctuations within the system and reduce the chance of possible system failure.

The rechargeable battery packs in electric vehicles can be made up of multiple cells, either connected in series or parallel, based on the requirement of the vehicular system. These battery cells can produce thousands of voltage to supply, which is to be controlled by BMS. The key purpose of the BMS is to make sure that there is no unbalance in supply voltage. Also, the BMS provides extra support to ensure zero supply failure.

The primary objective of using BMS is to ensure the safe action of the battery and also to provide its safety. Apart from this BMS checks the condition of the charge of the battery pack. BMS also manages cell balancing followed by its battery optimization which indirectly improves the longevity of the battery. This BMS also monitors the variations in voltage, different physical constraints like ambient temperature, atmospheric pressure, the flow of coolant, etc. Lithium-ion batteries possess a high charge density. Although these batteries are not so large, still they can be extremely unstable, which is not desirable. Therefore, it is recommended that these batteries should on no occasion be charged too much or be permitted to reach a circumstance of complete discharge on any occasion. When the current is supplied to charge the batteries to store the energy there may be a situation of rising in the cell temperature this phenomenon is called a Thermal Runaway. This is a highly undesired situation in electric vehicles because it reduces the life of the batteries drastically and sometimes results in the catching of fire. To avoid such hazardous circumstances an automatic arrangement to monitor its voltage and current is required. As this is a highly difficult process to control in which each cell needs to be monitored closely and the action needs to be taken based on the requirement of the vehicle for safe and effective operation the specially designed control system called BMS is employed. In addition, this BMS is required to compute various parameters of the system to such an extent that the vehicle can be monitored remotely with connected technologies and potentially unsafe situations can be avoided.

13.6   FUTURE SCOPE AND LIMITATION

India comes the fifth number in the car market in the world. It is estimated that by 2030 around 35 crores of consumers will need a promising solution for conveyance or transport. But considering the objectives set under the Paris treaty, the growing number of customers in the mobility domain shall not demand an increase in the ingestion of conventional fuels.

This indicates that to reduce the carbon footprint by reducing fossil fuel consumption, electric mobility is one of the promising technology for the future.

With more and more renewable energy being used for electricity generation across the globe in terms of wind, solar, etc., it has grasped the attention of many researchers and practitioners. As shown in Figure 13.5 the advancement in smart grid technology is helping to harness the renewable energy along with the conventiontional energy in much more effective way as the former is being the point of attention for all. A lot of effort is being put on distributed generation (DG) system with major focus on solar generaton in the countries where solar power is available for maximum time in the year.

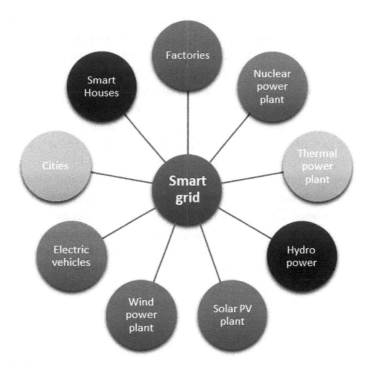

FIGURE 13.5   A schematic view of smart grid with distributed generation technology.
Source: Ref. [22].

To ensure achieving India's Net Zero Emissions by 2070 as per the government, a huge transport restructuring is required. The present situation is demanding superior and efficient transport for the public which includes roads, and railways. The efficient transport is already adopted as electrically driven by railways and now it's the time for the electric-powered cars. In this regard, the government of various countries has come up in support of growing the use of electric-powered vehicles by reducing duties and taxes. The changed and more encouraging faster adoption and manufacturing of electric vehicles (FAME II) scheme in India is an example of the same. With the promises of electric mobility as above there come some associated challenges which need to be addressed to make it successful.

The first challenge is the nonexistence of an industrial base for battery manufacturing in India, therefore to meet the rising demand of the electric vehicle market it is hugely dependent on the import from outside the country. The import bill related to electric vehicles, especially for batteries in the year 2021 was more than 1 billion dollars as per the government sources. As compared to the demand and future projections there is tiny penetration of electric vehicles. Moreover, this dependence will continue as the availability of lithium and cobalt which are the most important constituents for battery production. In addition, there are no clear guidelines to end of life of the batteries which will be a quite serious issue in the coming future.

The second challenge is charging infrastructure, which is quite less, than the immediate counterparts who already had a huge number of charging stations. The lack of charging stations in different states in India is a big distrust for the consumers to buy electric vehicles as they are not able to achieve the range presently. Further, it takes long hours up to 8 hours for a complete charge of a vehicle through the home socket using a company-supplied slow charger. And above this, the cost of cars running on conventional fuel is much less than the comparable electric cars of the dame segment. The cost factor is comparable for not only the cars but the electric two-wheelers also which come out to be nearly double the cost of conventional two-wheelers. Recently renewable energy-based charging systems especially solar are being installed in different areas to make the charging of electric vehicles independent of the conventional grid.

The third challenge is Policy Challenges, As EV production requires huge capital and long-term preparation to realize the profits. This lack of government on the policy front with the absence of clear guidelines, processes, and governance is resulting in the discouragement of new ventures to enter the market. The policies regarding electric mobility are still in nescent phase and are being developed with the time. Europian countries are ahead with clear policies and system for the electric vehicles as they have started adopting them early and contributing in reduction of carbon emission in a huge way.

The fourth challenge is the availability of Skilled Labor and technology. India needs self-sufficient production of electronics which still is a far-fetched dream. Electronics being in the heart to control and monitor the entire function of any electric vehicle, require huge and proactive initiatives for the in-house production of semiconductors. In addition, there is a huge demand for the skilled workforce in the EV domain but still, no dedicated course as of now is available in this segment. The only availability is conventional engineering courses which do not touch upon the electric vehicles in detail. Surprisingly there is a lack of laboratories related to hands-on learning about electric vehicle technologies even with the ranked and accreditated institutions of repute.

With the above discussion, it is evident that there is a enormous scope for electric mobility in the coming decades provided the challenges are met head-on to make it the mobility of the future. Also with such initiatives the major factor for the carbon emission that is transport sector accounting for 27% of total emission can very wellbe reduced as a result it will contribute towards the net zero emission goals of different nations.

13.7 CONCLUSION

With the above discussion on electric-powered vehicles being the possible solution for the new age mobility solution, there are still continuous enhancements being worked upon to make them more attractive to the masses in terms of features, efficiency safety, and handling. Such enhancements include ADAS, connected vehicle technology, etc.

Interestingly the electric mobility first was seen with the three-wheelers being used for the public transport in India which percolated to the remote areas owing to its advantages for the owners. Most of these three-wheelers are custom-made and use lead-acid batteries which results in a reduction in the price. Further, the electric power was seen with the four-wheelers that are cars and which aligned with the subsidy scheme of the government. Day by day the improvement is happening in electric cars and many automotive companies are considering switching and producing electric-powered cars with different capacities. Last but not least all the major two-wheeler (e-scooters) manufacturers are now developing different models due to rising demand as per the requirement of the masses which may result in reducing vehicular pollution to a large extent. The sales of electric two wheelers accounted for the 64.29% of the total electric vehicles in March 2022 [23]. Electric-powered buses are also on the card to manufacture in India for which many joint ventures are working.

The efforts are being made by the automotive industry and it will require some time for the consumer to get confidence in electric-powered vehicles and surpass the advantages of conventional vehicles in safety and efficiency. The government of India also taking proactive measures towards encouraging the transformation of the transport scenario with electric vehicles to reach 70% for commercial cars, 30% for private cars, 40% for buses, and 80% for two- and three-wheelers by 2030.

A great deal of the achieving the goals for a greener future by transforming the transport depends largely on the clear policies by the government, technological breakthroughs in research and development towards battery technologies, charging infrastructure, reuse of batteries, safety assurance, and at par vehicle cost with faster adoption by the masses. This can surely create a better future for all.

KEYWORDS

- **connected vehicle technology**
- **conventional vehicles**
- **electric mobility**
- **electric vehicles**
- **e-scooters**
- **safety assurance**
- **vehicular pollution**

REFERENCES

1. Jones, W. D., (2003). Hybrids to the rescue [hybrid electric vehicles]. *IEEE Spectrum*, *40*(1), 70, 71.
2. Curtis, D. A., & Judy, A., (2010). *Electric and Hybrid Cars: A History* (2nd edn., p. 257). McFarland & Company, North Carolina.
3. Jones, W. D., (2005). Take this car and plug it [plug-in hybrid vehicles]. *Spectrum, IEEE*, (Vol. 42, No. 7, pp. 10–13).
4. Wager, G., Whale, J., & Braun, T., (2016). Driving electric vehicles at highway speeds: The effect of higher driving speeds on energy consumption and driving range for electric vehicles in Australia. *Renew Sustain Energy Rev., 63*, 158–165. doi: 10.1016/j.rser.2016.05.060.
5. Al-Alawi, B. M., & Bradley, T. H., (2013). Review of hybrid, plug-in hybrid, and electric vehicle market modeling studies. *Renew Sustain Energy Rev., 21*, 190–203. doi: 10.1016/j.rser.2012.12.048.
6. Hannan, M. A., Azidin, F. A., & Mohamed, A., (2014). Hybrid electric vehicles and their challenges: A review. *Renew Sustain Energy Rev., 29*, 135–150. doi: 10.1016/j.rser.2013.08.097.
7. Arora, S., Shen, W., & Kapoor, A., (2016). Review of mechanical design and strategic placement technique of a robust battery pack for electric vehicles. *Renew Sustain Energy Rev., 60*, 1319–1331. doi: 10.1016/j.rser.2016.03.013.
8. Miller, J. M., (2009). Energy storage system technology challenges facing strong hybrid, plug-in, and battery electric vehicles. *Veh Power Propuls Conference 2009: VPPC '09* (pp. 4–10). IEEE. doi: 10.1109/VPPC.2009.5289879.
9. Offer, G. J., Howey, D., Contestabile, M., Clague, R., & Brandon, N. P., (2010). Comparative analysis of battery-electric, hydrogen fuel cell, and hybrid vehicles in a future sustainable road transport system. *Energy Policy, 38*, 24–29. doi: 10.1016/ j.enpol.2009.08.040.
10. Burke, A., Jungers, B., Yang, C., & Ogden, J., (2007). Battery electric vehicles: An assessment of the technology and factors influencing market readiness. *Public Interest Energy Res. Progr. Calif. Energy Comm.*, 1–24.

11. Miller, J. M., Bohn, T., Dougherty, T. J., & Deshpande, U., (2009). Why hybridization of energy storage is essential for future hybrid, plug-in, and battery electric vehicles. In: *2009 IEEE Energy Convers. Congr. Expo. ECCE 2009* (pp. 2614–2620). doi: 10.1109/ECCE.2009.5316096.

12. Li, Y., Yang, J., & Song, J., (2017). Nano energy system model and nanoscale effect of graphene battery in renewable energy electric vehicle. *Renew Sustain. Energy Rev., 69*, 652–663. doi: 10.1016/j.rser.2016.11.118.

13. Sabri, M. F. M., Danapalasingam, K. A., & Rahmat, M. F., (2016). A review on hybrid electric vehicles architecture and energy management strategies. *Renew Sustain. Energy Rev., 53*, 1433–1442. doi: 10.1016/j.rser.2015.09.036.

14. Deberitz, J., (2006). *Lithium Production and Application of a Fascinating and Versatile Element* (2nd edn.). Verlag Moderne Industrie.

15. Scrosati, B., & Garche, J., (2010). Lithium batteries: Status, prospects, and future. *J Power Sources, 195*, 2419–2430. doi: 10.1016/j.jpowsour.2009.11.048.

16. Riba, J. R., López-Torres, C., Romeral, L., & Garcia, A., (2016). Rare-earth-free propulsion motors for electric vehicles: A technology review. *Renew Sustain. Energy Rev., 57*, 367–379. doi: 10.1016/j.rser.2015.12.121.

17. Tahil, W., (2008). The Trouble with Lithium 2 Under the Microscope, *Meridian International Research*, Les Legers, 27210 Martainville, France, pp. 39–48. https://d1wqtxts1xzle7.cloudfront.net/60272860/Lithium_Microscope20190812-6534-jkxysn-libre.pdf?1565643520=&response-content-disposition=attachment%3B+filename%3DThe_Trouble_with_Lithium_2_Under_the_Mic.pdf&Expires=1689763676&Signature=K8xWiyOgSpfiCa2io0sdLz3BNSQAHBtHe5XttVbclWr2S7VMZ81kjbpWlfvByGAn4OAzjvZYTaGABP1~4sADfUoLRpgPR9DUt~VLZLJWVRARs3ut4-KLvgQox6pVvRAfubLOc~63bz0IPacMq02Ck~JMQIUTKnWpV6mrtTydApWsHzAv-BpZ8crfw9RHqs~k28kZ6PkVyAmkpJYrBAJ8v5eNF7q3URWHvxvaFCaMFCR3f23spS1sfO-AIG2ZFwGPya5PdFdHlVfxWRDIbqPIZcO11cU2ZAivXT1LAyl2moNL1H~lv8kTSTlcYYMW2RIXD4k40VoZ75rQL5CtC3wrkA__&Key-Pair-Id=APKAJLOHF5GGSLRBV4ZA.

18. Amiribavandpour, P., Shen, W., Mu, D., & Kapoor, A., (2015). An improved theoretical electrochemical-thermal modeling of lithium-ion battery packs in electric vehicles. *J. Power Sources, 284*, 328–338. doi: 10.1016/j.jpowsour.2015.03.022.

19. Li, Y., Yang, J., & Song, J., (2015). Electromagnetic effects model and design of energy systems for lithium batteries with gradient structure in sustainable energy electric vehicles. *Renew Sustain. Energy Rev., 52*, 842–851. doi: 10.1016/ j.rser.2015.07.155.

20. Bauer, S., Suchaneck, A., & Puente, L. F., (2014). Thermal and energy battery management optimization in electric vehicles using Pontryagin's maximum principle. *J. Power Sources, 246*, 808–818. doi: 10.1016/j.jpowsour.2013.08.020.

21. Zhang, X., Mi, C. C., & Yin, C., (2014). Active-charging based powertrain control in series hybrid electric vehicles for efficiency improvement and battery life extension. *J. Power Sources, 245*, 292–300. doi: 10.1016/j.jpowsour.2013.06.117.

22. Eolas Magazine, (2018). *Smart Grid Evolution*. Available online: https://www.eolasmagazine.ie/smart-grid-evolution/ (accessed on 12 June 2023).

23. https://evreporter.com/indias-electric-vehicle-sales-trend-march-2022/ (accessed on 12 June 2023).

CHAPTER 14

Power Loss Reduction and Voltage Improvement Through Capacitors and Their Optimization for the Distribution System

SHUBASH KUMAR,[1,2] CHANDAR KUMAR,[2] MUHAMMAD SUHAIL SHAIKH,[3] ANWAR ALI SAHITO,[4] and ZAHID ALI ARAIN[2]

[1]*School of Electrical Engineering, Yanshan University, Qinhuangdao, P. R. China*

[2]*Department of Sciences and Technology, Indus University, Karachi, Sindh, Pakistan*

[3]*School of Physics and Electronic Engineering, Hanshan Normal University, Guangdong, China*

[4]*Department of Electrical Engineering, Mehran University of Engineering and Technology, Jamshoro, Sindh, Pakistan*

ABSTRACT

To offer customers a constant power supply, a robust electrical power distribution system is necessary to be in place. The fluctuation in voltage is present in practically every distribution system, resulting in a significant voltage drop at the consumer's point of use. It is a prerequisite of a good distribution system that consumers be given a sufficiently consistent voltage for varied loads to operate satisfactorily. The modern usage of electronic ballast and energy conservation measures have modified the load pattern of business and residential users, which lowers the power factor to their desired

The Internet of Energy: A Pragmatic Approach Towards Sustainable Development. Sheila Mahapatra, Mohan Krishna S., B. Chandra Sekhar, & Saurav Raj (Eds.)

value, because the load on the distribution system is primarily inductive and requires a lot of reactive power, most power quality issues can be handled by controlling the reactive power. The installation of capacitors in the distribution system is widely used to accomplish reactive power compensation. According to studies, the distribution system wastes around 13% of the generated power due to Ohm's losses (I^2R) losses. High technical losses are being experienced, owing to an aging network and poor operating circumstances. The low power factor is one of the main causes of voltage variation and power loss in the distribution system. Voltages are given to consumers that are considerably below the lower permissible limits. In this chapter, the causes of the poor power factor, power losses, and voltage drop has been discussed. Different techniques have been used to minimize the above problems but optimal placement of capacitors in the utility network using the latest optimization technique. In this chapter, power capacitor compensation has been proposed and discussed briefly their advantages and optimal location in the distribution system. In the end comparison of the latest optimization techniques used for losses and voltage stability by optimal placement of the capacitor in the distribution system is also discussed. Installing capacitors and applying optimization techniques for appropriate placement in the distribution system, resulting in enhanced voltage profiles and reduction in power losses in the distribution system, can improve the stability and reliability of the power distribution system.

14.1 INTRODUCTION

The distribution system is the last step before electricity can be delivered to end-users, in any power system. Distribution feeders provide electricity from a distribution network to various end-users. It lowers the sub-transmission voltage to the distribution voltage level with this device (11 kV). The voltage of an 11 kV feeder is stepped down for different applications. Electricity is supplied to utility customers via the secondary side of the distribution transformer. Technical and non-technical losses can both occur in a distribution system. According to research, the distribution network wastes about 13% of electricity generated in the system. Because of the increased voltage drop along any line or transformer, as the system load grows, voltage drops at consumers' terminals [1]. Reduced consumer voltage has an impact on the performance and lifespan of consumer appliances [2]. It also raises the current required to operate loads, resulting in greater power losses. Distributor efficiency has a direct impact on both consumer appliances and power system economics [3]. A better understanding of voltage drop and power losses due to a low load power factor will enhance the system's overall performance and dependability various methods have been employed to solve these issues, including

Network reconfiguration, DG placement, and capacitor allocation. In the distribution system, shunt capacitors are used to reduce power loss, improve the voltage profile along with feeders, and increase the life of the equipment used for the distribution system. Power distribution networks often have shunt capacitors built to compensate for reactive power loss and decrease it. In distribution networks, however, shunt capacitors must be placed in the proper position and of the right size. When correctly installing shunt capacitors, capacitor location is critical to reducing loss as much as possible while keeping shunt capacitor prices low [4]. Power system architects and academics have had a difficult time figuring out where and how big to make shunt capacitors in distribution networks. There were several academic studies and publications on reactive power planning with optimization can be found in the literature. In Ref. [5], load flow analysis of the distribution feeder before and after installation of DG in the power loss reduction. In Ref. [6], a hybrid shuffling frog jumping and particle swarm optimization is given for the improvement of voltage profile and loss reduction in the radial distribution system. In Ref. [7], the author offers binary particle swarm optimization and shuffled frog leap (BPSO-SLFA) algorithms for optimal distributed generation (DG) placement in the radial distribution system to enhance voltage profile and power loss reduction. In Ref. [8], the proper placement of capacitors and PV systems with geographical location constraints is necessary for active power loss reduction in the radial distribution system. In Ref. [9], a hybrid optimization design was used to determine the best capacitor location and reconfiguration to improve the distribution system's performance. Firefly optimization is utilized in a radial distribution system to discover the ideal capacitor location and size [10].

This challenging combinatorial issue has been solved using a variety of methodological approaches. The problem of capacitor position and size is made considerably more complicated by imbalanced operating circumstances and the existence of harmonic sources in distribution systems. Many capacitor optimization studies have neglected to take these considerations into account. To solve capacitor placement problems, numerous different strategies have been developed, including analytical methodologies and numerical programming, as well as heuristic or AI-based solutions [11].

In this chapter, a variety of strategies have been used to tackle the optimum capacitor placement problem, including heuristics and artificial intelligence (AI). To deal with the nonlinearity of practical systems, AI is a potent knowledge-based technique. AI can simplify mathematical problems and respond quickly, making it useful for transitory analyzes like capacitor allocation and DG placement.

14.1.1   POWER LOSSES

Transformers, cables, overhead lines, and other infrastructure are used to transference electrical power from power plants to homes, businesses, and other consumers of electricity. To put it another way, the amount of electric energy created does not equal the amount of electric energy provided to consumers. There is a loss of a particular number of units. Transmission and distribution losses are defined as the difference between produced and dispersed units. The following categories can be used to categorize power system losses.

14.1.1.1   TECHNICAL LOSSES

The energy dissipated in the power system's conductors and equipment results in technical losses. The primary and secondary power lines of the electric grid have suffered significant technical losses. This means that the primary and secondary distribution networks must be well-designed or function within acceptable parameters. Technical losses have been higher than usual due to unanticipated increases in load. It's the system's current and Line characteristics that are to blame for the failure. Ohm's losses are the most common technical losses in transformers and conductors. Transformer losses also include losses from irons. Design maintenance and operation can help decrease technical losses, but they cannot be eliminated.

Technical failures can be caused by a wide range of circumstances. The overload, imbalance, and the lower power factor are the most important effects of a bad distribution system. Other factors to consider are the length of the feeder, the age of the distribution system, and the size of the conductor. The problem begins with poor design and planning, as well as an incorrect load estimate. Then there's the issue of mishandling and neglecting the system altogether. Penalties for new installations will be applied before those for existing ones, according to the current network load flow. Loads are carefully monitored, but they aren't stored. There's no regular upkeep going on. Due to the lack of a current maintenance schedule, burning transformers and line losses result in large losses. When a problem arises, the power supply must be restored and repair methods that are below par must be employed. It's the result of poor system planning and maintenance that the distribution network is in such poor shape, to begin with. There are a few instances of distribution system components that are in a terrible condition of repair including the old network, large feeders, an inappropriate size of conductors, improper earthing or grounds, and poor quality of the material.

14.1.1.2 NON-TECHNICAL LOSSES (NTLS)

NTLs occur in a power system when the computation of external actions and the load and circumstances are not taken into consideration. NTL's customer management systems and awareness of the program may contain several fraudulent automobiles. Theft of electricity is a primary cause of NTL.

14.2 VOLTAGE REGULATION

Transmission and distribution systems deliver energy to end-users in an electrical power system. Almost all distribution circuits have some degree of voltage fluctuation. Lighting, appliances, motors, and other loads require a somewhat steady voltage supply to operate properly [3]. As a result, the system as a whole operates at peak efficiency. Voltage regulation is defined as the difference between the voltages at the sending and receiving ends of a circuit. The variations should not exceed ±6% of the time.

14.2.1 CAUSES OF VOLTAGE VARIATIONS

There is a direct correlation between the amount of load placed on the supply system and voltage at the customers' terminal as well as the voltage drop along with the distribution network. When a system's load drops, the voltage at the consumer end rises because the voltage loss is reduced.

Voltage regulation requirements are described as: There is a ±6% need for a 120 V or 230 V system and a −2.5% to a + 6% proposed regulation for voltages more than 600 V [1].

Short intervals or unusual situations have an authorized upper limit of −8.3% to +5.8%. These circumstances are not explained in any way. Another definition of voltage regulation is "the difference between a device's no-load and full-load output voltage and the full-load output voltage."

$$Voltage\ Regulation(VR) = \frac{V_{nl} - V_{Rated}}{V_{Rated}} \times 100\% \tag{1}$$

where; V_{nl} is the voltage when no load is applied, and V_{Rated} is the voltage when the load is applied to the device. It is preferable if VR has a lower value. There will be less fluctuation in the voltage at the consumer terminals as a result of this. To keep the voltage differential between the transmitting and receiving ends under control, it is necessary to install voltage regulating equipment.

14.2.2   IMPORTANCE OF VOLTAGE REGULATION

The supply network's load changes cause the voltage to fluctuate at the consumer terminal. Voltage fluctuations can be dangerous, hence they must be controlled for the following reasons:

1. When the voltage increases by 6%, the illuminating power of a lamp increases by 50%, but when the voltage declines by 6%, the illuminating power of a lamp drops by 20% [12].
2. Low supply voltage might limit the starting torque of an induction motor. This can lead to inefficient operation and poor power factor as well as high magnetizing current.
3. Overvoltage produces extra heating and changes rated appliances like transformers, etc., to differing levels.

Because of this, voltage variations in power systems must be maintained to a minimum for better supply services. Transmission and distribution systems in electrical power systems include significant voltage drops, as well as varied circuits with load characteristics that differ from generation station to the consumer [13]. Each circuit must be able to be controlled independently. If the voltage drop exceeds permissible limits, voltage control equipment can be utilized at the generating station, substations, and feeders.

14.3   POWER FACTOR

In an AC circuit, the cosine angle between current and voltage is called the power factor (cos∅).

$$Power\ Factor\left(Cos\varnothing\right) = \frac{Active\ Power}{Apparent\ Power} = \frac{kW}{kVA} \qquad (2)$$

Inductive loads are primarily responsible for poor power factor, and poor power factor results in high apparent power need for continuous active power demand. To high apparent power demand, greater capacity transformers and transmission lines are required. Overall, the cost of the system components rises as a result of this. PF is stated as a percentage or as a unit of measurement.

14.3.1   REASONS FOR POOR POWER FACTOR

Knowing that a low power factor means that KW is less than KVA, this makes sense. Equipment having poor power factor includes pure inductive circuits, some lighting ballast, Induction motors, high-intensity discharge lights, transformers, and generators. As a result, inductive loads have higher reactive and apparent power and poor power factor (also known as efficiency) since the current is needed to create a magnetic field that does the intended job. In Figure 14.1, the power triangle is used to explain this impact.

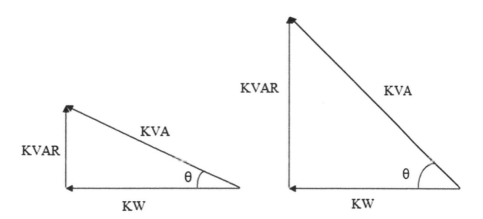

FIGURE 14.1   The impact of increasing the power angle.

14.3.2   IMPACTS OF A POOR POWER FACTOR

A low power factor has several detrimental consequences for the entire power system, which are detailed in subsections.

14.3.2.1   LARGE KVA RATING OF THE EQUIPMENT

$$KVA = \frac{kW}{Cos\emptyset} \tag{3}$$

KVA is the unit of measurement for electrical equipment. A higher KVA rating signifies that the KVA ratings of the equipment must be higher to fulfill the system requirements for the same KW demand [1]. It raises the price of the equipment and its associated accessories by a significant amount.

14.3.2.2   GREATER CONDUCTOR SIZE

As power consumption rises, the conductor size needs to be increased. Increased current flow in a system or piece of equipment due to a low power factor is not natural. Conductor size is therefore an additional factor that contributes to increased costs under these circumstances.

14.3.2.3 I^2R LOSSES

Higher I^2R losses result from increased current flow, which reduces the overall efficiency of the power system [14].

14.3.2.4   LOW VOLTAGE REGULATION

The low lagging power factor causes a big voltage drop, which in turn causes a larger voltage drop. Low lagging power factor To keep voltage drops within a certain range, voltage regulators must be installed.

14.3.3   TECHNIQUES OF IMPROVING SYSTEM POWER FACTOR

Power capacitors, one approach used to increase system power factor, will be addressed in subsections.

14.3.3.1   CAPACITORS

Using capacitors prevents inductive loads from causing trailing power. Figure 14.2 shows how this works quite clearly. Because inductive loads use lagging current and capacitance consumes leading current, adding capacitance balances out the effects of inductive loads overall.

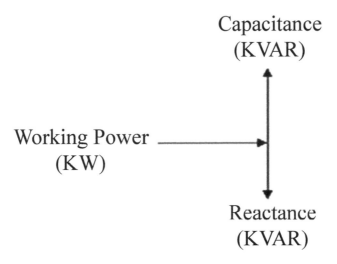

FIGURE 14.2   Impact of capacitors.

14.3.3.2   SHUNT CAPACITORS

The addition of shunt or parallel capacitor banks improves the power factor of the system [15]. Figure 14.2 illustrates how to shunt capacitors may enhance the power factor.

Let the feeder's impedance is, if $Z = R + jKL$

$$\text{Then Current } I_S = I_R$$

From the phasor diagram shown in Figure 14.3(b).

$$VD = I_R + I_X X_L \qquad (4)$$

here; X_L is the feeder reactance; and R is the feeder resistance. I_R and I_X represent the real and reactive parts of a current, respectively.

Referring to Figure 14.4(a), when shunt capacitors are installed, the leading capacitor current balances out the trailing inductive current, resulting in a reduction in total current. Figure 14.4(b) depicts this. The low voltage drop from permissible current indicates a higher power factor and better voltage management. However, a shunt capacitor has limitations and cannot boost the system power factor any further.

$$Z = R + jX_L$$

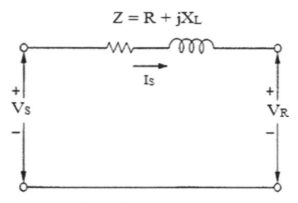

FIGURE 14.3(a)   The feeder circuit without capacitor.

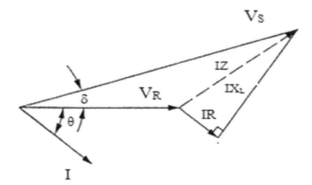

FIGURE 14.3(b)   The Phasor diagram from Ref. [16].

$$Z = R + jX_L$$

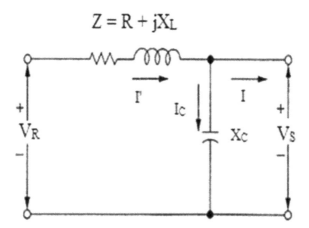

FIGURE 14.4(a)   The circuit of a feeder with a capacitor.

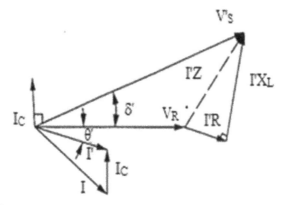

FIGURE 14.4(b) The phasor diagram.

Source: Ref. [16]

Here's how to figure out the voltage drop now:

$$VD = I_R R_R + I_X X_L - I_C X_C \tag{5}$$

here; the reactive component is I_C, and the voltage rise caused by the installation of a shunt capacitor may be expressed b.

$$VD = I_C X_L$$

14.3.3.3 BENEFITS OF SHUNT CAPACITORS

Shunt capacitors have the following benefits in a power system:

- High efficiency because of the little loss of energy during the operation.
- Because of the aging and insulation concerns, the reduced heating effect helps extend the life of machinery
- There is a reduction in demand (KVA) for a specific active power (kW).
- At greater power factors, energy dissipation (i.e., arcing time) is reduced, which reduces switchgear problems.
- Improved loading results and machine life.
- Improved control of voltage
- Generators with reduced excitation losses
- Generator capacity may be expanded without requiring further improvements to meet rising demand.

14.3.3.4   LOCATION OF SHUNT CAPACITORS

Capacitors must be installed everywhere reactive power needs are high to meet the system requirement. Capacitors should be installed close to the load since reactive power demand is higher there. Areas with high levels of low voltage are given precedence over those with low levels of low voltage. Transmission lines, distribution feeders, distribution and transmission substations, and the Long Term Network (LT Network) all benefit from the use of shunt capacitors. As a result, various compensation percentages are implemented on systems in different countries. For distribution feeders, shunt compensation is typically 60%, 30%, and 10% of the transmission system. A variety of techniques for locating the capacitor in an electrical system are discussed in the sections that follow.

14.3.3.4.1   Capacitor at Load

The capacitor in this manner is linked in such a way that it is energized when the load is operated [17]. The power factor correction capacitor (PFCC) is connected to the motor in the manner seen in Figure 14.5.

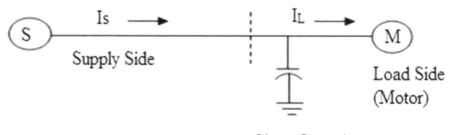

FIGURE 14.5   Separate compensation via a capacitor.

The following are the advantages of this method:

- It eliminates the need for a transformer and reduces loss;
- The most technically efficient and most adaptable solution;
- The decrease of voltage drop allows for better performance of the load;
- It is quite simple to install.

14.3.3.4.2 Fixed Capacitor Group

This method is commonly used when a huge number of powerful motors are readily available, allowing the business to make a valuable contribution to society. Controlling motors may be accomplished by adding a fixed capacitance value to the main bus bar control center. In most cases, the fixed capacitor bank is located near the service switch entrance, making it easy for customers to operate. To ensure the capacitor bank is safe from harm, a switch fuse or circuit breaker must always be positioned in front of it.

There will be an increase in the voltage value, which will lead to the failure of motors, lights, and other controls and devices when placing a fixed bank on a light load (for example, on weekends or holidays). Unstable loads or other related circumstances can cause harmonic distortion. In the study's conclusions, no more than 20% of the rated power of the kVA transformer should be in the fixed capacitance kVAR. Unexpectedly high numbers may have a harmful effect. Listed below are some advantages of this type of compensation:

- This method has the potential to enhance the power factor every month;
- Constant-load electrical facilities that are always on and drawing power;
- The correction of transformer reactive power;
- Motor compensation differs depending on the motor type;
- A fixed compensation amount is applicable.

14.3.3.4.3 Automatic Capacitor Bank

When reactive power requirements change dramatically over throughout operation, an automated system is a viable alternative for improving the power factor of the whole facility's inductive loads. The automated power factor capacitor system keeps track of reactive power needs and switches to the best switching capacitor whenever a high power factor is required. The capacitance is automatically altered to keep the right power factor in all working conditions, including fluctuating loads [14].

14.3.3.4.4 Combined Methodology

System integration may also serve as a compensating mechanism. Designing such a system necessitates thinking about things like cost and accessibility.

14.4 COMPENSATION FOR DISTRIBUTION FEEDER

When choosing a compensation scheme, it is vital to examine the following factors:

1. The fixed parallel capacitor's location about the mean reactive load should be considered.
2. The use of a big capacitor of the same size can be used to replace more than one capacitor group at the same time. Because of its economy, fixed capacitors of the same size may be used in a variety of locations.
3. Fixed capacitors with a high reactive load and the uniform load can be used across a distance of 2/3 of the feeder carrier length.

14.5 RESULT AND DISCUSSION

An 11 kV feeder from Hyderabad Electric Supply Company (HESCO), Sindh Pakistan, was used as a case study in Power System Simulator Siemens Calculation (PSSSINCAL) software to better understand the effect of voltage variation and power losses on the distribution system. In the program, three scenarios are created.

i.   Existing system;
ii. After installing capacitors on H.T. network; and
iii. After installing capacitors on L.T. network.

The simulation findings reveal that as the number of nodes in the present system grows, the voltage loss increases. After installing capacitors on HT and LT networks, the accompanying Figures 14.6(a) and (b) demonstrate a significant reduction in voltage dips.

Furthermore, the existing system has the highest possible technical losses, which must be minimized. Capacitors are used in H.T. and L.T. systems, and they reduce losses significantly. Table 14.1 presents a summary of losses in the distribution feeder.

The process for estimating economic advantages for existing network power losses of 724.555 kW, loss reduction, and yearly energy savings is shown in Table 14.2.

The present feeder of a distribution system is overloaded, as evidenced by the aforementioned figures and statistics, and has a maximum voltage drop in different nodes, as well as larger power losses. The voltage profile of the feeder is improved and power losses are reduced once capacitors are installed in the feeder (Figure 14.7).

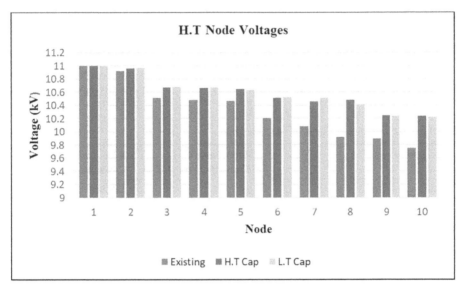

FIGURE 14.6(a)   H.T. node voltages.

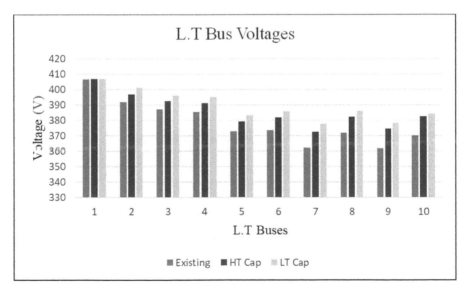

FIGURE 14.6(b)   L.T. bus voltages.

TABLE 14.1 Summary of Power Losses

Description	Existing Power Losses (kW)	Power Losses (kW) After Installing Capacitors on H.T.	Power Losses (kW) After Installing Capacitors on L.T.
Power losses (H.T.)	311.812	190.741	174.892
Power losses (transformer)	74.281	71.061	51.987
Power losses (L.T.)	338.462	301.459	221.962
Power losses (Total)	724.555	563.261	448.481

TABLE 14.2 Economic Benefits of Capacitors

Description	After Installing Capacitors on H.T.	After Installing Capacitors on L.T.
Power losses (kW)	563.261	448.481
Power loss reduction (kW)	161.294	276.074
Load factor	0.5	0.5
Loss factor	0.325	0.325
$FL = 0.3 * L.F + 0.7 * L.F^2$		
Annual energy savings (kwh)	459204.02	785982.68
$E_{S(Annual)} = P_L * 8760 * F_L$		

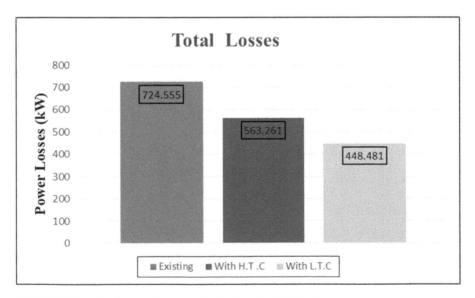

FIGURE 14.7 Total power losses in the feeder of a distribution system.

14.6 OVERVIEW OF OPTIMIZATION TECHNIQUES

The entire power system is divided into three divisions that are generation, transmission, and distribution. The generated power from different generating stations is supplied to the transmission line which is fed to a distribution system at substations. The fluctuation in voltage is present in practically every distribution system, resulting in a substantial voltage drop at the point of consumption. In line with the rising voltage drop and thus the rising reactive power demand, the amount of reactive power needed to maintain the same level of voltage at the consumer's terminal increases, leading to the poor load power factor and consequently increased reactive power consumption [19]. Approximately 70% of losses occur at the distribution system level, which comprises the primary and secondary distribution systems with the remaining 30% happening at the transmission system level, As a result, distribution systems are currently of primary concern, with losses targeted at the distribution level amounting to around 7.5% of total losses to reduce these losses optimal capacitors and DG are located in distribution system through optimization [20, 21].

Studies show that I^2R losses consume roughly 13% of generated power in the distribution system. Capacitors are commonly used for reactive power compensation in distribution systems due to their low cost, low repair, and maintenance requirements, and superior economic efficiency as compared to other techniques such as reactive power compensators [2]. The Inappropriate capacitor installation in any part of the power system increases losses and costs [23]. The distribution system's capacitor placement and size must be adjusted for optimal results.

Micro-grids, which are tiny, decentralized power systems, have emerged in recent years as a new way of thinking about the power system. Many dispersed and interconnected energy storage and generating units may be found in a typical micro-grid system. These systems can be islanding or grid-connected [24]. Poor capacitor placement can cause problems in systems that use capacitors. Optimal placement of capacitors in islanding and grid-connected modes for reactive power compensation is necessary for microgrids [25]. Based on electricity market pricing techniques, a unique hybrid modified grey wolf optimizer sine cosine and crow search algorithm (GWOSCACSA) is employed for optimal scheduling of distributed energy resources (DERs) in microgrid systems [26]. In Ref. [27], the author presented an oppositional gray wolf optimization (OGWO) for optimal reactive power when bus vulnerability was taken into account. In Ref. [28], author's suggestion for the solution of voltage constrained reactive power planning (VCRPP) in the power system, the Ameliorated Harris Hawks optimization (AHHO) and Harris Hawks optimization (HHO) algorithms were used. The nature-inspired different optimization algorithms proposed in Ref. [29] Oppositional Harris Hawk Optimization (OHHO) are found to deliver higher performance on the proposed test system for reactive power planning in power systems. In Ref. [30], author proposes a Sine-Cosine optimization algorithm for power loss reduction of transmission lines of the power system. In Ref. [31], the author provided a study on the HHOPSO algorithm's hybrid technique for voltage constrained reactive power planning. The simulation results demonstrate that active power loss and operating expenses are significantly reduced while voltage uniformity is maintained. In Ref. [32], author suggested an oppositional crow search optimization for optimal placement of capacitors for the reduction of transmission losses. In Ref. [33], author presented GWO-PSO hybrid optimization for reactive power planning in power systems, the presented system generates a reduction of operating cost in the IEEE 30 bus system and new England 39 bus system. In Ref. [34], the author devised a flux linkage approach for estimating transmission line parameters using multiple bundled conductors. The exact calculation of transmission-line properties, such as temperature correction resistance, using power-flow equations helps improve power-flow efficiency in the power system. In Ref. [35], the author develops and modifies a whale optimization technique for transmission line parameters estimation for various bundled conductors. Because of their high efficiency in discovering the optimal results, meta-heuristic optimization algorithms have gained favor in recent years.

To achieve this, many optimization algorithms are investigated, each with varying degrees of convergence power and precision. Researchers have studied the optimal placement of capacitors in distribution networks using a variety of optimization methodologies.

14.6.1 TEACHING-LEARNING-BASED OPTIMIZATION (TLBO)

Teachers have an impact on their students' performance [36] created a teaching-learning process-inspired algorithm called TLBO that considers it. The teacher phase (also known as algorithmic learning) and engaging with other learners are the two primary modalities of learning described by the algorithm (known as the learner phase). Using this optimization process, a population of students is treated as the optimization problem's distinct design factors, and the results of the students are compared to the optimization problem's 'fitness' value. Among all people, the teacher is thought to be the best answer. It is the objective function's best value that determines the optimum solution for a particular optimization issue, not the design variables [37]. The author used modified TLBO (MTLBO) with a constant and effective load to place capacitors where they will have the least influence on system performance.

14.6.2 PARTICLE SWARM OPTIMIZER (PSO)

Particles "fly" around a multidimensional search space modeling social activity, with each particle representing an intersection of all the search dimensions. Particles in a small region exchange memories of their "ideal" places for a brief time and utilize those memories to change their velocities and subsequent placements [38]. It offers the benefits of parallel processing and resilience, and it has a higher chance of finding the global optimum solution than other techniques. PSO is easy to implement, has a fast convergence rate, and has an intuitive design. PSO may be used in a variety of sectors, including scientific research and engineering. In Ref. [39], to reduce energy losses, the PSO algorithm creates and acquires capacitors in a radial distribution network [40]. The author proposes a multi-objective optimization technique for DG resources and capacitors in the distribution system. In Ref. [41], the capacitor bank is used to compensate for total reactive power, lowering the current and reducing power loss. In this study, for the best arrangement of a capacitor, energy storage system, and size in a microgrid, PSO and a generic algebraic modeling approach were applied.

14.6.3 THE GENETIC ALGORITHM (GA)

The genetic algorithm (GA) seeks the best solution utilizing evolutionary principles based on a string that is assessed and passed on to the next generation. The technique is meant to ensure that the "fitter" strings live and reproduce. GA's main benefit is that the answer is worldwide. GA may also find the global solution to functions that are not differentiable, linear or nonlinear, continuous or discrete, analytical or procedural. GAs is part of a wider family of evolutionary algorithms that solve optimization problems utilizing natural evolution mechanisms including inheritance, mutation, selection, and the crossover [42]. To reduce all functions simultaneously, a multi-objective technique is presented. Have recommended a site and size for the DG. The method is evaluated on an IEEE 69 bus system and compared to nonlinear optimization. According to the DG study, improper DG location and size would reduce system stability and dependability, increasing system losses. This strategy minimizes power losses and DG investment costs while optimizing DG placement and size [43]. A microgrid in both grid-connected and islanded mode is shown with the optimal capacitor allocation. Non-dispatch-able DGs account for the load. The imperialistic competitive algorithm (ICA) and GA are utilized in Ref. [44].

14.6.4   CUCKOO SEARCH (CS) ALGORITHM

Structural optimization problems are addressed using the cuckoo search (CS) approach. An initial benchmark nonlinear restricted optimization problem demonstrates the efficacy of the unique CS technique with Levy flights. The majority of structural engineering design optimization problems are nonlinear, and typically include a large number of diverse design variables under severe constraints. Material quality ranges, maximum stress, maximum deflection, minimum load capacity, and geometrical configuration are all examples of restrictions. Limited to basic limits or nonlinear interactions. This nonlinearity produces a multimodal response landscape. To improve the voltage profile while lowering the cost of reactive power, the CSA (cuckoo search algorithm) for optimal placement of shunt static capacitors in distribution networks is presented in Ref. [45].

14.6.5   ANT COLONY OPTIMIZATION (ACO)

Ant colonies are another popular evolutionary computing strategy that examines a large number of populations in parallel before assessing each population's competence using a cost function until convergence. The ant algorithm, which is based on ant behavior, calculates the quickest path from home to food. Pheromone trails are used by ants to communicate with one another and provide information about their journeys. Initially, a swarm of ants creates chaotic trails of constant density. As a result, denser trails for shorter courses are more relevant for later searches. These routes let the ants travel more efficiently [46]. The technology uses powerful ant colony optimization (ACO) to reconfigure feeders and insert capacitors in distribution networks. This research aimed to provide unique approaches to the best capacitor location, optimal feeder reconfiguration, and a combination of the two.

14.6.6   GREY WOLF OPTIMIZER (GWO)

The social (leadership) structure and hunting behaviors of wild Grey Wolves were replicated using a meta-heuristic algorithm. There are four subspecies of grey wolf optimizer (GWO) to consider (alpha, beta, delta, and omega). These models approximate the leadership hierarchy, which is based on the evolutionary struggle for the survival of the human brain. Assaulting your prey requires three separate tough maneuvers, thus planning a collective hunt will be necessary [47]. This chapter goes into considerable detail about these tactics. To reduce voltage loss and fluctuation, reactive power is also taken into consideration. To reduce network power loss, and voltage variation, and promote voltage stability, WCA and GWO are utilized to solve optimal capacitor placement problems in diverse distribution networks [48].

14.6.7   TABU SEARCH (TS)

Tabu exploration involves randomly searching the solution space. Fred Glover and colleagues created tabu search to find or regulate local optima. Using Tabu search effectively locates the history of the halt cycle and avoids being caught in local minima [49]. The higher-level heuristic algorithm is solved, and the answer is evaluated. Nara et al. [50] proposed a Tabu search to discover the ideal DG placement with the least amount of distribution network or system losses. The author of this study evaluated how much the power system's loss percentage may be decreased by optimum DG allocation on the demand side. This method assumes that the total number of DGs and total DG production capacity are equal to or greater than 1. A hybrid technique including TS has been used to position optimal capacitors [51]. TS currently incorporates components of practical heuristics and combinatorial techniques like GAs and simulated annealing. The method has been fully tested in a real-world 135-bus network as well as several other networks, with better quality and cost outcomes.

14.6.8   APPLICATION OF OTHER OPTIMIZATION TECHNIQUES FOR CAPACITOR ALLOCATION

It is proposed in Ref. [52] to use the artificial bee colony (ABC) optimizer to find the most optimum position and size possible in the distribution system to minimize the power losses and stability of voltage profile, besides lowering the lower of acquisition and installation capacitors. In Ref. [53], the shrimp straw method is used to determine the best position for the capacitor and the best way to reconfigure the distribution network to minimize distribution network losses under various load scenarios. For the optimal size of capacitive banks in the distribution system, the flower pollination algorithm (FPA) is utilized [54]. The biogeography-based optimization (BBO) is proposed in Ref. [55], to reduce the costs of power, as well as the quantity of energy collected from the upstream network. The harmony search method is used in Ref. [56] to offer the ideal position of capacitors to decrease losses and expenses (HSA). The ideal placement and size of shunt capacitors in the distribution network are found in Ref. [57] using a gravitational search algorithm (GSA) to reduce losses and increase financial advantage owing to the usage of capacitors (GSA). In Ref. [58], the allocation of capacitors is investigated using the sine–cosine algorithm (SCA) to maximize net savings while also improving dependability. In Ref. [59], the researcher proposed the modified cultural algorithm (MCA) to minimize the power loss of a distribution system by optimal placement of capacitors. For bus placement in RDNs, the author has used a combination of LSF and voltage stability index (VSI) in Ref. [60]. In Ref. [61], the authors show how they used shark smell optimization (SSO) to enhance OC allocation in small and large RDNs. Using a modified cultural algorithm (WOA), researchers in Ref. [62] were able to reduce the operational and energy expenses of RDNs. In Ref. [22], the authors employed a FPA to reduce the cost of the capacitor as well as the amount of energy lost. In Ref. [18], the author used a butterfly optimization approach to reduce the framework's generation costs to a minimum. Observation of four case test frameworks reveals that the proposed technique provides a superior arrangement regarding the speed of execution and the capacity to sustain regular costs. The literature review demonstrated that the meta-heuristic approach is extremely effective in determining the best location and capacity of reactive resources to obtain the best distribution network performance feasible. As a result, meta-heuristic techniques with high optimization power and low computing cost are necessary to achieve the best network performance.

14.7 CONCLUSION AND FUTURE WORK

In the entire power system approximately 70% of losses occur in distribution systems, as a result, the distribution system is the primary concern to minimize the losses through different techniques to provide reliable and stable power for consumption. In the field of optimization, also known as mathematical optimization, the process of picking the best element from a prospective collection of alternatives to meet particular criteria is known as optimization. This theory, which is concerned with finding the best and most optimum solution for a certain problem, applies to almost all areas of scientific and social sciences, and it has applications in a wide range of fields. The optimization domain may be used to maximize or reduce the value of a real function by using a systematic selection of input values from within a permissible set and calculation of the value of the function, respectively. Furthermore, for the challenges of power systems, a variety of optimization strategies are employed.

Several studies have demonstrated that line loss in low voltage and the weak distribution system wastes electrical power. However, distribution networks suffer the worst losses in rural areas of underdeveloped countries like Pakistan. As a result, reducing network loss in distribution networks is crucial. Distribution feeders in developing nations have a radial shape and are long. These issues cause poor voltage control in many feeds. As a result, it is employed in RDNs to minimize power losses and voltage variation while simultaneously improving the stability of voltage. However, the aforesaid advantages are largely dependent on the number, location, and size, of capacitors. As an optimal capacitor placement problem, it is a complex combinatorial problem to maximize net savings while reducing total expenses as an objective function. The classical approaches have been discovered to be easier, but they have drawbacks such as poor handling of qualitative constraints, poor convergence, and sluggish computing with many variables. Furthermore, for future work, AI approaches are quick and flexible. Designing capacitor banks with automated switching circuits for distribution networks and analyzing their cost-benefit through optimization techniques to attain the optimum network performance of the system.

This chapter will assist researchers in locating earlier work done on the topic of capacitor placement and reactive power planning for distribution system stability and reliability.

Furthermore, the chapter only discusses the optimal placement of capacitors and their optimization techniques for active and reactive power losses and voltage profile improvement, whereas the other devices (STATCOM, TSC, etc.) and distribution generations (PV, wind, fuel cell, etc.) may be considered for the same problem.

KEYWORDS

- **capacitors**
- **distribution system**
- **optimization techniques**
- **power factor**
- **power loss reduction**
- **voltage profile improvement**

REFERENCES

1. Umer, F., Noor, K. A., Anwar, A. S., & Agha, Z. P., (2013). Critical analysis of capacitors as a potential solution to achieve optimum level voltage regulation in HESCO network. In: *2013 The International Conference on Technological Advances in Electrical, Electronics and Computer Engineering (TAEECE)* (pp. 488–497). IEEE.
2. Sedghi, M., Aliakbar-Golkar, M., & Haghifam, M. R., (2013). Distribution network expansion considering distributed generation and storage units using modified PSO algorithm. *International Journal of Electrical Power & Energy Systems, 52*, 221–230.
3. Kumar, S., Chandar, K. F. Ur. R., Shoaib, A. S., & Anwar, A. S., (2018). Voltage improvement and power loss reduction through capacitors in utility network. In: *2018 International Conference on Computing, Mathematics and Engineering Technologies (iCoMET)* (pp. 1–5). IEEE.
4. Sirjani, R., Azah, M., & Hussain, S., (2012). Heuristic optimization techniques to determine optimal capacitor placement and sizing in radial distribution networks: A comprehensive review. *Organ, 7*, 12.
5. Farooq, U., Shahryar, Q., Sanaullah, A., & Fazal, W. K., (2017). Load flow analysis of 11 kV test feeder with and without the injection of DG. *International Journal of Computer Science and Information Security, 15*(3), 1.
6. Lotfi, H., Mahdi, S., & Ali, D., (2016). Optimal capacitor placement and sizing in radial distribution system using an improved particle swarm optimization algorithm. In: *2016 21st Conference on Electrical Power Distribution Networks Conference* (EPDC) (pp. 147–152). IEEE.
7. Hassan, A. S., Yanxia, S., & Zenghui, W., (2020). Multi-objective for optimal placement and sizing DG units in reducing loss of power and enhancing voltage profile using BPSO-SLFA. *Energy Reports, 6*, 1581–1589.
8. Nguyen, T. T., Bach, H. D., Thai, D. P., & Thang, T. N., (2020). Active power loss reduction for radial distribution systems by placing capacitors and PV systems with geography location constraints. *Sustainability, 12*(18), 7806.
9. Hussain, A. N., Al-Jubori, W. K. S., & Haider, F. K., (2019). Hybrid design of optimal capacitor placement and reconfiguration for performance improvement in a radial distribution system. *Journal of Engineering, 2019*.

10. Olabode, E., Ajewole, T., Okakwu, I., & Ade-Ikuesan, O., (2019). Optimal sitting and sizing of shunt capacitor for real power loss reduction on radial distribution system using firefly algorithm: A case study of Nigerian system. *Energy Sources, Part A: Recovery, Utilization, and Environmental Effects*, 1–13.

11. Eslami, M., Hussain, S., & Azah, M., (2011). Application of artificial intelligent techniques in PSS design: A survey of the state-of-the-art methods. *Przegląd Elektrotechniczny (Electrical Review)*, 87(4), 188–197.

12. Mehta, V. K., & Rohit, M., (2005). *Principles of Power System: Including Generation, Transmission, Distribution, Switchgear and Protection: For BE/B. Tech., AMIE and Other Engineering Examinations*. S. Chand Publishing.

13. Sahito, A. A., Jatoi, A. M., Memon, S. A., & Shah, S. A. A., (2015). Transmission system performance improvement through reactive power compensation. *Sindh University Research Journal-SURJ (Science Series)*, 47(3).

14. Tsong-Liang, H., Ying-Tung, H., Chih-Han, C., & Joe-Air, J., (2008). Optimal placement of capacitors in distribution systems using an immune multi-objective algorithm. *International Journal of Electrical Power & Energy Systems*, 30(3), 184–192.

15. Ahmadi, H., José, R. M., & Hermann, W. D., (2014). A framework for volt-VAR optimization in distribution systems. *IEEE Transactions on Smart Grid*, 6(3), 1473–1483.

16. Gonen, T., (2015). *Electric Power Distribution Engineering*. CRC press.

17. Almasoud, H., (2009). Shunt capacitance for a practical 380 kV system. *International Journal of Electrical and Computer Science (IJECS)*, 9(10), 23–27.

18. Ansari, M. M., Chuangxin, G., Muhammad, S., Nitish, C., Bo, Y., Jun, P., Yishun, Z., & Xurui, H., (2020). Considering the uncertainty of hydrothermal wind and solar-based DG. *Alexandria Engineering Journal*, 59(6), 4211–4236.

19. Reddy, M. D., & Kumar, N. V., (2012). Optimal capacitor placement for loss reduction in distribution systems using fuzzy and harmony search algorithm. *ARPN Journal of Engineering and Applied Sciences*, 7(1), 15–19.

20. Pazouki, S., Amin, M., Mahmoud-Reza, H., & Shahab, A., (2015). Simultaneous allocation of charging stations and capacitors in distribution networks improving voltage and power loss. *Canadian Journal of Electrical and Computer Engineering*, 38(2), 100–105.

21. Dinakara Prasasd Reddy, P., Veera Reddy, V. C., & Gowri Manohar, T. (2018). Ant lion optimization algorithm for optimal sizing of renewable energy resources for loss reduction in distribution systems. *Journal of Electrical Systems and Information Technology*, 5(3), 663–680.

22. Tamilselvan, V., Jayabarathi, T., Raghunathan, T., & Xin-She, Y., (2018). Optimal capacitor placement in radial distribution systems using flower pollination algorithm. *Alexandria Engineering Journal*, 57(4), 2775–2786.

23. Naderipour, A., Abdul-Malek, Z., Saber, A. N., Hesam, K., Amir, R. R., Saman, S., & Jiří, J. K., (2021). Comparative evaluation of hybrid photovoltaic, wind, tidal and fuel cell clean system design for different regions with remote application considering the cost. *Journal of Cleaner Production*, 283, 124207.

24. Ramakrishnan, K., Vijeswaran, D., & Manikandan, V., (2019). Stability analysis of networked micro-grid load frequency control system. *The Journal of Analysis*, 27(2), 567–581.

25. Baghaee, H. R., Mojtaba, M., Gevork, B. G., & Heidar, A. T., (2017). Nonlinear load sharing and voltage compensation of microgrids based on harmonic power-flow calculations using radial basis function neural networks. *IEEE Systems Journal*, 12(3), 2749–2759.

26. Dey, B., Saurav, R., Sheila, M., & Fausto, P. G. M., (2022). Optimal scheduling of distributed energy resources in microgrid systems based on electricity market pricing strategies by a novel hybrid optimization technique. *International Journal of Electrical Power & Energy Systems, 134,* 107419.

27. Babu, R., Saurav, R., Bishwajit, D., & Biplab, B., (2021). Optimal reactive power planning using oppositional grey wolf optimization by considering bus vulnerability analysis. *Energy Conversion and Economics.*

28. Swetha, S. G., Sheila, M., & Saurav, R., (2021). Voltage constrained reactive power planning by ameliorated HHO technique. In: *Recent Advances in Power Systems* (pp. 435–443). Springer, Singapore.

29. Gudadappanavar, S. S., & Sheila, M., (2021). Metaheuristic nature-based algorithm for optimal reactive power planning. *International Journal of System Assurance Engineering and Management,* 1–14.

30. Babu, R., Vishnu, K., Chandan, K. S., Saurav, R., & Biplab, B., (2022). Application of sine–cosine optimization algorithm for minimization of transmission loss. *Technology and Economics of Smart Grids and Sustainable Energy, 7*(1), 1–12.

31. Shekarappa, G. S., Sheila, M., & Saurav, R., (2021). Voltage constrained reactive power planning problem for reactive loading variation using hybrid Harris hawk particle swarm optimizer. *Electric Power Components and Systems, 49*(4, 5), 421–435.

32. Shiva, C. K., Swetha, S. G., Basetti, V., Rohit, B., Saurav, R., & Biplab, B., (2022). Fuzzy-based shunt VAR source placement and sizing by oppositional crow search algorithm. *Journal of Control, Automation and Electrical Systems,* 1–16.

33. Badi, M., Sheila, M., Bishwajit, D., & Saurav, R., (2022). A hybrid GWO-PSO technique for the solution of reactive power planning problem. *International Journal of Swarm Intelligence Research (IJSIR), 13*(1), 1–30.

34. Shaikh, M. S., Changchun, H., Munsif, A. J., Muhammad, M. A., & Aleem, A. Q., (2021). Parameter estimation of AC transmission line considering different bundle conductors using flux linkage technique. *IEEE Canadian Journal of Electrical and Computer Engineering, 44*(3), 313–320.

35. Shaikh, M. S., Changchun, H., Saurav, R., Shubash, K., Mannan, H., Muhammad, M. A., & Munsif, A. J., (2022). Optimal parameter estimation of 1-phase and 3-phase transmission line for various bundle conductor's using modified whale optimization algorithm. *International Journal of Electrical Power & Energy Systems, 138,* 107893.

36. Rao, R., & Vivek, P., (2012). An elitist teaching-learning-based optimization algorithm for solving complex constrained optimization problems. *International Journal of Industrial Engineering Computations, 3*(4), 535–560.

37. Bhattacharyya, B., & Rohit, B., (2016). Teaching learning-based optimization algorithm for reactive power planning. *International Journal of Electrical Power & Energy Systems, 81,* 248–253.

38. Bai, H., & Bo, Z., (2006). A survey on application of swarm intelligence computation to electric power system. In: *2006 6th World Congress on Intelligent Control and Automation,* (Vol. 2, pp. 7587–7591). IEEE.

39. Ramadan, H. S., Bendary, A. F., & Nagy, S., (2017). Particle swarm optimization algorithm for capacitor allocation problem in distribution systems with wind turbine generators. *International Journal of Electrical Power & Energy Systems, 84,* 143–152.

40. Zeinalzadeh, A., Younes, M., & Mohammad, H. M., (2015). Optimal multi objective placement and sizing of multiple DGs and shunt capacitor banks simultaneously

considering load uncertainty via MOPSO approach. *International Journal of Electrical Power & Energy Systems, 67*, 336–349.

41. Rajamand, S., (2020). Loss cost reduction and power quality improvement with applying robust optimization algorithm for optimum energy storage system placement and capacitor bank allocation. *International Journal of Energy Research, 44*(14), 11973–11984.

42. Shaikh, M. S., Changchun, H., Mannan, H., Saurav, R., Munsif, A. J., & Muhammad, M. A., (2021). Optimal parameter estimation of overhead transmission line considering different bundle conductors with the uncertainty of load modeling. *Optimal Control Applications and Methods*.

43. Naderipour, A., Abdul-Malek, Z., Mohammad, H., Zahra, M. S., Mohammad, A. F., Saber, A. N., & Iraj, F. D., (2021). Spotted hyena optimizer algorithm for capacitor allocation in radial distribution system with distributed generation and microgrid operation considering different load types. *Scientific Reports, 11*(1), 1–15.

44. Moradi, M. H., Arash, Z., Younes, M., & Mohammad, A., (2014). An efficient hybrid method for solving the optimal sitting and sizing problem of DG and shunt capacitor banks simultaneously based on imperialist competitive algorithm and genetic algorithm. *International Journal of Electrical Power & Energy Systems, 54*, 101–111.

45. El-Fergany, A. A., & Almoataz, Y. A., (2014). Capacitor allocations in radial distribution networks using cuckoo search algorithm. *IET Generation, Transmission & Distribution, 8*(2), 223–232.

46. Chung-Fu, C., (2008). Reconfiguration and capacitor placement for loss reduction of distribution systems by ant colony search algorithm. *IEEE Transactions on Power Systems, 23*(4), 1747–1755.

47. Shaikh, M. S., Changchun, H., Munsif, A. J., Muhammad, M. A., & Aleem, A. Q., (2021). Application of grey wolf optimization algorithm in parameter calculation of overhead transmission line system. *IET Science, Measurement & Technology, 15*(2), 218–231.

48. Kola, S. S., & Jayabarathi, T., (2020). Optimal capacitor allocation in distribution networks for minimization of power loss and overall cost using water cycle algorithm and grey wolf optimizer. *International Transactions on Electrical Energy Systems, 30*(5), e12320.

49. Sirjani, R., Azah, M., & Hussain, S., (2012). Heuristic optimization techniques to determine optimal capacitor placement and sizing in radial distribution networks: A comprehensive review. *Organ, 7*, 12.

50. Nara, K., Hayashi, Y., Yamafuji, Y., Tanaka, H., Hagihara, J., Muto, S., Takaoka, S., & Sakuraoka, M., (1996). A tabu search algorithm for determining distribution tie lines. In: *Proceedings of International Conference on Intelligent System Application to Power Systems* (pp. 266–270). IEEE.

51. Mori, H., & Shingo, T., (2005). Variable neighborhood tabu search for capacitor placement in distribution systems. In: *2005 IEEE International Symposium on Circuits and Systems* (pp. 4747–4750). IEEE.

52. El-Fergany, A. A., & Almoataz, Y. A., (2014). Capacitor placement for net saving maximization and system stability enhancement in distribution networks using artificial bee colony-based approach. *International Journal of Electrical Power & Energy Systems, 54*, 235–243.

53. Sultana, S., & Provas, K. R., (2015). Oppositional krill herd algorithm for optimal location of distributed generator in radial distribution system. *International Journal of Electrical Power & Energy Systems, 73*, 182–191.

54. Abdelaziz, A. Y., Ehab, S. A., & Abd, E. S. M., (2016). Flower pollination algorithm and loss sensitivity factors for optimal sizing and placement of capacitors in radial distribution systems. *International Journal of Electrical Power & Energy Systems, 78*, 207–214.

55. Ghaffarzadeh, N., & Hassan, S., (2016). A new efficient BBO based method for simultaneous placement of inverter-based DG units and capacitors considering harmonic limits. *International Journal of Electrical Power & Energy Systems, 80*, 37–45.

56. Ali, E. S., Abd, E. S. M., & Abdelaziz, A. Y., (2016). Improved harmony algorithm and power loss index for optimal locations and sizing of capacitors in radial distribution systems. *International Journal of Electrical Power & Energy Systems, 80*, 252–263.

57. Shuaib, Y. M., Surya, K. M., & Christober, A. R. C., (2015). Optimal capacitor placement in radial distribution system using gravitational search algorithm. *International Journal of Electrical Power & Energy Systems, 64*, 384–397.

58. Abdelsalam, A. A., & Hany, S. E. M., (2019). Optimal allocation and hourly scheduling of capacitor banks using sine cosine algorithm for maximizing technical and economic benefits. *Electric Power Components and Systems, 47*(11, 12), 1025–1039.

59. Haldar, V., & Niladri, C., (2015). Power loss minimization by optimal capacitor placement in radial distribution system using modified cultural algorithm. *International Transactions on Electrical Energy Systems, 25*(1), 54–71.

60. Devabalaji, K. R., Ravi, K., & Kothari, D. P., (2015). Optimal location and sizing of capacitor placement in radial distribution system using bacterial foraging optimization algorithm. *Int. J. Electr. Power Energy Syst., 71*, 383–390. https://doi.org/10.1016/j.ijepes.2015.

61. Gnanasekaran, N., Chandramohan, S., Kumar, P. S., & Mohamed, I. A., (2016). Optimal placement of capacitors in radial distribution system using shark smell optimization algorithm. *Ain. Shams Eng. J., 7*, 907–916. https://doi.org/10.1016/j.asej.2016.01.006.

62. Prakash, D. B., & Lakshminarayana, C., (2017). Optimal siting of capacitors in radial distribution network using whale optimization algorithm. *Alexandria Engineering Journal, 56*(4), 499–509.

CHAPTER 15

Solar Photovoltaic Powered Automatic Irrigation System for the Agriculture Sector

HIMANSHU SHARMA and PANKAJ KUMAR

Department of EEE, SRMIST, Delhi NCR Campus, Ghaziabad, Uttar Pradesh, India

ABSTRACT

The chapter was based on a self-contain home automatic system that was solar powered, providing plant watering of the necessary repeatability and moisture from a distinct water reservoir. It also able to alert the users in any kind of unusual situation. Inside a home is an extreme environment for many plants, as wetness was usually low, sunlight light levels vary significantly, and temperature can vary widely. This chapter contains an automatic self-watering system that sprinkles water to the plants when the soil moisture sensor, attached to the device, measures a small amount of soil moisture and overrides manual irrigation to keep the soil moisture level adequate. It was like hiring somebody to provide water for your plants.

15.1 INTRODUCTION

It was required to build an automatic solar power plant watering system which was irrigating the plant automatically after some time. This chapter would explain the hardware and software implementation of a solar power automatic power plant watering system. The plant watering system was a kind of automatic system which irrigates the plant after every time when sensor finds that soil was wet [1]. This system was a type of an embedded system,

---

The Internet of Energy: A Pragmatic Approach Towards Sustainable Development. Sheila Mahapatra, Mohan Krishna S., B. Chandra Sekhar, & Saurav Raj (Eds.)

as it was a combination of software and hardware aimed for a specific task. The circuit used here was electronic one that comes as a small electronic board which itself was an embedded system, the IC used here was "16F876," which was interfaced with a PCB containing a PIC micro-controller [2].

Block diagram in Figure 15.1 gives an overview of the system, the Automatic Solar Power Plant Watering System send the information to the micro-controller through a PWM signal from the sensor which was dig in the soil to check the moisture of the soil [3]. The signal to the micro-controller checks the signal and send signal according to that to the motor to run or not. For more details of micro-controller module refer to data-sheet of 16F876 [4]. The supply to the motor and the micro-controller was given by the battery. The battery was re-chargeable. The battery was charged by the solar panel. The solar panel gives the maximum voltage 17.5 V and the battery provides the 12 V. The difference in the voltage was adjusted by the resistance put in between.

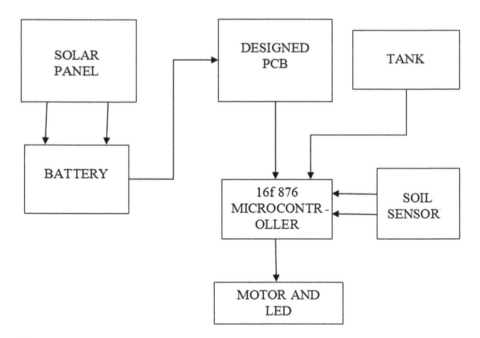

FIGURE 15.1   Block diagram of system.

This project was mostly suitable for those kinds of people who like to travel and want to go to for long trip. Also, such a system removes the need to water very often when you are at home. This device may be helpful for florists' people and other who like green environment at home but not remember to water them frequently to water them.

It was simple and easy to use for plants of various sizes and varieties. Similar devices are excellent for fields of vegetation in locations where the soil lacks water or where there has been a severe drought, in addition to potted plants. The device works with your plants as they are and so there was no need to repot. It was also convenient because it provides an easy way to water those hard to reach plants. This equipment may be used to automatically fertilize while watering and supports fruitful development since fertilizer can be added to the water supply [5].

15.1.1 AIMS

- To store the power supply in the battery from the solar panel from the solar energy.
- To use the power supply to run the system to irrigate the soil when required.
- Writing suitable assembly language program for the system to perform correct water supply operation.
- Design of output display unit, i.e., PCB with LED indicators and suitable driver circuit.

15.1.2 OBJECTIVES

- Theoretical study of an automated solar powered plant watering system.
- Construction of a working model of an automated solar powered plant watering system.
- Representation and explanation of system output.
- Exploration of advantages and disadvantages of the system.

15.1.3 DELIVERABLES

- Advance study and analysis of the automated solar powered plant watering system.
- Working model.
- Results achieved from the system.

15.2 BRIEF HISTORY OF SOLAR PANELS

15.2.1 *1880–1960: EARLY STAGE OF SOLAR ENERGY*

- ➤ **(1883) by the Charles Fritts:** An American inventor. According to his description the first solar cells made from Selenium wafers [5].
- ➤ **(1904) Wilhelm Hallwachs:** He explains that the photosensitivity can also be made by the combination of the copper and cuprous oxide [6].
- ➤ **(1932) Audobert and Stor:** Discover that cadmium sulfide (CdS) has the phovoltic effect after the nobel price given to Elbert einstine for his theory on photoelectric effect.
- ➤ **(1957) Hoffman Electronics:** Achieved 8% efficient photovoltaic cells and after one year later he got 9%.
- ➤ **(1960) Hoffman Electronics:** Achieves 14% efficient photovoltaic cells.
- ➤ **(1963) Japan:** That was the biggest achievement that the japans installs a 242-watt. That was the world's largest array at that time to collect the solar power from the sun.

15.2.2 *1970–1990: INTERMEDIATE STAGE*

- ➤ **(1977) World:** In that time total photovoltaic manufacturing production was not more then 500 kilowatts and that time it was first time it was increase more then 500 kilowatts.
- ➤ **(1982) World:** After five years the worldwide photovoltaic production was increase till 9.3 megawatt.

15.2.3 *1990–2007: ADVANCE STAGE*

- ➤ **(1991) America:** In this year the president of America redesignates Institute as the National Renewable Energy Laboratory.
- ➤ **(1996) World:** The most advance solar powered airplane was made on which 3,000 super solar cells was embedded from which it took the energy to fly.
- ➤ **(1998) World:** Remote control solar power aircraft.

15.3 LITERATURE SURVEY

Solar Photovoltaic technology is widely used to produce electricity [7] from the incident solar irradiation. This produced electricity can be widely used for the irrigation system. The various works has been performed based on the solar energized automatic [8–15] irrigation system. In this regards, Bolu et al. [8] implemented a solar photo-voltaic energized micro-controller based smart irrigation system which detects the moisture level and operates the pump when the moisture level goes below the threshold value. Al-Ali et al. [9] have produced a compendium of the usages of solar photo-voltaic [16]. According to the Abayomi-Alli et al. [10] smart solar photovoltaic system is also an viable alternative for the irrigation which can be taken as a pathway for the sustainable [17] agriculture. However, Mahalakshmi et al. [11] has focused on the distant monitoring and control of solar powered irrigation system for agricultural needs. The work presented by the author has used internet of things (IoT) technology for this purpose.

In a similar work, Tamoor et al. [12] has focused their study for remote rural area. The scholars investigated the socioeconomic as well as environmental impact after the integration of solar photovoltaic system [18] for remote agricultural needs. However, Rehman et al. [13] have implemented three different sensors namely moisture, temperature and humidity sensors to efficiently working of solar powered automatic irrigation system. Likewise, Rout et al. [14] have also used IoT technique to design solar photovoltaic energized smart watering system for agriculture purposes. The work can also be assessed from the existing literature in which role of hybridization [15] of solar power technology along with other technologies like diesel generator, battery [19–21], etc., are implemented to power the automatic irrigation system. This shows a wide scope of solar photovoltaic resource for powering a irrigation system across the globe.

15.4 HARDWARE STRUCTURE

15.4.1 MICROCONTROLLER

The microcontroller module as shown in Figure 15.2 will decode the incoming PWM signal and convert the information in electric form. To make the microcontroller to work according to the requirement a suitable program was required. The microcontroller used was PIC 18F1320 for more details refer to datasheet.

In this system embedded C was used for the development of the program, but microcontroller understands only machine codes for that some kind of tool was required that means a compiler, which will convert a high level language like embedded C into the required machine codes. Embedded C was used because as it was easy to understand and level of abstraction of program was much higher as compared with assembly language. For more information about embedded C look at reference.

MPLAB along with CCS C compiler was used to develop the program. ICD2 kit was used to transfer the machine codes into the flash memory of the microcontroller. To get familiar with CCS C compiler have a look at reference.

15.4.2   SOLAR PANEL PUMPING SYSTEM

Two basic components were used in the solar powered pumping system. The first component which was used was the combination of the solar panel and the chargeable battery. The solar cell was the smallest element of the solar panel. To the production of the direct current (DC) when the solar cell exposed to the light its two specially prepared layers do the special roles. Its gives the DC current to the battery which was then turns the DC pump on which give the supply of water to the soil. The rating of the voltage and the current at the output was calculated. The peak power was also calculated by multiplying of the voltage and the current of the solar panel. It was also depend on the irradiation and temperature from the sun on solar panel. The irradiation and temperature receives on the solar panel surface was 1,000 W/m^2 at 25°C. The amount of the DC produced by the solar panel was directly proportional to the light intensity striking on the solar panel [22, 23]. It could be explained like this if the intensity of the light was halved the output of the DC was half but the value of the direct voltage was reduce very little [24]. The ratings of the solar panel used are shown in Table 15.1.

TABLE 15.1   Characteristics Data of the Solar Panel

Maximum power	2.4 W
Maximum voltage	17.5 V
Maximum current	137 mA

15.4.3 BATTERY SYSTEM

The battery which was selected was rechargeable lead acid battery in the Hardware. There are many benefits of this battery such as easy to maintain, cheap, available in the market in the different variety of the size. They can also withstand for one day with their only 80% of the rated capacity. There are other batteries also available such as gel cell that have gel instead of the electrode. They are portable and easy to use. They can be use upside down and in any way because they are seal packed. Another type of the battery which could be use was nickel cadmium plates used with the solar panel system. Although, the lead acid battery has high initial cost but the life cycle cost was very low. There were some advantages of the NiCd batteries like it is long lasting but it needs less maintenance. It can withstand in the extreme conditions also. NiCd could tolerate more to complete discharge. Car batteries was also an option but car batteries were not use to store the power from the solar panel because the design of the car batteries were made like that they produce cold cracking ampere for short period of time. Battery banks were also the good option while charging the battery by the solar panel. In the battery banks series of 12 V batteries could be connect in series to get the desire voltage at the end. They are rechargeable. In the end after going through all the available batteries the lithium ion battery was selected due to its much advantage over other available batteries like it was easy to use, portable and easy to handle. The battery 2 Ah storage capacities was selected.

15.4.4 PUMP

The pump was the essential component of the project. The pump which was used was especially selected to use the solar power efficiency. The convention pumps need alternative current that generates the supply. The pump which was use was the solar pump which was especially design for the DC supply input from the battery. It was especially selected to work effectively during low-light conditions, at reduced voltage, without stalling or overheating. There were various types and size of pumps was available. The especially 12 V DC pump was selected to get the application for the low volume. The low volume pumps keep the cost of the system down and use the less electricity from the chargeable battery. Some pumps are fully submarine but the pump which was used was not. The pump which was used was used two pipes connected and takes charge from the battery. If the pump which was used under water the benefit would be it eliminates potential priming and freezing problems. The range of the solar panel used would use between 12 and 36 V DC. Some pumps could be use in the project could be positive displacement pumps. It seals the water in cavities inside the pump and forces it up. They design in such a manner that they maintain their lift capability all through the solar day slowly if they were connected directly with the solar panel. If the lighting conditions were fine it runs fast but if the lighting condition were not good it run slow. The speed of the pump if connected directly depends upon the lightning conditions available. The required system was not required on the pumping vertically. So the pump which was used was the simple 12 V DC pump.

Pumping was the mechanical work. The amount of water pump by the pump and the length of the pipe used before and after that were the two main factors which determine the rating of the pump. If the length of the pipe was increase the required power of the pump was increase. By the appropriate measurement it was found that the used pump of 12 V was alright. The friction of the pipe with the water was also reducing by the use of the smooth pipe to increase the efficiency. The choice of pump depends on water volume needed, efficiency, price and reliability.

15.5   PROBLEM ANALYSIS

15.5.1   THE SYSTEM DIAGRAM

As shown in Figure 15.2, the "Solar Powered Automatic Irrigation System" can detect and sense the moisture level in the soil and water the plants when needed. When the moisture level drops below the set level, the water pump will automatically turn on and water the plant until the set moisture level is reached. The sensor's probes in the ground record the moisture content through the conductivity established between the probes. When soil moisture and mineral levels are high, conductivity is established and the corresponding conductivity level is indicated by the LEDs. The energy to pump the water is produced from the solar panel which stores in the battery system. The irrigation system, which is commonly used in open fields, can fully exploit the sun's resource and use solar energy to power itself and even the water pump.

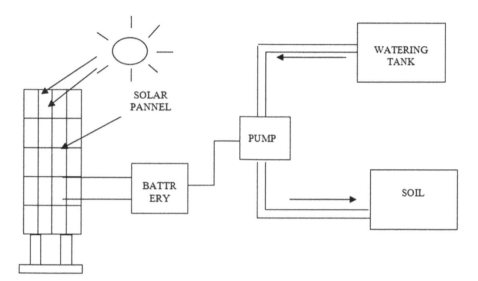

FIGURE 15.2 The system block diagram.

15.5.2 SPECIFICATION OF THE PROJECT (FIGURE 15.3)

15.5.2.1 INPUT

- Solar power energy.

15.5.2.2 OUTPUT

- Plant watering after every time when sensor finds that soil was wet (for simplicity LED display board used).

15.5.2.3 PROCESS

- Solar Batter charger charges the battery, and the battery gives the power supply to the system.
- The sensor senses the moisture in the soil and supply the water supply when required.
- LED glow up when water supply required and stay off when not required.

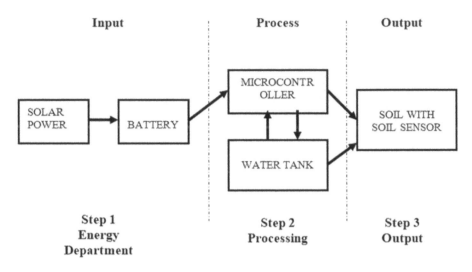

FIGURE 15.3   The input, output, and the process.

They were two types of connection which could be done. The first type was when the solar panel and the pump directly connected with each other and the second method was when the battery was intermediate between them. They are called battery coupled and direct couple. A variety of factors must be considered in determining the optimum system for a particular application.

There were some drawbacks for the using of the battery system. First was the efficiency of the overall system can be reduced because the voltage on which the whole system works was dictated by the battery directly not with the solar panel? It depends upon the amount of the battery charged. The voltage supplied by the batteries can be one to four volts lower than the voltage produced by the panels when the sun is at its maximum. This reduced efficiency can be minimized by using an appropriate pump controller that boosts the battery voltage supplied to the pump. But the connection was straight and forward. There was no pump controller was used.

15.5.3   BATTERY-COUPLED SOLAR PUMPING SYSTEMS

Battery-couple water pumping system required the direct coupling of the solar panel and the battery. The battery was connected with the project then. The hour charges the batteries and the batteries in turn power the pump whenever water is needed. Using batteries spreads pumping over a longer period of time by providing a constant operating voltage to the pump's DC motor. Thus, the system can provide a constant source of water for the project even during the night and in low light.

15.5.4 DIRECT-COUPLE SOLAR PUMPING SYSTEMS

This system was not used in the project but could be used if the system not required the battery and the amount of energy coming from the sun was good it means there would be constant source of the sunlight. It could be used straight in the sunlight.

15.6 PROBLEM IMPLEMENTATION

Because it combines hardware and software components, users can change requirements depending on seasons, plant types or water requirements. In addition, it does not require a highly skilled person to control the system. All that is required is knowledge about the plant, which must be provided in simple steps. This system finds many uses on farms owned by large corporations that produce crops for the world market. Installation requires very little cost and minimizes labor costs for the company as no labor is required to water the plants by testing soil moisture. Only one person is needed to operate the machine in the control room. The use of solar energy saves a large amount of electrical energy.

15.6.1 HARDWARE SPECIFICATION

- Port B0 was connected with the 12 V DC pump.
- Port B1 was connected with the electrodes.
- Port B2 was connected with the Led

In the hardware implimentation as shown in Figure 15.4,

- If the soil is wet then signal goes to microcontroller and PIN B becomes 1. Then eventually pump stops running and the led glow.
- If the soil is dry then signal goes to microcontroller and PIN B becomes 0. Then eventually pump start working and the led stop.

The flow chart of the above working has been shown in Figure 15.4.

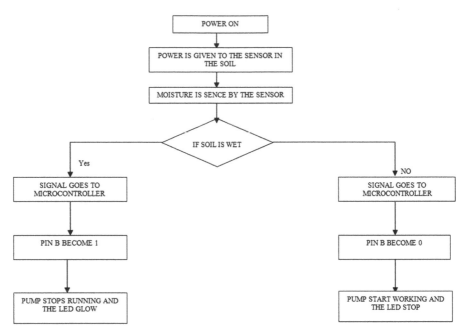

FIGURE 15.4   The flow chart of working.

15.6.2   PROGRAM USED

#include <16F877A.h>
#use delay(clock=4000000)
// macro-definition
#define motor PIN_B0
#define sensor input (PIN_B1)
#define led PIN_B2
#define on 0x00 // for on condition
#define off 0xff // for off condition

// main part of the program
void main()
{
int count;
// setup_oscillator(OSC_4MHZ);

```
count=0;
output_high(motor); // to turn on the pump
// to flash LED
output_low(led);
delay_ms(500);
output_high(led);
while(1) // infinite loop
{
while(input(sensor)==0); // to checking the moisture of the soil
output_high(motor); // to start the motor
output_high(led);       // turn on LED
delay_ms(250);          // delay
// if soil was wet
while(input(sensor))
{
output_low(led);        // led was stop glowing
output_low(motor);      // turn off motor
delay_ms(250);          // delay
}
}
}
```

15.6.3 CIRUIT DIAGRAM

In Figure 15.5, the ciruit diagram has been shown. The power has been through by the transister "7805." The 5 V supply from outside has been given to the microcontroller. The PDS was used to use the microcontroller and the power was given from outside. The optocoupler used to transfer the signal from one electronic equipment to another with out making any connection. The microcontroller and the optocoupler both connected with the 5 V supply. The transister was used to drive the motor properly. The base of the transister connected with the optocoupler and the collector was connected with the pump. The emitter was connected with the voltage of 12 V. The transister was giving the voltage and the current to the motor properly. The leads are labeled base (B), collector (C), and emitter (E). These terms refer to the

internal operation of a transistor, but they are not much help in understanding
how a transistor was used, so just treat them.

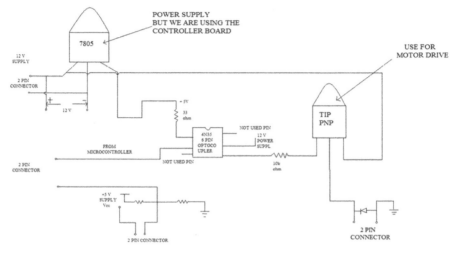

FIGURE 15.5   The flow chart of working.

15.7   CONCLUSION

The proposed work undertakes the feasible solution to a problem of the
unsustainable plants irrigation system. There is a rising need for technically
advanced irrigation practices to save thousands gallons of water that would or
else go to waste. The proposed hardware system uses solar panels to power
itself, making it a self-contained independent hardware model. The use of
clean renewable energy in engineering events cannot be underestimated if
hydrocarbons are not used in a sustainable manner because of the energy
disaster we are heading towards. The working model also eliminates the need
for manpower to irrigate slowly and intensively; the opportunity cost saved
allows the farmer to focus on other technical improvements and maximizing
crop yield.

Due to the cheap building materials and further cost optimization when
the model comes in the marketplace, it finds lots of applications in a broad
range. The scalability of the proposed work would be significantly easier
since the model can be used on both small and large farms. So, this model
would not only confine the market of larger and more affluent farms, but also
minor farms with limited resources. The system is also easy to use, and no

multifaceted training is necessary on the part of farmers to use this device. In addition to the environment, this benefit from lower water consumption in the fields, the farmer also makes profit from a higher yield through optimized irrigation. In addition to the positive environmental impact that this hardware model would have, there will be huge economic benefits for farmers.

Given the needful demand for such a project, a win for all the key stakeholders is obvious when using such a project. Aside from detecting soil moisture, this device can also be used to test the difference between mineral water and distilled water, the power of acids and any material that involves the ionization thereof, making it a multi-purpose tool for measuring the levels of various substances. As a responsible citizen of this planet, it is our individual responsibility to take an active job in water conservation. And so, this project is a result of that responsibility.

KEYWORDS

- **cost optimization**
- **hardware specification**
- **optocoupler**
- **PIC micro-controller**
- **solar PV**
- **stakeholders**
- **transistor**

REFERENCES

1. Balaji, V. R., & Sudha, M., (2016). Solar powered auto irrigation system. *International Journal of Emerging Technology in Computer Science & Electronics (IJETCSE), 20*(2), 203–206.
2. Barman, A., Neogi, B., & Pal, S., (2020). Solar-powered automated IoT-based drip irrigation system. In: *IoT and Analytics for Agriculture* (pp. 27–49). Springer, Singapore.
3. Priya, P., & Madhumitha, R., (2021). Modeling and simulation for automatic irrigation system with PV solar tracking. *Materials Today: Proceedings.*
4. Khemissi, L., Khiari, B., Andoulsi, R., & Cherif, A., (2012). Low cost and high efficiency of single phase photovoltaic system based on microcontroller. *Solar Energy, 86*(5), 1129–1141.

5. Zhai, Z., Chen, X., Zhang, Y., & Zhou, R. (2021). Decision-making technology based on knowledge engineering and experiment on the intelligent water-fertilizer irrigation system. *Journal of Computational Methods in Sciences and Engineering* (pp. 1–20). (Preprint).

6. Morison, W. L., (2004). Photosensitivity. *New England Journal of Medicine, 350*(11), 1111–1117.

7. Kumar, P., Pal, N., & Sharma, H., (2020). Performance analysis and evaluation of 10 kWp solar photovoltaic array for remote islands of Andaman and Nicobar. *Sustainable Energy Technologies and Assessments, 42,* 100889.

8. Bolu, C. A., Azeta, J., Alele, F., Daranijo, E. O., Onyeubani, P., & Abioye, A. A., (2019). Solar powered microcontroller-based automated irrigation system with moisture sensors. In: *Journal of Physics: Conference Series* (Vol. 1378, No. 3, p. 032003). IOP Publishing.

9. Al-Ali, A. R., Rehman, S., Al-Agili, S., Al-Omari, M. H., & Al-Fayezi, M., (2001). Usage of photovoltaics in an automated irrigation system. *Renewable Energy, 23*(1), 17–26.

10. Abayomi-Alli, O., Odusami, M., Ojinaka, D., Shobayo, O., Misra, S., Damasevicius, R., & Maskeliunas, R., (2018). Smart-solar irrigation system (SMIS) for sustainable agriculture. In: *International Conference on Applied Informatics* (pp. 198–212). Springer, Cham.

11. Mahalakshmi, M., Priyanka, S., Rajaram, S. P., & Rajapriya, R., (2018). Distant monitoring and controlling of solar driven irrigation system through IoT. In: *2018 National Power Engineering Conference (NPEC)* (pp. 1–5). IEEE.

12. Tamoor, M., ZakaUllah, P., Mobeen, M., & Zaka, M. A., (2021). Solar powered automated irrigation system in rural area and their socio economic and environmental impact. *International Journal of Sustainable Energy and Environmental Research, 10*(1), 17–28.

13. Rehman, A. U., Asif, R. M., Tariq, R., & Javed, A., (2017). Gsm based solar automatic irrigation system using moisture, temperature and humidity sensors. In: *2017 International Conference on Engineering Technology and Technopreneurship (ICE2T)* (pp. 1–4). IEEE.

14. Rout, K. K., Mallick, S., & Mishra, S., (2018). Solar powered smart irrigation system using internet of things. In: *2018 2ⁿᵈ International Conference on Data Science and Business Analytics (ICDSBA)* (pp. 144–149). IEEE.

15. Tapiceria, P. A., & Magwili, E. G., (2021). Hybrid solar-hydrokinetic powered automated irrigation system. In: *2021 IEEE 13ᵗʰ International Conference on Humanoid, Nanotechnology, Information Technology, Communication and Control, Environment, and Management (HNICEM)* (pp. 1–5). IEEE.

16. Sharma, H., Kumar, P., Pal, N., & Sadhu, P. K., (2018). Problems in the accomplishment of solar and wind energy in India. *Problemy Ekorozwoju, 13*(1).

17. Kumar, P., Sharma, H., Pal, N., & Sadhu, P. K., (2019). Comparative assessment and obstacles in the advancement of renewable energy in India and China. *Problemy Ekorozwoju, 14*(2).

18. Kumar, P., Pal, N., & Sharma, H., (2022). Optimization and techno-economic analysis of a solar photo-voltaic/biomass/diesel/battery hybrid off-grid power generation system for rural remote electrification in eastern India. *Energy, 247,* 123560.

19. Kumar, P., Pal, N., & Kumar, M., (2021). Hybrid operational deployment of renewable energy—A distribution generation approach. In: *Design, Analysis, and Applications of Renewable Energy Systems* (pp. 627–643). Academic Press.

20. Kumar, P., Pal, N., & Sharma, H., (2021). Techno-economic analysis of solar photo-voltaic/diesel generator hybrid system using different energy storage technologies for isolated islands of India. *Journal of Energy Storage, 41*, 102965.
21. Kumar, P., Kumar, M., & Pal, N., (2021). AN efficient control approach OF voltage and frequency regulation IN an autonomous microgrid. *Revue Roumaine Des Sciences Techniques-Serie Electrotechnique Et Energetique, 66*(1), 33–39.
22. Hosenuzzaman, M., Rahim, N. A., Selvaraj, J., Hasanuzzaman, M., Malek, A. A., & Nahar, A., (2015). Global prospects, progress, policies, and environmental impact of solar photovoltaic power generation. *Renewable and Sustainable Energy Reviews, 41*, 284–297.
23. Kumar, P., Chandra, K. A., Patel, S., Pal, N., Kumar, M., & Sharma, H., (2020). Operational challenges towards deployment of renewable energy. *Resources, Challenges and Applications,* 129.
24. Kumar, P., Sikder, P. S., & Pal, N., (2018). Biomass fuel cell based distributed generation system for Sagar Island. *Bulletin of the Polish Academy of Sciences: Technical Sciences,* 5.

CHAPTER 16

Optimal Parameter Estimation of 3-Phase Transmission Line Using a Grey Wolf Optimization Algorithm

MUHAMMAD SUHAIL SHAIKH,[1] ABDUL LATIF SHAH,[2]
SHAFIQ UR REHMAN MASSAN,[2] RABIA ALI KHAN,[2] MUNSIF ALI JATOI,[4]
SHUBASH KUMAR,[3] and MANNAN HASSAN[5]

[1]*School of Physics and Electronic Engineering, Hanshan Normal University, Guangdong, China*

[2]*Newports Institute of Communications and Economics, Karachi, Pakistan*

[3]*School of Electrical Engineering, Yanshan University, Qinhuangdao, Republic of China*

[4]*Salim Habib University, Karachi, Pakistan*

[5]*School of Electrical Engineering, Southwest Jiaotong University, Chengdu, Republic of China*

ABSTRACT

A optimization, selects the best option from a list of possible options to meet a predetermined set of criteria. In pursuit of the best and most optimal solution for a particular issue, this theory is relevant to virtually all areas of natural as well as social sciences, and it has applications across the board. It's possible to utilize the optimization domain to either maximize or decrease the value of an actual function by selecting input values carefully and computing their values. A significant field of applied mathematics is devoted to the application of minimization and maximization with different restrictions on a wide variety of objective functions. Many different optimization methods may be used

The Internet of Energy: A Pragmatic Approach Towards Sustainable Development. Sheila Mahapatra, Mohan Krishna S., B. Chandra Sekhar, & Saurav Raj (Eds.)

constructed on the cost function and restrictions that are existing. As a result, optimization methods are generally classified into the following categories: discrete and continuous optimization, Unconstrained and Constrained optimization, and none or multi-objective functions. Swarm intelligence is a kind of artificial intelligence (AI), which mimics the intelligent behavior of swarms, flocks, herds, and other groups of animals seen in nature. As a result, these algorithms make use of the collective behavior of animals' groupings. The whale optimization algorithm (WOA), dolphin echolocation (DE), CS algorithm, bat algorithm (BA), Harmony Search (TS), artificial bee colony (ABC) algorithm, grey wolf optimizer (GWO), fruit fly optimization algorithm (FOA), the, hill-climbing, iterative local search, and other SI algorithms are available. (This chapter will cover the idea of optimization, as well as mathematical derivations and explanations of their applications in real-world situations). Following that, it will examine the specific application of optimization methods to one of the electrical engineering issues, namely, the parameter estimation of overhead transmission lines, in more detail. It is the transmission line, which is composed of inductance, capacitance, and resistance, that is the most critical component of the power system. During the transmission line design phase, these characteristics are very important. The overhead transmission line parameter is computed using a new optimization method known as gray wolf optimization, which is described in more detail in the literature. GWO is a meta-heuristic algorithm that was recently created based on natural inspiration. For testing, 3-phase transmission line test systems are used. The command structure and hunting technique of gray wolves served as inspiration for the suggested algorithm. The technique is used for several optimization functions that differ in size and a lot of search agents participating. The GWO algorithms provide optimal results that outperform those produced by other algorithms previously in use. For the vast majority of these statistically verified functions, the suggested method provided the most optimum solutions that could be found. Following the results, it has been discovered that the suggested method is more computationally efficient than the conventional algorithm and exceeds it in terms of accuracy, resilience, and convergence time. Moreover, the best optimal obtained using GWO for 3-phase capacitance and inductance are 0.22436, 0.022935, 0.65915, and 0.34938.

16.1 INTRODUCTION

It is important to estimate line parameters since they are used in a variety of power applications. The techniques of calculating line parameters may be divided into two categories: off-line methods and online approaches. Off-line techniques measure the line characteristics based on the design of the tower and the kind of wire used. However, because of the fluctuating nature of the load and the changing weather conditions throughout the operation, the measured line characteristics may prove to be incorrect. As a consequence of the noise introduced during PMU measurements, estimate results that differ from their real value have been seen [1]. The capacity of the transmission line must be adequate to meet power flow changes. Transmission line parameter estimation models are also commonly employed in a range of network applications. Previously, power transmission parameters were computed using physical transmission line features such as cable geometry, tower structure, and length. These measures are subject to mistakes because they overlook constantly changing operational elements including age, skin effects, air temperature, and other environmental influences, which are all taken into account. According to a recent trend, physicist-based strategies are progressively being supplanted by measurement-based approaches. Such measurements may be performed using a supervisory control and data acquisition (SCADA) device, a phasor measuring unit (PMU) fault record. Failure records are used in certain techniques for calculating parameter estimates. To categorize parameters from reported fault transient recordings, a phase domain strategy is provided, as well as relying on prony predicting methods to detect fault transient features. Using state growth methods necessitates a significant degree of redundancy in the computation as well as subjective standards regarding parameters, which makes them impractical for practical use. Furthermore, the computation of the parameter may be done independently of the estimate of the current state of affairs. It is proposed to use numerous measurement snapshots to do off-line parameter estimation, including iterations for parameter estimate and states utilizing multiple measurement snapshots. To estimate the parameters, a PMU technique is created that uses the estimated condition of many measurement snapshots to determine their values. As previously stated, several of the techniques make use of numerous measurement snapshots to enhance repetition. To build a well-conditioned coefficient matrix, the snapshot number must be maintained as low as practicable. The limitations of parameter estimating strategies based on distinct measurement snapshots stem from these competing factors. Using dynamic parameter approximation, which is useful for future grids, can alleviate these concerns [2].

Any power system's principal goal is to provide high-quality, reliable, and cost-effective energy to its customers. Furthermore, the energy system must achieve transmission, production, and supply of power systems, since meeting changing consumer demand is difficult [3–5]. The use of mathematical simulation is critical in the study of power systems. In the domain of power systems, there are multiple complex problems with no single solution. So, optimization approaches are employed to solve these undetermined challenges. Multiple artificial intelligence (AI) optimization methods have been adapted for different settings and issues. The main advantage of AI is computation accuracy [6, 7]. There are several different AI algorithms, including the ant lion optimizer (ALO), Hirschberg–Sinclair algorithm (HS), dragonfly algorithm (DA), Tabu search (TS), compare and swap algorithm (CAS). The overall costs and computing durations incurred by the GWO method were lower than those incurred by other optimization algorithms, and the algorithm demonstrated better convergence rates and superior performance when compared to other optimization algorithms to demonstrate the effectiveness.

To optimize engineering design issues, a gray wolf optimization method (GWO) is being suggested. It all starts with creating an initial position for the gray wolf population using a tent map, which distributes the population fairly and prepares for a massive global search and rescue effort. Second, to prevent the method from being trapped in a local optimal state, Gaussian mutation perturbation is employed to perform various actions on the present optimum solution.

The swarm intelligence method is capable of dealing with a wide range of issues that traditional optimization approaches are unable to handle effectively or efficiently. As a result, optimization methods have become more popular in a broad range of fields of study. In other words, current algorithms are capable of addressing certain problems well, but not all of them do so consistently. Some novel heuristic algorithms are suggested each year as a consequence of this, and ongoing research is conducted in this area. Actual monetary worth is a novel population-based meta-heuristic technique called grey wolf optimizer (GWO). Because the GWO method is straightforward, versatile, and efficient, it may be employed efficiently in a wide variety of real-world situations. There are many variants of the GWO algorithm available for download. GWO algorithms were combined with other investigative optimization approaches by several authors. As all of these approaches depend on binary encoders to track down the gray wolf, the information stored in genetic information is limited in its utility. In every evolutionary method, the convergence rate is given precedence over the quality of solutions while addressing an optimization issue, regardless of the technique. Generalized evolutionary computing technology (GWO) produces superior outcomes in comparison to previous evolutionary computer technologies. Real-time applications, on the other hand, are primarily concerned with the convergence time [8–12].

A GWO is a recently developed competent meta-heuristic process that mimics the public pyramid behavior of gray wolves [13]. Grey wolves live in a group with an average number of participants ranging from 5 to 12 individuals. Grey wolves are regarded as excellent hunters, and their survival and reproduction are dependent on the leadership chain and hunting etiquette. Section 16.2 provides a thorough description of the GWO algorithm's internal structure. The GWO algorithm has been utilized to tackle issues in a variety of engineering fields, including energy storage system sizing [17], optimum dispatch of actual and reactive power [15, 16], fuzzy control system tuning [18], and load frequency management [14]. Each new system is described in its simplest form, after which it may be tweaked and hybridized with existing algorithms to solve global or specialized functions [19–21]. Among other things, the suggested MGWO algorithm is distinguished by its simplicity, adaptability, and globalism. Scientific experts have suggested many changes to the GWO algorithm to avoid slipping into the local optima trap. Ref. [22] presents the GWO technique, which combines the GWO with Lévy Flight (LGWO) to avoid local goals stagnation and increase performance. However, this incorporation increased the complexity of the original GWO technique. To be assessed utilizing different bundle conductor designs, transmission line parameters must be analyzed using an optimization technique that meets the following requirements. However, the complexity of the original GWO method was increased as a result of this integration. Transmission line parameters must be evaluated using an optimization method that has the following criteria to be evaluated using various bundle conductor configurations.

The following are the chapter's major contributions:

- A detailed review of different optimization techniques is presented.
- The 3-phase transmission line's parameters are estimated using GWO on a separate bundle conductor.
- Four cases are studied while performing transmission line problems, two bundle capacitance, 3 bundle capacitance, two bundle inductance, and finally 3 bundle inductance.
- Best optimal solution considering the cases.
- Proposed GWO was compared with WOA available in the literature.

16.2   LITERATURE REVIEW (DIFFERENT OPTIMIZATION TECHNIQUES)

The field of computational intelligence (CI) includes meta-heuristic algorithms based on nature. Mathematics CI (MathCI), Physicist CI (PhyCI), bioinformatics CI (BioCI), and chemistry CI (ChmCI) are the four main types of processes in CI [23]. Bioinformatics-based algorithms are those that are based on biological data. The two main types of biology-based meta-heuristic algorithms are evaluation-based algorithms and swarm-based algorithms. Evaluation-based algorithms are the most common kind of biology-based meta-heuristic algorithm. It has become widely accepted in several engineering fields because of its benefits over other selections of nature-based algorithms. Swarm meta-heuristic Algorithms have gained widespread acceptance in a variety of engineering sectors due to their benefits on classifications of nature-based algorithms. It has benefits over estimation-based algorithms in that it keeps track of the search space throughout each iteration and needs minimal operators for successful implementation, resulting in rapid execution. For their part, the knowledge retained by evaluation-based algorithms does not hold up whenever a new population is created, and this results in a higher number of operators being required [24]. With exponential growth in the issue size, swarm-based algorithms have been demonstrated to be more efficient than other algorithms in addressing high-dimensional across a wide area, combinations and nonlinear optimization problems search region [25–27]. The academic community has been fascinated with meta-heuristic algorithms for almost two decades, and it is no surprise that they remain so. Particle swarm optimization (PSO) [28, 29], ant colony optimization (ACO) [30, 31], artificial bee colony optimization (ABC) [32], CS algorithm [33, 34], krill herd (KH) algorithm [35], bat-inspired (BA) algorithm [36], firefly algorithm [37], genetic algorithm (GA) [38, 39], The, evolution strategy (ES) [40], probabilistic incremental learning (PBIL) [41], Genetic programming (GP) [42, 43], and biogeography-based optimizer (BBO) [44, 45] are examples of another generation of evaluation-based algorithms.

16.2.1   GWO

Recently [46], The GWO is a meta-heuristic algorithm that was created to imitate the social configuration and Grey Wolves in the Wild Hunting

Behavior. The GWO is composed of four types of Grey Wolves, each of which has its unique hunting behavior (alpha, beta, delta, and omega) are the four types of the wolf. Moreover, they utilized to approximate the leadership hierarchy which is evolved as a result of the group's struggle for survival as a whole, which is modeled after the human brain. To carry out a collective hunt, it is frequently necessary to use 3 complex movements: searching for prey, surrounding of prey, and attacking of prey. These maneuvers have been described in detail in Ref. [14]. When it comes down to it, gray wolf optimization has the following 3-distinct advantages: When associated with other standard heuristic methods using a gravitational search method, Thus, a greater rise in global optimization is possible since it allows for effective exploitation and inquiry, as well as proactive local optimization prevention and promising implementation in the face of uncertainty. A multi-objective optimization technique based on differential evolution (DE) was used to improve the size of a solar water pumping system which has since been recommended [16]. Evolution programming (EP) is used to improve the viable path supplied by a set of local activities, and the ABC approach is also utilized as a local search method [17]. Many other areas have benefited from GWO, for example, preventing a real power system from becoming dark owing to failures in generating units or major transmission lines [18]. Furthermore, in Ref. [19], the optimal reactive power is determined to reduce loss and voltage variation to the bare minimum.

This part models hunting methods as well as the social hierarchy to create Grey Wolf Optimization and carry out work in the field.

16.2.1.1 INSPIRATION

Researchers believe gray wolves are top-of-the-food-chain predators, making them top hunters. They would rather live in groups, and a distinct group of 5 to 12 candidates almost invariably follows a strict social leadership structure. Because the alpha wolf should be followed by the rest of the pack, he is often referred to as the dominating wolf. In the Mathematical model of hunting, the GWO mechanism is outlined as below:

- Moving towards prey;
- Follow the prey till he moves, surrounding him, and harassing him until he moves;
- Finally, attacking the prey.

16.2.1.2   SOCIAL HIERARCHY

During mathematical modeling, make certain that the Grey Wolf Optimization is designed in such a way that the optimal solution in the alpha is accepted. As a result, beta () is the second-best solution to the problem. The delta () wolf is the third-best, while the remaining wolves are omega () wolves. The alpha, beta, and delta algorithms are responsible for the hunting process.

16.2.1.3   TRACKING

As the first stage in Grey Wolf Optimization, they take down the gray wolf and catch up with it during the hunting process. The mathematical Eqns. (1) and (2) may be used to explain the wolf's behavior.

$$\vec{X}(t+1) = \overrightarrow{X_p}(t) - \vec{A}.\vec{D} \tag{1}$$

$$\vec{D} = \left| \vec{C}.\overrightarrow{X_p}(t) - \vec{X}(t) \right| \tag{2}$$

where; $\vec{X}$ is the gray wolf position, t shows the number of iterations; $\overrightarrow{X_p}$ indicates the location of the prey; D denotes the distance between both prey and wolf; A denotes the number of coefficients, which are computed as follows:

$$\vec{A} = 2\vec{a}.\overrightarrow{r_1} - \vec{a} \tag{3}$$

where; r_1 and $r2$ is the vector's two random values within the range $[0,1]$.

16.2.1.4   ENCIRCLING

The technique of GWO while attacking agents is shown in this attacking prey illustration. As a consequence of the transmission line features a study, the attacker's behavior, which is dependent on leaders and their colleagues, encircles the agent. Because capacitors and inductors are submissive to the transmission line parameter, delta wolves compute it and assault the agent farther down the transmission line parameter chain. The transmission line parameter, also known as omega and scapegoat, is defined as the effects of agents on end-users. We may update their positions by computing alpha, beta, and delta, which are $(\overrightarrow{X_\alpha})$, $(\overrightarrow{X_\beta})$, and $(\overrightarrow{X_\delta})$, respectively.

$$\vec{X}(t+1) = \frac{\vec{X_1} + \vec{X_2} + \vec{X_3}}{3} \tag{4}$$

$$\vec{X_1} = \left| \vec{X_\alpha}(t) - \vec{A_1}.\vec{D_\alpha} \right| \tag{5}$$

$$\vec{X_2} = \left| \vec{X_\beta}(t) - \vec{A_2}.\vec{D_B} \right| \tag{6}$$

$$\vec{X_3} = \left| \vec{X_\delta}(t) - \vec{A_2}.\vec{D_B} \right| \tag{7}$$

$$\vec{D_\alpha} = \left| \vec{C_1} - \vec{X_\alpha}(t).\vec{X_t} \right| \tag{8}$$

$$\vec{D_\beta} = \left| \vec{C_2} - \vec{X_\beta}(t).\vec{X_t} \right| \tag{9}$$

16.2.1.5 ATTACKING

Grey wolves complete their hunt by attacking their prey after it stops moving. The value of "a" in Eqn. (3) will drop linearly as a result of these kinds of activities. This time also includes a change in the quantity of A, which goes from [–a] to [a]. As an alternative, whenever |A| <1, the wolf's location becomes closer to the prey's position. When the value of |A| >1, the pack of wolves will run away from their target. As a result, in gray wolf optimization, the update equation of "a" is the same as in Eqn. (10).

$$a = 2 - 2 * t / \text{Max iter} \tag{10}$$

where; 't' is the current iteration.

16.2.2 WOA

WOA is a continuous optimization problem-solving method based on swarm intelligence. When utilized in conjunction with current meta-heuristics techniques, it has been demonstrated to improve performance [52, 53]. It is simple to create and durable, for example, when compared to other swarm intelligence systems, making it comparable to other nature-inspired algorithms in terms of performance. The algorithm requires a lesser number of control parameters; in reality, only one parameter (the time interval) needs to be fine-tuned. The researchers have built a multi-dimensional search habitat for the humpback whale population in WOA forages. It's important to note that the distance between humpback whales and their food is connected to the amount of objective cost in this model, hence individual whale locations are option variables. Three operational strategies are employed to determine a whale's time-dependent position: shrinking around prey, bubble-net assault technique (exploitation phase), and seeking for prey (exploration phase). The next subsections explain various operating procedures, as well as a mathematical expression for them.

16.2.2.1   ENCIRCLING PREY

When in their native habitat, humpback whales can recognize and circle their prey. Because the ideal design is situated in the ideal location. Considering that the search space is unknown at the onset, the WOA considers that the current best answer is either the intended prey or something that is close by. Other search agents are updating their locations around the top search agent while the best search agent is being sought. The equations shown below reflect the behavior.

16.2.2.2   SEARCHING AND ENCIRCLING PREY

Eqn. (11) may be used to demonstrate the process of looking for food (prey)

$$\vec{D} = \left| \vec{C}.\vec{X}_{rand} - \vec{X} \right| \tag{11}$$

$$\vec{X}(t+1) = \vec{X}_{rand} - \vec{A}.\vec{D} \tag{12}$$

where; A and C are vectors of coefficients, as shown:

$$\vec{A} = 2\vec{a}.\vec{r} - \vec{a} \tag{13}$$

$$\vec{C} = 2.\vec{r} \tag{14}$$

The value of a decreases linearly from 2 to 0, while the value of r is within the range [0,1].

$$\vec{D} = \left| \vec{C}.\overrightarrow{X^*}(t) - \vec{X} \right| \tag{15}$$

where; $\overrightarrow{D}$ is the prey update location.

$$\vec{X}(t+1) = \overrightarrow{X^*}(t) - \vec{A}.\vec{D} \tag{16}$$

If A≥1 then Eqn. (11) and (12) are used. It means that the whale is searching for prey. If A<1, then Eqns. (15) and (16) are used, which means that the whale already searched for food and now attacking the prey.

Where; 't' is the number of iterations currently being performed; 'X' denotes the vector position; and X* denotes the best possible value.

16.2.2.3 POSITION OF SPIRAL UPDATE

Eqn. (17) shows the new position:

$$\vec{X}(t+1) = \begin{cases} \overrightarrow{X^*}(t) - \vec{A}.\vec{D} \, if p < 0.5 \\ D.e^{bl}.\cos(2\pi l) + \overrightarrow{X^*}(t) \, if p \geq 0.5 \, \#\# \end{cases} \tag{17}$$

where; p is a number between [0,1], l is between [−1,1], and b is a constant.

16.2.3 MFO

It was Mirjalili [54] who first suggested the moth–flame optimization (MFO) method. Specifically, it falls under the category of population-based meta-heuristic methods. A random night butterfly is created in the solution space, the fitness values of each moth are calculated, and then the moth with the greatest fitness value is marked with a flame using MFO. A spiral movement function is used to update the moths' locations to attain flame-decorated ideal positions, As a result, each moth's location is updated, and new positions are generated, all of which are repeated until the termination requirements have been reached. Table 16.1 lists the MFO's features.

TABLE 16.1 Features of MFO

Algorithm Description	MFO Elements
Decision variable	Moths position in every dimension
Solutions	Moth position
Initial solutions	Moth random position
Current solutions	Moth current position
New best solution	New position
Best optimal solution	Flames position
Fitness function	The distance between moth and flame
New generating process for solution	A spiral path towards the flame

Creating the first population of Moths: As previously stated in Ref. [55], Mirjalili believed that each moth can fly in one of four dimensions:

one-dimensional, two-dimensional, 3-dimensional, and hyperdimensional space. The following is an expression for the collection of moths:

$$M = \begin{bmatrix} MO_{1,1} & MO_{1,2} & \cdots & \cdots & MO_{1,d} \\ MO_{2,1} & MO_{2,2} & \cdots & \cdots & MO_{2,d} \\ \vdots & \vdots & \vdots & \vdots & \vdots \\ \vdots & \vdots & \vdots & \vdots & \vdots \\ MO_{n,1} & MO_{n,2} & \cdots & \cdots & MO_{n,d} \end{bmatrix} \tag{18}$$

where; 'n' is the number of moths; and 'd' is the number of columns of matrix 'M.'

The MFO algorithm's remaining components are mostly flames. Each flame is represented by a 3-dimensional matrix, with fitness function vectors connected with each flame.

$$F = \begin{bmatrix} F_{1,1} & F_{1,2} & \cdots & \cdots & F_{1,d} \\ F_{2,1} & F_{2,2} & \cdots & \cdots & F_{2,d} \\ \vdots & \vdots & \vdots & \vdots & \vdots \\ \vdots & \vdots & \vdots & \vdots & \vdots \\ F_{n,1} & F_{n,2} & \cdots & \cdots & F_{2,d} \end{bmatrix} \tag{19}$$

where; 'n' is the number of moths and d is the dimension of the moth matrix.

It should be mentioned that both moths and flames are effective remedies in this situation. It is the manner we handle and modifies them for each iteration that distinguishes them from one another. A search agent is a creature that moves throughout the search area, While flames are the greatest habitat for moths thus far. To put it another way, think of flames as flags or pins that moths deposit in their search area as they fly around looking for food. It's because of this that each moth searches for an alternative to the current solution near a flame (and changes the flag accordingly). Because of this procedure, a moth's best option is always within reach.

16.2.3.1   UPDATING THE MOTHS' POSITIONS

MFO uses three-types of functions for convergence of the global optimization of the optimization problems which are defined below:

$$MFO = (I, P, T) \tag{20}$$

where; I is a function; 'P' is the flight of the moth in search of space; and 'T' is the stopping criteria.

$$M_i = S(M_i, F_j) \tag{21}$$

where;

- Mi is the number of ith moths;
- Fj indicates the number of jth flames;
- S is the twisting function and can be expressed as follows:

$$S(M_i, F_j) = D_i.e^{bt}.\cos(2\pi t) + F_j \tag{22}$$

where;

- Di is the distance between moth and flame;
- b is a constant value [1];
- t is a random number within the range of $[-1,1]$;
- Di is the distance can be calculated as:

$$Di = |F_j - M_i| \tag{23}$$

The moth's position is obtained by Eqn. (23) and the distance between moth and flame is controlled by t. The twisting motion of the moth describes how to update its position around the flame. Updating the position about diverse locations to contain this situation, Eqn. (26) is utilized in this regard.

$$\text{Flame number} = \text{round}\left(N - I * \frac{N-1}{T}\right) \tag{24}$$

where;

- I is the number of iterations;
- N is the total amount of flames;
- T represents the number of iterations.

Another kind of meta-heuristic is a first-generation method that may operate in conjunction with the top secondary heuristics to efficiently find and provide the best or just slightly better than optimal solutions for optimization issues. All meta-heuristics must search in both depth and breadth throughout their search process. Developing an appropriate diversification strategy is critical to dealing with this situation effectively. Meta-heuristic algorithms are used to find excellent results to optimize issues in situations when the complexity of the problem being addressed or the amount of search time accessible prevents the usage of precise optimization algorithms from being used.

16.2.4   TABU SEARCH (TS)

Tabu exploration is a meta-heuristic method that involves randomly exploring the entire solution space. Tabu search was developed by Fred Glover and colleagues as a way of locating or controlling local optima. Using the Tabu search has the advantage of effectively locating the history of the stop cycle and avoiding getting caught in local minima using customizable remembers [56]. The higher-level heuristic algorithm is solved, and efforts are made to determine whether the solution is good. Nara et al. [57] proposed a Tabu search to find the ideal position in terms of DG with the fewest losses in distribution networks or systems. According to the conclusions of this study, the author investigated how much the power system's loss percentage may be reduced following optimal DG allocation on the demand side of the system. This technique is based on the premise that the total number of DGs and total DG producing capacity are both equal to or greater than 1. When the NSGA II is employed as a pointing out technique, a multi-objective Tabu search is compared to the NSGA II, and MOTS delivers a much superior result in terms of processing time, an enticing aspect, especially as the investigation progresses and the passage of time becomes more critical [4].

16.2.5   PARTICLE SWARM OPTIMIZER

It is one of the most frequently employed stochastic optimization methods, and it is known as particle swarm optimization (PSO). Developed by Dr. Eberhart and Dr. Kamedy in 1995, this method is especially well suited for convivial breeding or nesting in bird colonies or gregarious breeding in bird colonies, or gregarious breeding in bird colonies. To increase the voltage profile and decrease power losses as well as Total Harmonic Distortion (THD), a PSO-based technique is employed as a distribution network solution tool. It also reduces THD. The primary objective of this technique is to decrease total harmonic distortion (THD) while concurrently improving the voltage profile. Smart PSO is, by its nature, very comparable to GA in terms of functionality. A random population will be used to start the system and subsequently, an upgraded population will be used to achieve optimal outcomes, based on the outcome of PSO and GA. The Proposed approach has provided a solution of better quality than the GA technique, which was previously used. Contrasting PSO to GA, the time required to complete the task is much less, and the method can be implemented to a real network with a high degree of ease. By focusing on bus voltage limitations as the major constraints, the recommended technique is one way to reduce system power losses. To reduce total power losses, the optimal placement and capacity of distributed generation (DG) must be determined by comparing the new network to the old one. As well as this, the upgraded power system's line voltage consistency is later connected, and the DG is recognized by the line voltage consistency signal. Applications for the real-time power system utilize the system.

16.2.6   PARTICLE SWARM OPTIMIZER

The ant colony transmits the algorithm for ant interbreeding with a single inhalation. The ants disperse pheromones on the earth's surface to choose a suitable path and then operate following the colony's social organization. The ant colony method, often known as a beginning algorithm, is an optimization approach. At the start of each reiteration of a procedure, new artificial ants are utilized to recognize or construct the process. Every single person participating in inventing a solution for performing the traveling action from the lowest point to the highest point on a plot chart while avoiding a barrier that does not tour the highest point that has already been visited. At the end of each iteration, the pheromone price is altered based on the quality of the ants' solutions, to bias the ants' future time iterations to construct the best feasible solution in comparison to the previous. An ACO approach for locating DG in the distribution network. The suggested ACO algorithm has been tested and shown to minimize operational expenses while improving electricity quality and customer service dependability. DG increases electricity quality and customer service dependability while cutting operational costs. To reach the distributed system's goal, an ant colony employs optimization dependability to execute a distributed system's optimal distribution generation location plan. To solve the problem, the ACO method was adopted. They discovered that the ACO technique outperforms both the external genetic algorithm (GA) and the need for normal computing time addition. This has the potential to cause ACS to persist for a long time, and it should be taken carefully since it may lead to much more solved possibilities in the future. However, because of the frequent response complication size and the computational efficiency of the system, extra time is not needed for the procedure under investigation to be completed.

16.2.7   SIMULATED ANNEALING (SA)

Meta-heuristics are a global optimization problem of finding a decent approximation to the global ideal of a job and useful when the location and range are distinct. Annealing guarantees that the natural optimization process happens with the least energy. And in 1985, Vlado Cerny described it to Scott Kirkpatrick, C. Daniel Gelatt, and Mario P. Vecchi. The well must be solved using simulated annealing cold. What are the beginning heat intensity and cooling procedures for the great significance of simulated annealing? It is possible to achieve more than standard solutions effectively and similarly as well as simply one goal using simulated annealing.

16.2.8   GENETIC ALGORITHM (GA)

A kind of optimization technique is the GA, which uses natural progression mechanisms including mutation, option crossover, and inheritance to solve an optimization problem. When an encoding candidate is solved to the ideal difficulty supplied for a great solution, the genetic approach first builds randomly residents (or chromosomes) of the string. The general health of the residents is assessed at the end of each cycle to build a new dweller. The algorithm's next or upcoming iteration is used in the new modified population. A good health plane or flat is obtained by repeating this approach until an extreme measure or number of generations is reached. DG placement and sizing GA The DG is normally nearer the substation. Moving DG closer to the load center minimizes power loss and has additional advantages, according to the suggested algorithm. By reducing power losses and improving power quality, this research endeavor provides the greatest answer for a group's size or upgrade. The suggested multi-objective technique minimizes all functions simultaneously. Have recommended a site and size for the DG. The method is evaluated on an IEEE 69 bus system and compared to nonlinear optimization. According to the DG study, improper DG location and size would reduce system stability and dependability, increasing system losses. This strategy minimizes power losses and DG investment costs while optimizing DG placement and size.

16.2.9   HYBRID OPTIMIZATION TECHNIQUES

Generating GAs is very useful for figuring out the optimal size and location for distributed generating systems. The GA, on the other hand, can sometimes produce low-quality results, making it inefficient when exact results are required. The evolutionary algorithm is employed in concert with other heuristic optimization approaches to overcome this constraint. The equivalence along traditional GA was solved by integrating GA with SA to uncover the method's ability to construct a fantastic class. A solution based on GA and fuzzy logic that takes into consideration the different forces of the objective function is one of the applications of multi-objective issues. To get a better result, researchers used additional optimization methods with GA. The transmission and distribution of energy from generators to consumers is the primary function of the electrical transmission and distribution system Many power applications need an accurate estimation of line parameters [49–58]. In a deregulated environment, the overhead transmission line system (OHTL) is critical [47, 48].

16.3 RESULTS AND DISCUSSION

This section will present, analyzes, and discusses the results. A Computer with a 64-bit Intel(R)-Core (TM) i7-4610M CPU @ 3.00 GHz and 12 GB RAM is set up with the Windows 10 operating system. Finally, the following section is organized:

16.3.1 CASE 01: THE CAPACITANCE WITH A TWO-BUNDLE CONDUCTOR FOR A 3-PHASE TRANSMISSION LINE

Initially, the suggested GWO was used to calculate 3-phase transmission line characteristics, and a value of 0.22436 was determined as the optimal capacitance per unit length for use with the two bundle conductors. Figure 16.1(a) shows the convergence curve between the GWO and WOA for transmission line parameter estimation when the iteration number is 100. As you can see in Table 16.2, the optimum solution to the transmission line problem may be found.

16.3.2 CASE 02: THE CAPACITANCE WITH A 3-BUNDLE CONDUCTOR FOR A 3-PHASE TRANSMISSION LINE

In this case, we have analyzed capacitance per unit length considering 3 bundle conductors. Figure 16.1(b) displays the suggested algorithm's convergence characteristic, with the ideal value for this example being 0.022935 when the number of search agents is 20 and the number of iterations is 100. As shown in Table 16.2, using GWO provides the best possible outcomes when compared to those produced by using WOA. An algorithm's reliability and superiority over an existing one must be demonstrated.

16.3.3   CASE 03: THE INDUCTANCE WITH A TWO-BUNDLE CONDUCTOR FOR A 3-PHASE TRANSMISSION LINE

For the two bundle conductors, the recommended GWO was utilized to compute 3-phase transmission line inductance. Figure 16.1(c) shows the convergence curve for the estimate of transmission line parameters using GWO and WOA at 100 iterations. As shown in Table 16.2, the transmission line issue may have an optimal solution and the best optimal value is 0.65915.

16.3.4   CASE 04: THE INDUCTANCE WITH A 3-BUNDLE CONDUCTOR FOR A 3-PHASE TRANSMISSION LINE

Here, we looked into inductance per unit length for 3 bundle conductors. Figure 16.2(d) illustrates the suggested algorithm's convergence characteristic, with the most optimum value being 0.34938 for 20 search agents and 100 iterations. Figure 16.1(d) shows the suggested method's convergence. Table 16.2 compares the results acquired using GWO to those obtained using WOA. To show the GWO algorithm's supremacy over the WOA algorithm's reliability.

16.4   CONCLUSION

This chapter covered a variety of optimization techniques for increasing the size and position of DG in the distribution system, including analytical, heuristic, and hybrid methodologies. In the appendix, there is a brief comparison of different optimization strategies. The research found that analytic methods perform poorly when dealing with complicated network issues. While heuristic methods provide a simple answer. However, it may occasionally result in low-quality outputs and more processing time. A combined strategy that incorporates two or more heuristic techniques may provide a high-quality result. This chapter discusses the parameter computation of the overhead transmission line (OHTL) power system using the suggested GWO algorithm. Three-phase

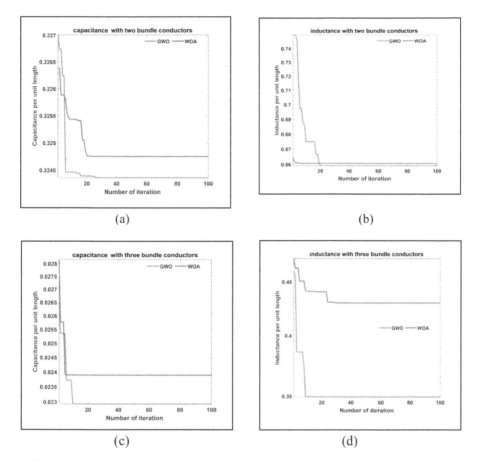

FIGURE 16.1 Three-phase transmission line convergence characteristics – (a) the capacitance with the two-conductor bundle; (b) the capacitance with the 3-conductor bundle; (c) the inductance with the two-conductor bundle; and (d) the inductance with a 3-conductor bundle.

transmission lines models, the method was applied to 06 functions, respectively. The total costs and calculation durations encountered while using the GWO method were smaller than those attained when utilizing different optimization procedures with a greater convergence rate, such as WOA. The collected results confirmed that the suggested GWO algorithm outperforms the other methods in terms of exploitation. For power systems, the suggested method is both quick and robust. Grey Wolf Optimization's convergence feature shows its ability to tackle a tough issue. The suggested method's findings are compared to the traditional approach for 3-phase transmission line systems. The findings

TABLE 16.2 Comparative Analysis of 3-Phase Transmission Line

Parameters	Capacitance per Unit Length (µF)		Parameters	Inductance per Unit Length (mH)	
	GWO	WOA		GWO	WOA
	Control variable obtained by GWO	Control variable obtained by WOA		Control variable obtained by GWO	Control variable obtained by WOA
Case 01	-0.013912	-5.4542e-07	Case 03	0.0051506	0.00031172
	0.00098618	-1.4406e-07		1.05	1.01
	0.881	0.82		0.34048	0.008144
	-0.074601	0.82		-0.46989	1.01
	0.2162	0.82		-0.011444	1.01
	Best Optimal Value	**Best Optimal Value**		**Best Optimal Value**	**Best Optimal Value**
	0.22436	0.22475		0.65915	0.66063
Case 02	0.94	0.89	Case 04	-0.00052943	0.019337
	0.94	0.89		2.5	0.078428
	-0.42275	0.31239		-0.047056	0.077129
	0.94	0.89		0.084181	0.02629
	-0.39658	0.27428		0.012898	0.089032
	Best Optimal Value	**Best Optimal Value**		**Best Optimal Value**	**Best Optimal Value**
	0.022935	0.023897		0.34938	0.42958

demonstrate Grey Wolf Optimization's efficacy in resolving the transmission line parameters issue. Finally, the GWO method outperforms other algorithms in terms of resilience, computing effort, and avoidance of premature convergence. The GWO algorithm's convergence accuracy has been significantly improved by evaluating 06 functions and comparing the results with ALO and DA. With 100 maximum iterations and 20 populations, it achieves a 90% success rate for all optimization functions, demonstrating GWO's efficacy in addressing difficult problems. The best optimal values for 3-phase capacitance and inductance are 0.22436, 0.022935, 0.65915, and 0.34938 are the best ideal values for 3-phase capacitance. For the time being, we shall concentrate on two problems. On the one hand, we will broaden GWO's applicability to additional real-world engineering issues. On the other hand, to tackle optimization issues more efficiently, we shall create new meta-heuristic optimization techniques.

KEYWORDS

- **algorithms**
- **hybrid optimization**
- **objective function**
- **optimization**
- **parameter estimation**
- **simulated annealing**
- **transmission line**

REFERENCES

1. Zhang, Y., & Yuan, L., (2020). Kalman filter approach for line parameter estimation for long transmission lines. In: *2020 IEEE Power and Energy Conference at Illinois (PECI)*. IEEE.
2. Ren, P., Ali, A., & Lev-Ari, H., (2019). Tracking transmission line parameters in power grids observed by PMUs. In: *2019 IEEE Milan PowerTech*. IEEE.
3. Ansari, M. M., et al., (2020). Planning for distribution system with grey wolf optimization method. *Journal of Electrical Engineering & Technology, 15*(4), 1485–1499.
4. Ansari, M. M., et al., (2019). A review of technical methods for distributed systems with distributed generation (DG). In: *2019 2nd International Conference on Computing, Mathematics and Engineering Technologies (iCoMET)*. IEEE.
5. Ansari, M. M., et al., (2020). Considering the uncertainty of hydrothermal wind and solar-based DG. *Alexandria Engineering Journal, 59*(6), 4211–4236.

6. Wolpert, D. H., & William, G. M., (1997). No free lunch theorems for optimization. *IEEE Transactions on Evolutionary Computation, 1*(1), 67–82.

7. Zheng-Ming, G., & Juan, Z., (2019). An improved grey wolf optimization algorithm with variable weights. *Computational Intelligence and Neuroscience, 2019.*

8. Luo, Q., et al., (2016). A novel complex-valued encoding grey wolf optimization algorithm. *Algorithms, 9*(1), 4.

9. Zhang, S., & Yongquan, Z. H. O. U., (2018). Grey wolf optimizer with ranking-based mutation operator for IIR model identification. *Chinese Journal of Electronics, 27*(5), 1071–1079.

10. Zhang, S., Qifang, L., & Yongquan, Z., (2017). Hybrid grey wolf optimizer using elite opposition-based learning strategy and simplex method. *International Journal of Computational Intelligence and Applications, 16*(2), 1750012.

11. Zhang, S., et al., (2016). Grey wolf optimizer for unmanned combat aerial vehicle path planning. *Advances in Engineering Software, 99*, 121–136.

12. Raj, S., & Biplab, B., (2018). Reactive power planning by opposition-based grey wolf optimization method. *International Transactions on Electrical Energy Systems, 28*(6), e2551.

13. Mirjalili, S., Seyed, M. M., & Andrew, L., (2014). Grey wolf optimizer. *Advances in Engineering Software, 69*, 46–61.

14. Muro, C., et al., (2011). Wolf-pack (*Canis lupus*) hunting strategies emerge from simple rules in computational simulations. *Behavioral Processes, 88*(3), 192–197.

15. Xu, Y., et al., (2016). An adaptively fast fuzzy fractional order PID control for pumped storage hydro unit using improved gravitational search algorithm. *Energy Conversion and Management, 111*, 67–78.

16. Muhsen, D. H., Abu, B. G., & Tamer, K., (2016). Multiobjective differential evolution algorithm-based sizing of a standalone photovoltaic water pumping system. *Energy Conversion and Management, 118*, 32–43.

17. Contreras-Cruz, M. A., Ayala-Ramirez, V., & Hernandez-Belmonte, U. H., (2015). Mobile robot path planning using artificial bee colony and evolutionary programming. *Applied Soft Computing, 30*, 319–328.

18. Mahdad, B., & Srairi, K., (2015). Blackout risk prevention in a smart grid based flexible optimal strategy using grey wolf-pattern search algorithms. *Energy Conversion and Management, 98*, 411–429.

19. Sulaiman, M. H., et al., (2015). Using the gray wolf optimizer for solving optimal reactive power dispatch problem. *Applied Soft Computing, 32*, 286–292.

20. Jayabarathi, T., et al., (2016). Economic dispatch using hybrid grey wolf optimizer. *Energy, 111*, 630–641.

21. Yang, B., et al., (2017). Grouped grey wolf optimizer for maximum power point tracking of doubly-fed induction generator-based wind turbine. *Energy Conversion and Management, 133*, 427–443.

22. Nishmitha, B., et al., (2018). Computation of transmission line parameters using MATLAB. *Int. J. Res. Sci. Innovation, 5*(4), 2321–2705.

23. Xing, B., & Wen-Jing, G., (2014). *Innovative Computational Intelligence: A Rough Guide to 134 Clever Algorithms,* 22–28.

24. Mirjalili, S., & Andrew, L., (2016). The whale optimization algorithm. *Advances in Engineering Software, 95*, 51–67.

25. Beheshti, Z., & Siti, M. H. S., (2013). A review of population-based meta-heuristic algorithms. *Int. J. Adv. Soft Comput. Appl., 5*(1), 1–35.

26. Hazir, E., Emine, S. E., & Küçük, H. K., (2018). Optimization of CNC cutting parameters using design of experiment (DOE) and desirability function. *Journal of Forestry Research* 29(5), 1423–1434.

27. Song, M., & Dongmei, C., (2018). An improved knowledge-informed NSGA-II for multi-objective land allocation (MOLA). *Geo-spatial Information Science, 21*(4), 273–287.

28. Zeugmann, T., et al., (2011). Particle swarm optimization. *Encyclopedia of Machine Learning, 1*(1), 760–766.

29. Shi, Y., & Russell, C. E., (1999). Empirical study of particle swarm optimization. *Proceedings of the 1999 Congress on Evolutionary Computation-CEC99 (Cat. No. 99TH8406).* (Vol. 3). IEEE.

30. Stützle, T., et al., (2011). A concise overview of applications of ant colony optimization. *Wiley Encyclopedia of Operations Research and Management Science, 2*, 896–911.

31. Dorigo, M., & Gianni Di, C., (1999). Ant colony optimization: A new meta-heuristic. *Proceedings of the 1999 Congress on Evolutionary Computation-CEC99 (Cat. No. 99TH8406)* (Vol. 2). IEEE.

32. Karaboga, D., & Bahriye, B., (2007). A powerful and efficient algorithm for numerical function optimization: Artificial bee colony (ABC) algorithm. *Journal of Global Optimization 39*(3), 459–471.

33. Chiroma, H., et al., (2017). Bio-inspired computation: Recent development on the modifications of the cuckoo search algorithm. *Applied Soft Computing, 61*, 149–173.

34. Xin-She, Y., & Suash, D., (2014). Cuckoo search: Recent advances and applications. *Neural Computing and Applications, 24*(1), 169–174.

35. Singh, G. P., & Abhay, S., (2014). Comparative study of krill herd, firefly and cuckoo search algorithms for unimodal and multimodal optimization. *International Journal of Intelligent Systems and Applications in Engineering, 2*(3), 26–37.

36. Xin-She, Y., (2010). A new metaheuristic bat-inspired algorithm. *Nature Inspired Cooperative Strategies for Optimization (NICSO 2010)* (pp. 65–74). Springer, Berlin, Heidelberg.

37. Xin-She, Y., & Xingshi, H., (2013). Firefly algorithm: Recent advances and applications. *International Journal of Swarm Intelligence, 1*(1), 36–50.

38. Deb, K., et al., (2002). A fast and elitist multiobjective genetic algorithm: NSGA-II. *IEEE Transactions on Evolutionary Computation, 6*(2), 182–197.

39. Dasgupta, D., & Zbigniew, M., (2013). *Evolutionary Algorithms in Engineering Applications.* Springer Science & Business Media.

40. Zhao, C., et al., (2009). Independent tasks scheduling based on genetic algorithm in cloud computing. In: *2009 5th International Conference on Wireless Communications, Networking and Mobile Computing.* IEEE.

41. Knowles, J., & David, C., (1999). The pareto archived evolution strategy: A new baseline algorithm for pareto multiobjective optimization. *Proceedings of the 1999 Congress on Evolutionary Computation-CEC99 (Cat. No. 99TH8406)* (Vol. 1). IEEE.

42. Khandelwal, M., et al., (2017). Function development for appraising brittleness of intact rocks using genetic programming and non-linear multiple regression models. *Engineering with Computers, 33*(1), 13–21.

43. Koza, J. R., Forrest, H. B., & Oscar, S., (1999). Genetic programming as a Darwinian invention machine. *European Conference on Genetic Programming.* Springer, Berlin, Heidelberg.

44. Bhattacharya, A., & Chattopadhyay, P. K., (2010). Application of biogeography-based optimization to solve different optimal power flow problems. *IET Generation, Transmission & Distribution, 5*(1), 70–80.
45. Sun, S., et al., (2018). Fast object detection based on binary deep convolution neural networks. *CAAI Transactions on Intelligence Technology, 3*(4), 191–197.
46. Mirjalili, S., Seyed, M. M., & Andrew, L., (2014). Grey wolf optimizer. *Advances in Engineering Software, 69*, 46–61.
47. Shaikh, M. S., et al., (2022). Optimal parameter estimation of 1-phase and 3-phase transmission line for various bundle conductor's using modified whale optimization algorithm. *International Journal of Electrical Power & Energy Systems, 138*, 107893.
48. Mastoi, M. S., et al., (2021). Research on power system transient stability with wind generation integration under fault condition to achieve economic benefits. *IET Power Electronics.*
49. Shaikh, M. S., et al., (2021). Optimal parameter estimation of overhead transmission line considering different bundle conductors with the uncertainty of load modeling. *Optimal Control Applications and Methods.*
50. Shaikh, M. S., et al., (2021). Parameter estimation of AC transmission line considering different bundle conductors using flux linkage technique estimation des paramètres d'une ligne de transmission à courant alternatif en tenant compte de différents conducteurs de faisceau à l'aide de la technique de liaison de flux. *IEEE Canadian Journal of Electrical and Computer Engineering.*
51. Shaikh, M. S., et al., (2021). Application of grey wolf optimization algorithm in parameter calculation of overhead transmission line system. *IET Science, Measurement & Technology, 15*(2), 218–231.
52. Mirjalili, S., & Andrew, L., (2016). The whale optimization algorithm. *Advances in Engineering Software, 95*, 51–67.
53. Rana, N., Muhammad, S. A. L., & Haruna, C., (2020). Whale optimization algorithm: A systematic review of contemporary applications, modifications and developments. *Neural Computing and Applications*, 1–33.
54. Mirjalili, S., (2015). Moth-flame optimization algorithm: A novel nature-inspired heuristic paradigm. *Knowledge-Based Systems, 89*, 228–249.
55. El-Ghazali, T., (2009). *Metaheuristics: From Design to Implementation* (Vol. 74). John Wiley & Sons.
56. Nara, K., et al., (2001). Application of tabu search to optimal placement of distributed generators. In: *2001 IEEE Power Engineering Society Winter Meeting; Conference Proceedings (Cat. No. 01CH37194)* (Vol. 2). IEEE.
57. Maciel, R. S., & Padilha-Feltrin, A., (2009). Distributed generation impact evaluation using a multi-objective tabu search. In: *2009 15ᵗʰ International Conference on Intelligent System Applications to Power Systems.* IEEE.
58. Shaikh, M. S., et al., (2020). Analysis of underground cable fault techniques using MATLAB simulation. *Sukkur IBA Journal of Computing and Mathematical Sciences, 4*(1), 1–10.

CHAPTER 17

Network Reconfiguration-Based Outage Management for Reliability Enhancement of Microgrid: A Hardware in Loop Approach

SHRUTI PRAJAPATI, SONAL, SOURAV KUMAR SAHU, and
DEBOMITA GHOSH

Department of Electrical and Electronics Engineering, BIT Mesra, Ranchi, Jharkhand, India

ABSTRACT

The demand of energy with the sprawl of power system network has posed a challenge for reliable power. Also, the rising fuel cost adds pressure to power demand. With increasing attention to reliable power supply and distributed generation (DG), the concept of microgrids (MGs) has become increasingly important. MG is an aggregation of distributed energy resources (DERs) and connected loads, facilitating remote applications during abnormal scenarios. Hence, MGs are providing reliable, affordable, and secure power supply. However, the external uncertainties may lead to sudden outages of components leading to mal-operation. Therefore, the outage management and reliable operation of MG should be a priority for enhancing the quality power. Reliability is coined as the probability of a system or component to perform its intended role adequately for a specified time period. Thus, to know the practicality of these MGs with optimal injections, firstly steady state modeling is presented using distribution load flow (DLF) analysis. Load outage management using network reconfiguration to enhance the network reliability is further proposed. The proposed method estimates and

The Internet of Energy: A Pragmatic Approach Towards Sustainable Development. Sheila Mahapatra, Mohan Krishna S., B. Chandra Sekhar, & Saurav Raj (Eds.)

compares the reliability of both IEEE 33-bus base and reconfigured network with underground cables and overhead lines for variable weather conditions. Finally, the optimal topology that minimizes the customer based reliability indices is selected as the best fit topology. Careful contemplation of both the networks tested in MATLAB and Typhoon hardware in loop (HIL) for healthy and possible outage conditions gives an incredibly comprehensible depiction of reconfigured network efficacy in terms of enhanced outage management and reliability indices.

17.1   INTRODUCTION

Microgrids are evolving with the advent of power system and integration of DERs. The major objective of these MGs is to meet the ever increasing energy demand. Consequently, the orientation of network plays a vital role to meet the demand in an uninterrupted and reliable manner. Reconfiguration of network is very essential in wide aspects of power system application. The reconfiguration of any power system network depends on the specific objective such as voltage profile improvement, loss reduction in network leading to improvement of local power injection, minimizing the line current, fault current minimization, etc. Junlakarn et al. [1] showcased the importance of reconfiguration in improving voltage profile and reducing the losses. Zhao et al. [2] and Ali et al. [3] presented issues related to fault management of DC microgrids, including fault detection, location, identification, isolation, reconfiguration, and comprehensive fault management of MGs. Song et al. [4] established the importance of establishing small power grids through distributed generation (DG) in the accident on the power grid. Liao et al. [5] analyzed the influence of line impedance, load demand and network topology on voltage unbalance caused by DERs. Alam et al. [6] explained the operation of MGs in co-ordination with DERs for enhancing the reliability and resiliency of the power supply. Ahmed et al. [7] classified MGs in terms of their intended application and the control in terms of their operating principle and performance.

Munzo-Delgado et al. [8] established the fact that reconfiguration done only for network loss reduction may cause detrimental effect on voltage profile and voltage stability. Penuela Meneses et al. [9] prioritized reconfiguration to reduce the power quality issues through amalgamation of multi-objective function into a single objective function. Although authors in the above literatures improved the network parameters in distinct ways, which indirectly improved the reliability of network, but Li et al. [10] used network reconfiguration to improve the availability of power in case of any abnormalities in the network. Farzin et al. [12]; Xin et al. [13]; and Nazmul Huda et al. [14] detailed the reliability evaluation methods based on the customer precedence and also analyzed the reliability for line switching operation in the power system. Arefi et al. [15] explained an analytical reliability evaluation technique for reliability assessment of the network, and network reconfiguration is considered as a post-fault scenario. Esmaeilian et al. [16] proposed a hybrid method for reconfiguration of the network to boost the network performance and to reduce the run time of the distribution network which resulted in decrease in power loss due to DG integration. Abdul Rahim et al. [17] presented an optimal network reconfiguration technique considering limitations of the protection system to reduce the impact of the integration of distributed generators. Wu et al. [11]; and Paterakis et al. [18] enhanced the service reliability, by analyzing the distribution system reconfiguration problem by utilizing the mixed-integer linear programming method to reduce the power losses and to improve the reliability indices of the system. de Quevedo et al. [19] introduced a framework that incorporated the possibility of creating intentional islanding in case of faults and highlights the benefits of network reconfiguration in reliability improvement of the network.

To analyze the practical implementation of these MGs with optimal injections, steady state modeling and analysis is significant. Thus, DLF is an effective method. Sahu et al. [20] coined the DLF method and presented a very effective way to handle the high R/X ratio by topological matrices. This method showcased very effective results for unbalanced 3-phase systems as well. Unlike the traditional load flow, the 'Y' matrix is not required in this method. Due to all these topological matrix advantages and non-repeated use of the matrices, the convergence of this method is faster as compared to the previously used load flow methods.

Real-time modeling environment is very effective and essential for the various power system application such as fault detection, power management and power injection studies. Ghosh et al. [21] verified the power system parameters values for a microgrid with highly penetrated DERs under of real-time environment. Sahu et al. [22] also showcased the importance of real-time environment for verifying the proposed fault detection technique within microgrid. Kumar et al. performed the the unbalance studies in microgrid for highly penetrated solar photovoltaic (SPV) in the Typhoon HIL based real-time environment. Osama et al. [23]; Selim et al. [24] cross verified the application of Typhoon HIL based real-time environment with the effectiveness of inhouse developed prototype of PMU. Due to the high proximity to the real-time scenarios, the application of Typhoon HIL real-time environment is very crucial for any power system result validation.

This chapter addresses the network outage management for the reliability enhancement of microgrid. In this chapter, the DLF technique is used for load flow analysis of distribution system. It uses two matrices bus injection to branch current (BIBC) and branch current to bus voltage (BCBV). BIBC matrix consists of bus injection and branch current relationship matrix and BCBV matrix comprising of branch current and bus voltage matrix. Both the matrices are used for solving the load flow of a system, and hence there is no requirement to build the admittance matrix, i.e., the Jacobian matrix, and LU which is usually required in the conventional load flow technique. Hence, the proposed method is better than the conventional load flow solution method. The DLF method is thus highly tailor-fitted for the steady-state analysis of microgrid. Typhoon HIL real-time simulation platform is used for the validation of DLF. For this work, a real-time hardware configuration of 602+ is used for the modeling of both normal and reconfigured IEEE-33 bus network. Load outage management using network reconfiguration is analyzed for MG reliability enhancement. To deal with the uncertainties involved in MG operability condition, the Monte Carlo simulation is performed. Several case studies are carried out and the analysis is done for both underground cables and overhead lines under variable weather conditions for the considered IEEE 33 bus base and reconfigured network. Careful contemplation of both the networks tested in Typhoon HIL for healthy as well as all possible outage conditions gives an incredibly comprehensible depiction of reconfigured network efficacy in terms of enhanced outage management and reliability indices.

The chapter is organized as follows. Section 17.2 details steady-state analysis of microgrid, Section 17.3 provides modeling methodology of microgrid using Typhoon HIL for real-time analysis, Section 17.4 presents an overview for reliability assessment of the modeled microgrid, Section 17.5 reflects case studies, simulation, and result analysis. Section 17.6 concludes the work.

17.2 STEADY-STATE ANALYSIS OF MICROGRID

Steady-state modeling is the primary stage to know the practicality of the power system network. Selim et al. [25]; Baran et al. [26]; Sahu et al. [20]; and Schweitzer et al. [28] performed the load flow analysis for possible power injection.

To proceed further with the above-mentioned injections, it is very essential to know if the network parameters, such as voltage and currents are within the specified limit or not. Below are the steps that are involved in calculating the currents and the voltages of the network required to be analyzed for injection.

> **Step 1:** This step involves writing the KCL equations for each branch considering every node current down the radial network, starting from branch 1 to the last branch. These KCL equations for the branches can be expressed as in Eqn. (1) [18]:

$$[B]_{((n-1)\times1)} = [S]_{((n-1)\times(n-1))} \times [I_i]_{((n-1)\times1)} \tag{1}$$

where; $[S]_{((n-1)\times(n-1))}$ represents square matrix containing ones and zeros. For a continuously connected radial distribution system, the elements in the main diagonal and above it is ones and below the main diagonal, all elements are zeros. This matrix is called BIBC matrix. $[I_i]_{((n-1)\times1)}$ represents the line current variable of $(n-1)$ branches.

> **Step 2:** In this step voltage deviation of each node with respect to the reference node is calculated by using the branch currents and the corresponding impedances of the lines. Eqn. (2) represents the voltage deviation matrix.

$$[\Delta V]_{((n-1)\times1)} = [Z]_{((n-1)\times(n-1))} \times [B]_{((n-1)\times1)} \tag{2}$$

Now, the value of [B] can be put in Eqns. (2) and (3) can be obtained as follows:

$$[\Delta V]_{((n-1)\times1)} = [Z]_{((n-1)\times(n-1))} \times [S]_{((n-1)\times(n-1))} \times [I_i]_{((n-1)\times1)} \tag{3}$$

Again, $[Z]_{((n-1)\times(n-1))} \times [S]_{((n-1)\times(n-1))} = [DLF]$, thus Eqn. (3) can be represented as in Eqn. (4)

$$[\Delta V]_{((n-1)\times1)} = [DLF]_{((n-1)\times(n-1))} \times [I_i]_{((n-1)\times1)} \tag{4}$$

The currents in the network can be found out by Eqn. (5), as follows:

$$[I_i^t]_{((n-1)\times1)} = \left[\left(\frac{P_i + jQ_i}{v_i^t} \right)^* \right]_{((n-1)\times1)} \tag{5}$$

where; I is node current at i^{th} node and t^{th} itration and V is the voltage at i^{th} node and t^{th} iteration.

From Eqns. (6) and (7), the voltage can be found as follows:

$$\left[\Delta V^{t+1}\right]_{((n-1)\times 1)} = \left[DLF\right]_{((n-1)\times (n-1))} \times \left[I_i^t\right]_{((n-1)\times 1)} \tag{6}$$

The effective change in voltage, i.e., ΔV as that of the previous iteration can be calculated from Eqn. (6) and the final voltage at i^{th} node can be calculated from Eqn. (7).

$$\left[V^{t+1}\right]_{((n-1)\times 1)} = \left[V^t\right]_{((n-1)\times 1)} + \left[\Delta V^{t+1}\right]_{((n-1)\times 1)} \tag{7}$$

The difference in voltage 'D' between two consecutive iterations can be calculated from Eqn. (18), as follows:

$$D = V^{t+1} - V^t \tag{8}$$

The LF convergence can be achieved by Eqn. (9), where the convergence criteria 'E,' should be set according to the user.

$$D > E \tag{9}$$

This process will continue until the preset convergence 'E,' is achieved.

Upon calculating the voltages and currents in the network and comparing them to the predefined standards, the realization of the network can be done. The load flow analysis values are considered to be optimistic boundaries for the real-time analysis. Further, Typhoon HIL-based real-time simulation platform is used for validation.

17.3   MICROGRID MODELING METHODOLOGY USING TYPHOON HIL FOR REAL TIME ANALYSIS

Real-time analysis of network aids in immediate precautionary measures during the time of outages and guide in proper decision making. Since the recent past, power networks have grown exponentially in terms of size and technology due to the incorporation of active microgrids. As a result, utility companies must strive to ensure that the customer's reliability considerations are satisfied.

To ensure the reliability to the customers, an accurate result analysis needs to be performed. To collect the required result, in this work HIL 602+ real-time simulator is used. Typhoon HIL 602+ real-time simulator is a six core (FPGA) device with two ARM Cortex A53 64 bit processors. The ARM processors are used to handle the I/O signal of the simulator and FPGA processes the plant part for higher speed real-time analysis. The 602+ is also having 16 analog I/O capable of transferring signals at a rate of 2 MS/s. The

real-time simulator is also having 32 digital I/O with the capability of transferring 280 MS/s. Most importantly, the simulator runs on its own software
platform to avoid compatibility issues.

Due to the mentioned advantages, Typhoon HIL real-time simulation
platform is used for the modeling of both the normal configuration and the
reconfigured IEEE-33 bus network. Both the networks are modeled in the
real-time environment for the analysis. For further calculation of the reliability parameters, the voltage, currents, and the power delivered must be
calculated with the atmost accuracy. The followed procedure as provided in
Figure 17.1, ensure industry-standard reliability in network dynamic parameter measurement. It depicts the overall work flow for accurate calculation
of the network parameters.

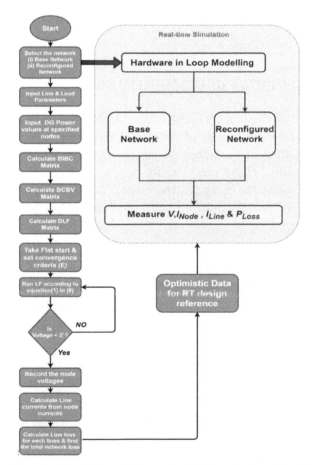

FIGURE 17.1 Flowchart for HIL modeling and validation of microgrid network.

17.4 OVERVIEW FOR RELIABILITY ASSESSMENT OF MICROGRID

The main challenge faced by electric power utilities in developing countries today is that the power demand is increasing at a rapid rate. Also, electric supply is dependent on the component health, weather condition, unpredictable DERs nature, and the related concerns. This has resulted in the need for more extensive justifications for the MG modeling, planning, and improvements in its operation with respect to customer demands. The customer failure statistics reveal the system's behavior, by depicting the gap in power supply to the customer. Reliability aims to maximize the performance by minimizing interruptions of the customers and providing a continuous power supply. This section presents the overview of reliability evaluation of MG considering the effect of outages and managing its reliable operation under various outage scenarios and selecting the topology which is robust to these disturbances.

17.4.1 COMPUTATION OF RELIABILITY INDICES

The reliability indices SAIFI, SAIDI, CAIDI, and AENS reflect MG performance under the impact of outages. These reliability indices include frequency of outages, measures of outage duration, response time, and system availability. These are vital for any corrective action planning to be taken in response to outages experienced. Incorporation of appropriate strategies protect the MG from further degradation and keeps it intact for a longer duration.

1. **System Average Interruption Frequency Index (SAIFI):** It is one of the most basic reliability indices; SAIFI indicates the average number of interruptions encountered by the customers over the span of one year. SAIFI can be enhanced by reducing the number of interruptions. It also depicts the recurrence of fault. Its unit is (interruption/system customer/year) [3, 5, 29].

$$SAIFI = \frac{\sum_{j \in L_T} \lambda_j * L_j}{\sum_{j \in L_T} L_j} \text{ failure/customer/year} \tag{10}$$

where; λ_j denotes the failure at node j; Lj is the number of customers being served at load point j; and L_T is the total load points of the considered network.

2. **System Average Interruption Duration Index (SAIDI):** It is the ratio of the annual duration of interruptions (sustained) to the number of consumers. It doesn't show the state of the system; while it shows how quickly utility can rejuvenate the supply after an outage takes place in the network. Its unit is (hours/system customer/year) [3, 5].

$$SAIDI = \frac{\sum_{j \in L_T} U_j * L_j}{\sum_{j \in L_T} L_j} \quad \text{hour/customer/year} \tag{11}$$

where; U_j denotes the duration of outage per year at load point j; Lj is the number of customers being served at load point j; and L_T is the total load points of the considered network.

3. **Consumer Average Interruption Duration Index (CAIDI):** It is the ratio of the total duration of interruptions to the number of customer interruptions encountered over the span of a year. CAIDI indicates how much longer do an average interruption may last. Its unit is (hours/customer interruption) [3, 5, 27].

$$CAIDI = \frac{\sum_{j \in L_T} U_j * L_j}{\sum_{j \in L_T} \lambda_j * L_j} \quad \text{hour/failure} \tag{12}$$

where; U_j denotes the duration of outage per year at load point j; λ_j is the failure at node j; Lj is the number of customers being served at load point j; and L_T is the total load points of the considered network.

4. **Average Energy Not Supplied (AENS):** It is an index that provides information regarding energy not supplied to the customers.

$$AENS = \frac{\sum_{j \in L_T} P_{aj} * U_j}{\sum_{j \in L_T} L_j} \quad \text{kW/hour} \tag{13}$$

where; U_j denotes the duration of outage per year at load point j; P_{aj} is the average load in kW at node j Lj is the number of customers being served at load point j; and L_T is the total load points of the considered network.

Based on the above discussed reliability metrics, Monte-Carlo simulation is used to find the distribution of reliability indices for the considered MG and assess the effectiveness of the network to supply power to the customers.

17.4.2 MONTE-CARLO SIMULATION FOR RELIABILITY EVALUATION

Monte-Carlo simulation (MCS) is a computational technique that estimates quantitative analysis using randomness of parameters in a range of values that might be deterministic in principle. This allows proper decision-making and reduces the risk of failure owing to external or environmental factors.

When any line interrupts, a part of the network will be out of service, as a result the customers will experience disruption. Hence, reliability indices AENS, SAIFI, SAIDI, and CIADI can be appropriately estimated to reflect the loss of energy and customer-related impact using Eqns. (10)–(13). Hence, it gives the performance measure of MG in terms of reliability indices. MCS creates random operational history based on the average component failure rate per unit time, and average repair time and with the different outage cases with different probability of occurrences under normal and abnormal weather conditions. Thus, the probability of failure can be estimated by using the expression of conditional probability P, as given by Eqns. (14)–(16) [25, 26].

$$P(l,w) = \{OH, UG, aw, nw\} \tag{14}$$

The conditional probability of line failure with respect to weather variation is given by Eqn. (15).

$$P(l|w) = \frac{P(l|nw)*P(aw)}{P(nw)} \tag{15}$$

The conditional probability under normal weather with respect to line type is given by Eqn. (16).

$$P(nw|l) = \frac{P(nw|OH)*P(UG)}{P(OH)} \tag{16}$$

Here; $P(l|w)$ denotes the conditional probability of line type with respect to weather, $P(l)$ and $P(w)$ are the probabilities of failure of line and occurrence of abnormal weather scenario, respectively. The updated joint probability of line with respect to weather scenarios, $P(l|ws)$ is given by Eqns. (17)–(18).

$$P(l|ws) = \frac{P(l, ws)}{P(ws)} \tag{17}$$

$$P_{new} = \frac{P(l|ws)}{\sum_{(l|ws)} P(l|ws)} \tag{18}$$

The value of joint probability of line with variation of weather condition can be constructed from the combined probabilities, which satisfies the condition that the probability values should be greater than 0, as given by Eqn. (19).

$$For\ any\ event, P(l|ws) \geq 0 \tag{19}$$

This conditional probability is capable of estimating the system parameters of failure and repair rate according to the weather as well as line type for a specific network under consideration. Figure 17.2 represents the flowchart for reliability analysis of MG taking into consideration MCS method.

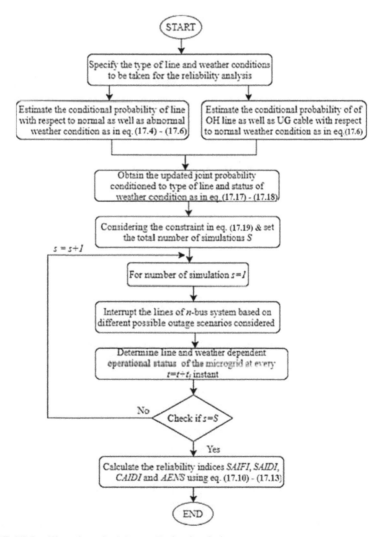

FIGURE 17.2 Flowchart for Monte-Carlo simulation.

For different outage scenarios, the average duration, its frequency, average interruption time as well as system availability are calculated. After initializing the beginning of MCS the consequetive simulation steps are referred to as $S=s+1$. For this, the reliability indices SAIFI, SAIDI, CAIDI, and AENS are evaluated. Scenario-wise different values of indices are calculated for variable weather conditions and different types of lines tested on both IEEE 33 bus base and reconfigured network. This assessment can be used for MG planning and operation utilizing the obtained reliability indices indicating a comparison between normal and reconfigured network topologies.

17.5 ANALYSIS AND RESULTS

This section summarizes and analyzes the results for reliability-oriented outage management concerning the description of the previous sections and its related mathematical expressions.

To claim the aforesaid concept, the IEEE 33-bus network with normal configuration and reconfiguration is tested with DLF initially. This is followed by modeling and testing the two networks in Typhoon HIL software as in Figure 17.3 with optimal DG location for real power integration as reflected in Table 17.1.

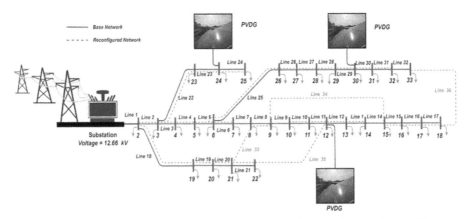

FIGURE 17.3   Base and reconfigured IEEE 33-bus model with DGs in typhoon HIL.

Initially, both the networks are tested under healthy conditions, followed by interruptions at all possible locations in both the network to analyze in which network the customers are least affected. This is also extended by varying the type of lines and the lines exposed to variable weather conditions.

TABLE 17.1   IEEE 33-Bus Optimal DG Location and Capacity [13]

SL. No.	Bus No.	Real Power (kW)
1.	12	913.05
2.	24	882.86
3.	30	1079.05

When the rated power is injected in both the base and reconfigured IEEE 33-bus networks, the reconfigured network results in low power consumption without derating the load, which is a considerable amount, i.e., 192.69 kW. The reduction in network loss, inherently improved network efficiency. This improvement is about 5.18% of the total network load and this difference in power can be observed from Figure 17.4. With the injections as per Table 17.1, there is a possibility of an increase in the voltage profile in the microgrid. To ensure that the voltage profile remains within the acceptable operational range, simulation is performed to find the bus voltages.

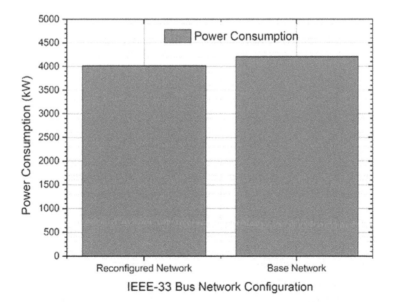

FIGURE 17.4   Power consumption for IEEE 33 bus base and reconfigured network without derating the loads.

Figure 17.5 shows the real-time voltage profile results for the base as well as reconfigured IEEE 33-bus distribution system. The radius in Figure 17.5 shows the voltage in volts and the labels on circumference denote the nodes of the network. It can be observed from Figure 17.5, that the voltage profile for reconfigured as well as for base network is within acceptable operational range for the smooth operation, with some additional improvement for the reconfigured network.

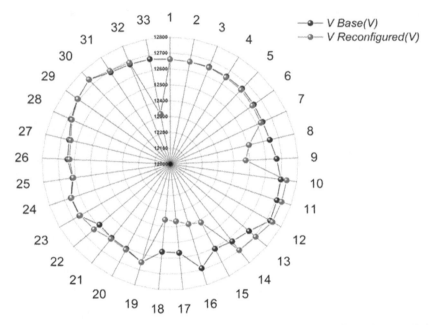

FIGURE 17.5 Real-time voltage profile results for base and reconfigured IEEE-33 bus system.

This is followed by creating different possible outages at all possible locations in both IEEE 33-bus base and reconfigured networks. For creating outages it is necessary to partition the core of the simulator. The six cores of the 602+ is partitioned according to the placement of the SPV location outage. Due to the effective core partition, the sampling rate can be made high. In this work, a high sampling rate, i.e., 1.6 μsec is chosen for capturing all the dynamic variations involved. Table 17.2 shows the result for different possible outages due to different interruptions created in both the base and reconfigured network for comparative analysis of kW outage.

From Table 17.2 and also from Figure 17.6 it can be observed that in most of the interruption conditions, reconfigured network kW outages are less than the normal network kW outages with OH lines under both normal and abnormal weather conditions. In the case of underground cables also the reconfigured configuration proved to be a better approach.

TABLE 17.2 HIL based Comparative Assessment for Different Possible Outages in Base and Reconfigured IEEE 33-Bus Network

Line Outages	Outage of Base Network (kW)	Outage of Reconfigured Network (kW)	Remark
Line2	3710.56049	3737.1665	Base network is better
Line3	3254.76077	2123.35	Reconfigured network is better
Line4	2238.2668	1305.65147	Reconfigured network is better
Line5	2119.5091	1187.01253	Reconfigured network is better
Line6	2056.5201	1124.08553	Reconfigured network is better
Line7	1997.4417	1065.06603	Reconfigured network is better
Line8	876.60251	581.97634	Reconfigured network is better
Line9	685.9533	386.47352	Reconfigured network is better
Line10	629.2668	328.66186	Reconfigured network is better
Line11	572.95688	77.44923	Reconfigured network is better
Line12	531.07485	122.89062	Reconfigured network is better
Line13	423.31029	170.64403	Reconfigured network is better
Line14	364.48119	112.70026	Reconfigured network is better
Line15	248.8984	381.96678	Base network is better
Line16	192.03078	371.37474	Base network is better
Line17	153.97354	417.07935	Base network is better
Line18	80.78684	158.90791	Base network is better
Line19	355.45252	1315.0686	Base network is better
Line20	265.92161	1225.4435	Base network is better
Line21	176.67179	1136.09866	Base network is better
Line22	87.97258	466.62614	Base network is better
Line23	928.56059	927.44207	Reconfigured network is better
Line24	840.28199	838.76545	Reconfigured network is better
Line25	404.59669	406.41387	Base network is better
Line26	931.62789	869.92182	Reconfigured network is better
Line27	874.738	811.84028	Reconfigured network is better
Line28	818.0863	754.00655	Reconfigured network is better
Line29	761.61187	696.44164	Reconfigured network is better
Line30	650.88921	583.05862	Reconfigured network is better
Line31	398.57522	347.88171	Reconfigured network is better
Line32	253.19203	200.58135	Reconfigured network is better
Line33	55.22984	76.83586	Base network is better

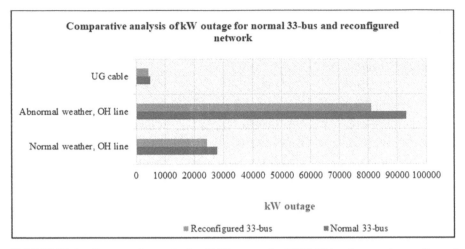

FIGURE 17.6   Comparative analysis of kW outage for IEEE 33-bus base and reconfigured networks.

Further, a comparative analysis is done for the IEEE 33-bus base and reconfigured system in terms of reliability indices (SAIFI, SAIDI, CAIDI, and AENS). The absolute failure and repair rates considered are as shown in Table 17.3 and are used for conditional probability computation pertaining to normal and wind-related weather conditions for different types of lines. These conditional probabilities obtained for different scenarios are further used as the input parameters for MCS for the computation of reliability indices considering outage cases in the network. Analyzing the reliability indices of both the networks with OH lines and UG cables for variable weather conditions, the observations are as presented in Figures 17.7–17.9, respectively.

TABLE 17.3   Failure and Repair Rate Considering Normal and Wind Related Weather Conditions for Different Types of Lines [20, 21]

Weather Condition	Type of Lines	Failure Rate (f/yr.)	Repair Rate (hrs.)
Normal	Overhead	0.79	4
	Underground	0.15	6
Abnormal	Overhead	2.23	15
	Underground	0.15	6

Figure 17.7 represents a graphical comparison obtained for different OH line disruptions created in IEEE 33-bus base and reconfigured system for

analyzing the reliability indices under normal weather conditions. It can be observed that reliability indices value evaluated for OH lines under normal weather conditions shows improvement in kW outage as well as improvement in AENS, SAIFI, SAIDI, and CAIDI values with the reconfigured network. Also, these values are superior as compared to the OH line exposed to abnormal weather conditions as in Figure 17.8. UG cables are on the other hand are not exposed to the external environment so they are less susceptible to disruption. However, it can be observed that reliability indices value evaluated even for UG cables show an improvement in the reconfigured network as in Figure 17.9.

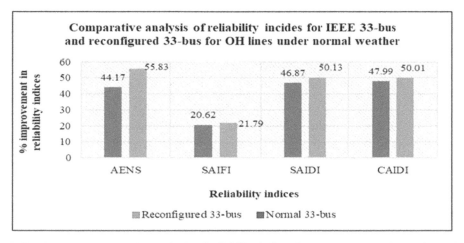

FIGURE 17.7 Comparative analysis of reliability indices for IEEE 33-bus and reconfigured 33-bus for OH lines in normal weather.

17.6 CONCLUSION

The ever-increasing demand for electrical energy led to the development of strategies for reliable outage management and enhanced quality power. In this proposed work, the practicality of MG with optimal injections is presented using its steady-state modeling. This is estimated by the DLF method and results are supported by Typhoon HIL. The quick convergence of DLF in case of a high R/X ratio makes it a very effective tool for this analysis. The real-time data with very high resolution from Typhoon HIL simulator includes all possible variations for effective deployment in the MG and ensures high reliability of the scheme. Load outage management using network reconfiguration to enhance the network reliability is further analyzed. The reliability assessment of both the network validates the efficacy of the network. The results for SAIFI, SAIDI, CAIDI, and AENS substantiates that in the IEEE 33-bus reconfigured network, for different possible outage scenarios, as compared to the base network for its efficacy.

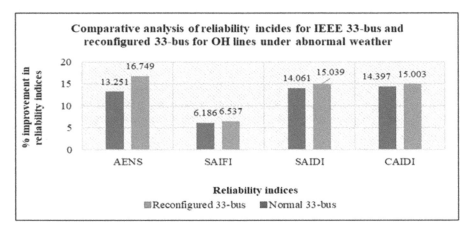

FIGURE 17.8   Comparative analysis of reliability indices for IEEE 33-bus and reconfigured 33-bus for OH lines in abnormal weather.

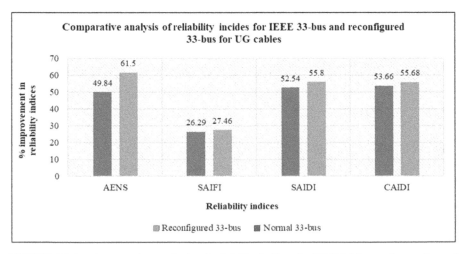

FIGURE 17.9   Comparative analysis of reliability indices for IEEE 33-bus and reconfigured 33-bus for UG cables.

KEYWORDS

- average energy not supplied (AENS)
- customer average interruption duration index (CAIDI)
- distributed load flow (DLF)
- distribution energy resources (DERs)
- microgrid (MG)
- Monte Carlo Simulation (MCS)
- network reconfiguration
- system average interruption duration index (SAIDI)
- system average interruption frequency index (SAIFI)
- typhoon hardware in loop (HIL)

REFERENCES

1. Junlakarn, S., & Marija, I., (2014). Distribution system reliability options and utility liability. *IEEE Transactions on Smart Grid, 5*(5), 2227–2234.
2. Zhao, S., & Chanan, S., (2017). Studying the reliability implications of line switching operations. *IEEE Transactions on Power Systems, 32*(6), 4614–4625.
3. Ali, Z., Yacine, T., Syed, Z. A., Mustafa, A. H., Muhammad, S., Chun-Lien, S., & Josep, M. G., (2021). Fault management in DC microgrids: A review of challenges, countermeasures, and future research trends. *IEEE Access.*
4. Jing-Hui, S., Xue-Rong, J., Hong-Kai, L., Hong-Bo, G., & Qing-Xin, L., (2008). Distributed power generation technology and the effect of it on the accident of large-scale power grids. In: *2008 China International Conference on Electricity Distribution* (pp. 1–5). IEEE.
5. Liao, H., & Jovica, V. M., (2017). Methodology for the analysis of voltage unbalance in networks with single-phase distributed generation. *IET Generation, Transmission & Distribution, 11*(2), 550–559.
6. Alam, M. N., Saikat, C., & Arindam, G., (2018). Networked microgrids: State-of-the-art and future perspectives. *IEEE Transactions on Industrial Informatics, 15*(3), 1238–1250.
7. Ahmed, M., Lasantha, M., Arash, V., & Manoj, D., (2020). Stability and control aspects of microgrid architectures–A comprehensive review. *IEEE Access, 8*, 144730–144766.
8. Muñoz-Delgado, G., Javicr, C., & José, M. A., (2016). Reliability assessment for distribution optimization models: A non-simulation-based linear programming approach. *IEEE Transactions on Smart Grid, 9*(4), 3048–3059.
9. Meneses, C.A. P., & Jose, R. S. M., (2013). Improving the grid operation and reliability cost of distribution systems with dispersed generation. *IEEE Transactions on Power Systems 28*(3), 2485–2496.

10. Li, Z., Wenchuan, W., Xue, T., & Boming, Z., (2020). Optimization model-based reliability assessment for distribution networks considering detailed placement of circuit breakers and switches. *IEEE Transactions on Power Systems, 35*(5), 3991–4004.

11. Li, Z., Wenchuan, W., Boming, Z., & Xue, T., (2019). Analytical reliability assessment method for complex distribution networks considering post-fault network reconfiguration. *IEEE Transactions on Power Systems, 35*(2), 1457–1467.

12. Farzin, H., Fotuhi-Firuzabad, M., & Moeini-Aghtaie, M., (2017). Role of outage management strategy in reliability performance of multi-microgrid distribution systems. *IEEE Transactions on Power Systems, 33*(3), 2359–2369.

13. Xin, S., Cheng, Y., Zhang, X., & Wei, C., (2017). A novel multi-microgrids system reliability assessment algorithm using parallel computing. In: *2017 IEEE Conference on Energy Internet and Energy System Integration (EI2)* (pp. 1–5). IEEE.

14. Huda, A. S. N., & Rastko, Ž., (2018). Study effect of components availability on distribution system reliability through multilevel Monte Carlo method. *IEEE Transactions on Industrial Informatics, 15*(6), 3133–3142.

15. Arefi, A., Gerard, L., Ghavameddin, N., & Behnaz, B., (2020). A fast adequacy analysis for radial distribution networks considering reconfiguration and DGs. *IEEE Transactions on Smart Grid, 11*(5), 3896–3909.

16. Esmaeilian, H. R., & Roohollah, F., (2014). Energy loss minimization in distribution systems utilizing an enhanced reconfiguration method integrating distributed generation. *IEEE Systems Journal, 9*(4), 1430–1439.

17. Rahim, M. N. A., Hazlie, M., Abdul, H. A. B., Mir, T. R., Ola, B., & Nurulafiqah, N. M., (2019). Protection coordination toward optimal network reconfiguration and DG sizing. *IEEE Access, 7*, 163700–163718.

18. Paterakis, N. G., Andrea, M., Sergio, F. S., Ozan, E., Gianfranco, C., Anastasios, G. B., & João, P. S. C., (2015). Multi-objective reconfiguration of radial distribution systems using reliability indices. *IEEE Transactions on Power Systems, 31*(2), 1048–1062.

19. De Quevedo, P. M., Javier, C., Andrea, M., Gianfranco, C., & Radu, P., (2017). Reliability assessment of microgrids with local and mobile generation, time-dependent profiles, and intraday reconfiguration. *IEEE Transactions on Industry Applications, 54*(1), 61–72.

20. Sahu, S. K., & Debomita, G., (2019). Hosting capacity enhancement in distribution system in highly trenchant photo-voltaic environment: A hardware in loop approach. *IEEE Access, 8*, 14440–14451.

21. Sahu, S. K., & Debomita, G., (2021). Spectral kurtosis-based fault detection for a highly penetrated distributed generation: A real-time analysis. In: *Advances in Smart Grid Automation and Industry 4.0* (pp. 649–656). Springer, Singapore.

22. Sahu, S. K., & Debomita, G., (2021). Photovoltaic hosting capacity increment in an unbalanced active distribution network. In: *2021 1ˢᵗ International Conference on Power Electronics and Energy (ICPEE)* (pp. 1–5). IEEE.

23. Kumar, R., Sourav, K. S., Debomita, G., & Sarbani, C., (2021). A multi HIL-based approach for real-time phasor data monitoring using phasor measurement unit. In: *Advances in Smart Grid Automation and Industry 4.0* (pp. 623–630). Springer, Singapore.

24. Osama, R. A., Ahmed, F. Z., & Almoataz, Y. A., (2019). A planning framework for optimal partitioning of distribution networks into microgrids. *IEEE Systems Journal, 14*(1), 916–926.

25. Selim, A., Salah, K., Ali, S. A., & Francisco, J., (2020). Optimal placement of DGs in distribution system using an improved Harris hawk's optimizer based on single-and multi-objective approaches. *IEEE Access, 8*, 52815–52829.

26. Baran, M. E., & Felix, F. W., (1989). Network reconfiguration in distribution systems for loss reduction and load balancing. *IEEE Power Engineering Review, 9*(4), 101, 102.
27. Dalsgaard, M. T., & Guangya, Y., (2016). *Reconfiguration of Power Distribution Networks by Evolutionary Algorithm for Reliability Improvement*. Technical report, special course, DTU.
28. Schweitzer, E. O., David, W., Héctor, J. A. F., Demetrios, A. T., David, A. C., & David, S. E., (2011). Line protection: Redundancy, reliability, and affordability. In: *2011 64th Annual Conference for Protective Relay Engineers* (pp. 1–24). IEEE.
29. Jae-Han, K., Ju-Yong, K., Jin-Tae, C., Il-Keun, S., Bo-Min, K., Il-Yop, C., & Joon-Ho, C., (2014). Comparison between underground cable and overhead line for a low-voltage direct current distribution network serving communication repeater. *Energies, 7*(3), 1656–1672.

CHAPTER 18

A Net Energy Meter-Based Approach for Islanding Detection in Modern Distribution Systems

SOHAM DUTTA,[1] AKASH KUMAR PANDEY,[2] SOURAV KUMAR SAHU,[3] and PRADIP KUMAR SADHU[4]

[1]Department of Electrical and Electronics Engineering, Manipal Institute of Technology, Manipal Academy of Higher Education, Manipal, Karnataka, India

[2]Design Engineer, Larsen and Toubro, Chennai, India

[3]Department of Electrical and Electronics Engineering, BIT Mesra, Ranchi, Jharkhand, India

[4]Department of Electrical Engineering, IIT(ISM) Dhanbad, Jharkhand, India

ABSTRACT

The era of smart grid has provided unprecedented opportunities, which has taken the power industry to a new level assuring better reliability, continuity and productivity. However, the grid system has to face challenging uncertainties due to faults, voltage dip, islanding, etc. Islanding can be defined as a situation during which distributed generation (DG) continue to feed some portion of loads even after being disconnected from the main grid. Island situations can be dangerous for the utility workers and users end equipments. Net energy meter (NEM) has the ability to record flow of energy in two directions. To feed the household loads, consumer utilizes power from two sources i.e. from the grid and different DGs. If there is a surplus power

The Internet of Energy: A Pragmatic Approach Towards Sustainable Development. Sheila Mahapatra, Mohan Krishna S., B. Chandra Sekhar, & Saurav Raj (Eds.)

generation by DGs, the excess power will be fed back to the grid through a NEM. To deal islanding events with NEM, this chapter recommends an advance technique of detecting islanding for doubly-fed induction generator (DFIG) wind turbines based microgrid. Walsh Hadamard transform (WHT) is used as the signal processing tool to generate the WHT coefficients of the signals derived from net energy meter (NEM) and artificial neural network (ANN) has been employed for the examination of these coefficients and classification/detection of islanding events. The results obtained for various islanding and non-islanding scenarios show that the presented method is faster and effective. The simulation is performed in MATLAB/SIMULINK environment.

18.1   INTRODUCTION

The era of smart grid has provided unprecedented opportunities, which has booked the power business to a fresh level assuring better reliability, continuity and productivity [1]. However, the grid system has to face challenging uncertainties due to faults, voltage dip, islanding, etc. Islanding can be defined as a situation during which distributed generation (DG) carry on to nourish some percentage of loads even after being disconnected from the main grid. Island situations can be risky for the service workers and users end equipments [2].

Net energy meter (NEM) has the ability to record flow of energy in two directions [3]. To feed the household loads, consumer utilizes power from two sources, i.e., from the grid and different DGs. If there is a surplus power generation by DGs, the excess power will be fed back to the grid through a NEM.

To deal islanding events with NEM, this chapter recommends an advance technique of detecting islanding for doubly-fed induction generator (DFIG) wind turbines based microgrid. Walsh Hadamard transform (WHT) is used as the signal processing tool to generate the WHT coefficients of the signals derived from NEM and artificial neural network (ANN) has been employed for the examination of these coefficients and classification/detection of islanding events. The results obtained for various islanding and non-islanding scenarios show that the presented method is faster and effective. The simulation is performed in MATLAB/ SIMULINK environment.

18.2 MODERN DISTRIBUTION SYSTEM

18.2.1 ADVANCEMENT OF ELECTRICAL TECHNOLOGY

Unlike any other developing country, India cannot ignore the importance of power industry in pillaring its overall development. Moreover, to be a developed nation with increased industrial productivity and large job creation, India has to rely on power that can fuel the factories and provide the opportunity to be a developed nation. As the productivity of all other sectors mainly depends upon the energy fueling, every power industry have to increase its output for supplying the required demand. Thus, there is a need of shift in paradigm from the point of view of energy production and energy transmission by using the latest power electronics and smart grid technology [4]. It can result in effective and efficient generation and utilization of energy that can result in development of power sector in any country. The excessive usage of fossil fuels can result in high amount of carbon dioxide (CO_2) production that consistently results in global warming increasing the average global temperature. This proves to be very dangerous for the living beings and our animal planet. Hence, the Government has taken various steps for the improvement in the field of renewable energy sources (ES) [5].

18.2.2 DG

Power networks are becoming smarter progressively and concurrently, the prominence of DGs is growing. The notion of DG has contributed the power business to a fresh elevation that has guaranteed improved dependability and stability [6]. The word DG implies power produced by numerous decentralized sources for definite engagements. DGs have decreased the dependence on main grid source because it performs as a source in a micro-grid possessing the capability to cater small amounts of loads like household loads. Nevertheless, as the infiltration of these DGs has augmented in the grid system, several stimulating worries have surfaced like faults, voltage dip and islanding that have a difference in pattern from previously recognized conventional faults [7]. Thus, they need to be identified and resolved.

18.2.3   DG AS A FEASIBLE ENERGY SOURCE

Electricity production and its lossless grid circulation is mostly adminis-
trated by the Government which is the regulating authority of this matter.
Likewise, the electrical power providers have to adhere to guideline and
stability of demand-supply ratio. Nevertheless, there is always a chance of
power catastrophe and abnormalities of supply owing to several causes like
faults, natural disaster, etc. To deal with them, the power system needs an
alternative but unfailing sources like DG. Thus, the Government and public
sectors recently has concentrated on deregulating the power supply process
and DGs are the paramount response for the necessities. However, each coin
has two sides. The use of DG comes with its own hassle. For instance, its
ability to cope in the high competition power market [8]. Contemporary
growth in the area of technology, e.g., ultra-fast switching and competence
to cope up with the huge voltage level, has pressed the power industry to use
DG as a feasible energy source [9].

18.2.4   DG CLASSIFICATION

The DGs are broadly classified into two categories traditional and nontraditional as shown in Figure 18.1. Traditional DGs are classified as sources
which work on combustion engines and whose whole output depends upon
the internal combustion. Some examples are gas microturbine and low speed
turbine. They are quite small in size and area but have a unique advantage of
being not dependent on geographical resources for their operation. Moreover,
they can be installed at any location [10]. Nontraditional are the one that
have the ability to produce electricity with assurance of zero emission and
eco-friendly. However, they have a disadvantage of producing DC power
which cannot be connected to the main grid directly. Thus, this dc power
is converted into ac by utilizing power electronics devices (for example,
rectifier or inverter) before connecting it to the main distribution system.
Due to this drawback, power sources such as photovoltaic and fuel cell uses
converters before getting connected to the grid.

18.2.5   DISTRIBUTION UNIT WITH MULTIPLE DGS

Since, DG or decentralized sources are known for producing electrical energy
apart from the main grid supply, they are placed nearer to the load center in a

single unit or multiple unit as per suitability for maintaining the continuity of supply. One such arrangements is shown in Figure 18.2. It basically provides an alternate power generation unit for the upgradation and modernization of previously running grid supply system [8–10]. Also, for the utility purposes, multiple DGs can be connected simultaneously for supplying different kinds of complex load demand. In the figure, loads are supported by main grid as well as by the sources of DG through transformers and circuit breakers. Here the purpose of DGs is to provide backup power during any uncertainties like faults and islanding, etc. Hence, from above figure the advantages of DGs can be better understood in case of faulty operation [7–10].

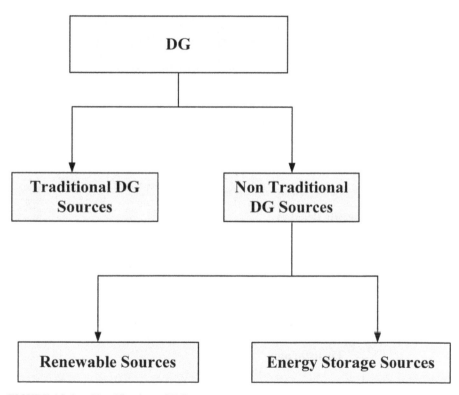

FIGURE 18.1 Classification of DG.

18.2.6 USEFULNESS OF DGS

The usefulness of the DGs can be can be summarized as follows:

1. **Better Reliability:** DG are load located power resources that are placed nearer to the demand site as a alternate power resources. Due to this, DG provides increased power availability which is unaffected from the impacts of transmission and distribution system faults which further helps in decreasing the peak demand overload. Furthermore, utility of multiple DG distribute the demand requirement equally on all the ES and hence reduce the overload on main grid [8].

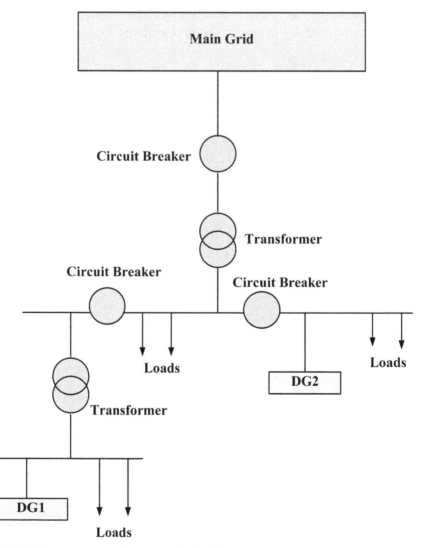

FIGURE 18.2   System consisting of multiple DGs.

2. **Continuity:** As DG are located nearer to the load center and are flexible in its area of application, an uninterrupted production in output of DGs can be obtained. This is useful when there is any disturbance like faults or by an act of terror. This gives DGs an additional advantage over the main grid, i.e., despite being any fault, DGs continue to feed power to the load center [9].

3. **Flexibility:** DG are much versatile when it comes to their location of application as they can be placed anywhere close or a bit far from the load center as per the requirement without affecting the quality and continuity of output that are being obtained from the micro-grid [10].

4. **Reduction in Peak Demand:** DGs, by generating power, helps the main grid in reducing its overall load demand during the peak hours [11].

5. **Improved Power Quality:** DG reduces the power degradation by supplying the additional support to the main grid and hence improves the power quality and profile of voltage [11].

6. **Reduced Transmission Losses:** DGs being located near the load center, the requirement of transmission and distribution lines reduce to a much smaller extent as there is small distance between the source and demand center. As a result of this, the transmission losses associated with it is lesser than the main grid [12].

Apart from the various advantages associated with DG as listed above, DGs also suffer some shortcomings. There are few challenges that must be noted for DGs which will have impact on the whole working unit and the output that is obtained from the DGs. Such issues are discussed in the next section.

18.2.7 VARIOUS ISSUES ASSOCIATED WITH DGS

DGs have some particular negative aspects as followed [9, 11, 12]:

1. **Setup Cost:** Capital investment, i.e., the cost for setting up a DG system is quite higher when compared to large central plants as DGs requires advanced technologies. Moreover, in a hybrid system, the flow of power is bi-directional which makes the whole connected unit costlier due to varios controllers.

2. **Reduction of Power Quality:** Though, with large number of benefits related to power and voltage stability, the idea of utilizing multiple DGs as a substitute resource can work in a negative manner because

of many factors. DGs affects the system working frequency. Thus, for maintaining a constant frequency, suitable load frequency control systems are needed with the potential of maintaining frequency constant for multiple DG system.

3. **Connection Issues:** Connecting multiple DGs with the main transmission line in hybrid grid mode requires a great effort and technological advancement to maintain the stability of voltage and frequency.

18.2.8 MICRO-GRIDS

As the power scientists have progressed in the field of power engineering, the production of power mainly relies on non-renewable energy resources like diesel, coal, etc., as they constitute for more than 70% of total energy production. However, on the other side, these non-renewable sources causes huge expanse of pollution that contributes to numerous environmental issues like global warming and healthiness related issues. Therefore, to provide mankind the superior substitutes, the power researchers and engineers have engrossed on renewable and non-conventional energy segment. As the usage of nontraditional ES are on hike in today's world, power engineers have the responsibility and requirement to cope up with these changes in a very positive manner and embrace the ways of renewable energy generation techniques. In hunt for improved and quality power generation, one have to adopt a clever and faster tactic that would be fulfilling the existing energy obligation in a eco friendly means. It has to be done in such a way that after shifting from one resources to the other, we can uphold the continuousness of power supply in an efficient manner. For this requirements, micro-grid were developed to attend the drive of generating and supplying the power mainly by utilizing the natural resources. Micro-grids [13, 14] possess the ability to work both as minor power grids those can operate in association with several other minor grids as well as with the main-grid. Decentralized sources mainly constitute of micro-grids that can operate on small scale for energy production that can serve the purpose of operating various small loads besides sending back excess power to the main grid. Decentralized source planted for any specific purpose and having the capacity of small scale energy production and which can also be stored for future usage are known as micro-grids [13].

Micro-grids predominantly exploits the renewable form of energy like wind and solar power for their input energy demand that can be converted into electrical energy. The electrical energy that are produced by micro-grid can be deployed for small load demand and also can act as reserve storage during the peak demand [15]. Thus, micro-grids are the conclusive answers for our environment friendly energy requirement that can provide additional support to the main grid in an efficient way for maintaining the continuity of supply specially during the localized fault. In this chapter, for making a hybrid micro-grid, wind turbine based decentralized source has been taken in the form of a micro-grid through a 30 km transmission line, fastened to the grid of 120 kV. Thus, from above-mentioned, the advantages of the micro-grid can be stated as following [15]:

- Maintaining the steadiness in power supply during any localized fault which disturbs the main supply of the power grid;
- Supplying the required reactive and active power demand during island condition;
- Collaborating with the main grid to supply the peak load demand;
- Supplying the generated surplus power to the main grid.

18.3 CONCEPT OF ISLANDING IN POWER SYSTEM

18.3.1 ISLANDING PHENOMENON

Islanding is considered as a situation in which a DG or group of DGs remain to feed some part of load in spite of being disconnected from the main source as shown in Figure 18.3. For avoiding islanding, the concept of islanding condition must be understand thoroughly and its impact on the system [16, 17]. System parameters, like voltage and frequency, vastly fluctuate after islanding. Thus, islanding condition must be properly taken care of for the protection of system. In some cases, islanding also affects the power quality and operation of system. Islanding can be of two types intentional and unintentional [16]. In intentional islanding, a section of unit is willingly isolated from the main grid for some maintenance or removal of faults. This type of islanding generally does not pose any problem as everything is preplanned. However, the main concern is unintentional islanding, where a section is isolated unwillingly which can be dangerous for the whole electrically islanded system.

18.3.2 DISADVANTAGES OF ISLANDING

There are negligible disadvantages of intentional islanding. However, inadvertent islanding welcomes various drawbacks. Some of them are as follows [18]:

1. **Dangerous for Utility Workers:** Islanding is a sudden phenomenon that adversely affects not only the isolated unit but it is life threatening for the utility staff. This is because, during islanding, loads in the islanded grid are energized from the DGs even when the main grid supply is cut off which can be unsafe for the service workers. Thus, the workers may get electrocuted.

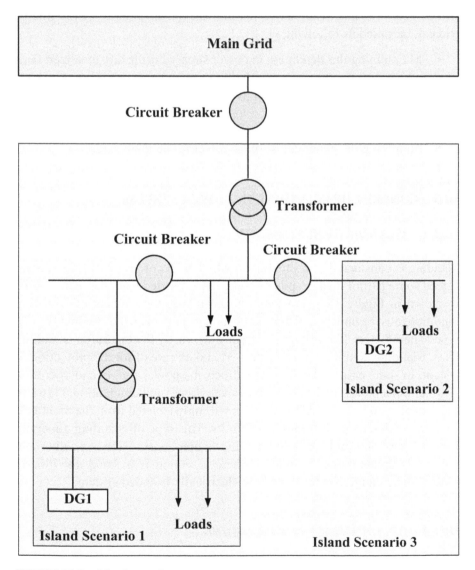

FIGURE 18.3   Island scenarios.

2. **Instability of Voltage:** In islanding instants, main grids are cut out suddenly due to any of the reasons like circuit breaker opening or faults, etc. The islanded part is unable to cope with the required reactive power and that fluctuates the voltage of islanded section. This instability in voltage of the working unit can have very negative effect on the connected loads or user based appliances.

3. **Fluctuation in Frequency:** Grid connected systems operate at a constant frequency but frequency of islanded section fluctuates due to the inability of the islanded part to cope up with the required active power. This fluctuation in frequency can be dangerous for the connected loads as the loads are designed to operate on a constant frequency.

18.3.3 ISLANDING DETECTION TECHNIQUES

Islanding detection is vital for the security of service workers and utility system connected to the grid. There are several approaches already been given by scientists in the past that are largely categorized into classical methods and modern methods [18]. Classical methods are divided into active, passive hybrid and local methods as shown in Figure 18.4.

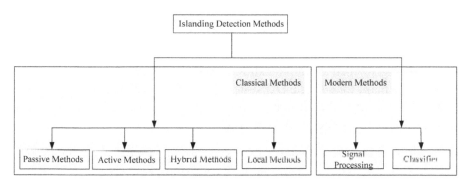

FIGURE 18.4 Classification of islanding detection methods.

18.3.3.1 CLASSICAL METHOD OF ISLANDING DETECTION

Classical methods of islanding detection are directly or indirectly related with the measurement of magnitude of signal like voltage and frequency, etc. Classical method is further divided into passive, active hybrid and local methods [19].

Passive approaches deal with assessment of parameter changes like harmonic distortion, frequency, voltage, etc., with respect to the threshold rate. These parameters are largely affected by islanding situation. Passive methods have found advantage when there is a high power difference between generation and load demand [20]. Passive techniques are reliable, swift and do not create any system variation, but they have high non-detection zone (NDZ).

Classical method of islanding detection has active technique as their second classification [21]. Active method completely depends upon injecting a minor disturbance or perturbation signal at point of common coupling (PCC) and this perturbation signal will give noteworthy alteration in system parameters in islanded system. The range of fluctuations is checked due to disturbance signal and if this range lies outside the predefined value, then islanding is detected otherwise, the whole process for detecting islanding is repeated. Under the scenario of matched load and generation, this technique becomes superior to the passive technique. Active technique, by using disturbance signal, can detect islanding with high efficacy and lesser NDZ [18, 22]. Nevertheless, they also reduce the power quality and upsurge the total harmonic distortion.

Hybrid technique is a synthesis of passive and active methods of islanding detection [23]. Hybrid technique firstly apply the technique discussed in passive technique and then apply the active method for detecting islanding condition [24]. Local approaches like transfer trip scheme and power line carrier communication, etc., [25] utilizes a communication path between the DG and utility section of islanded portion. Local methods are a bit costly owing to their communication channel prerequisite rendering local approaches inapt for a single DG system [18, 26]. Local approaches possess extreme advantage when it is used in multiple system consisting of several DGs as these approaches have zero NDZ.

18.3.3.2   MODERN METHODS OF ISLANDING DETECTION

Modern methods are relatively newer than the traditional methods of this category. It detects islanding with signal processing tools and intelligent classifiers [27]. Modern methods are very advantageous as they have zero NDZ and detect islanding very effectively. Island detection by signal processing uses numerous signal processing methods for obtaining the features of signals that are attained from the islanded section for detection of islanding instants. Some of these techniques are Stockwell transform, wavelet transform, time-time transform, mathematical morphology, etc. [18, 28–30]. Intelligent techniques are the latest techniques in the field of islanding detection method. Classifiers like fuzzy logic, ANN and decision tree, etc., are used for detecting islanding in an effective and efficient manner [31–35]. Classifier methods have additional advantage of zero NDZ.

In all the above methods, a set of hardware and software needs to be installed in the grid for detecting islanding conditions. It takes a lot of time for installation of these devices. Besides time, these installments also requires a lot of capital. Since, NEM are already present in modern day grid, its functions can be leveraged for island detection. The hardware of NEM can be employed for the acquisition of signals and its software can be used for processing of these signals. However, islanding signals involves transients that are not detectable by time analysis. Thus, an advanced signal processing tool is required for analyzing these signals. In the present method, WHT is used which is explained in the next section.

18.4 ISLANDING DETECTION EMPLOYING WALSH HADAMARD TRANSFORM

18.4.1 NEED OF FAST ISLANDING DETECTION METHOD

Connecting DGs in the main stream transmission networks is the key for providing continuous and efficient power for energy demand in the present scenario [3–5]. DGs have been the most significant component in the field of energy market and modern infrastructure. DG, on one side are acting as a pillar to the energy demand and supply, but on the other side faults related with the DGs like islanding could be proved fatal for the utility workers and safety equipments connected to the system [7]. So, for the protection of whole working unit, it is compulsory that islanding phenomenon is detected and corrected in smallest possible time. Unable to detect islanding situation in time could be dangerous for the working personnel, reduce power quality for consumer loads or can cause false triggering of protective devices [18]. Hence, for providing protection against islanding now a day's islanding detection relays employing advance techniques are being used.

18.4.2 BLOCK DIAGRAM OF SIMULATION MODEL

In this chapter, the block diagram of the simulation model developed is presented in Figure 18.5. It is important that the model reflects a real system in all vital parts. In the model main grid is connected to DFIG wind turbine based micro-grid which is rotating at 15 m/sec through a 30 km transmission line and grounding transformer. The model also contains of a NEM so that bidirectional energy flow can be measured which also serves as an additional advantage for collecting samples of signals that can processed through signal processor and trained by intelligent classifiers for the purpose of islanding detection. A circuit breaker separates the micro-grid from the main grid to demonstrate the islanding phenomenon. At PCC, the main grid at a voltage of 120 kV is connected to a micro-grid to form a hybrid grid system. Hence, the behavior of the simulated system represents a real situation.

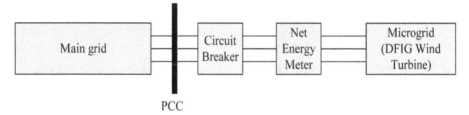

PCC

FIGURE 18.5   Block diagram of simulation model.

18.4.3   TEST SYSTEM CONSIDERED

The test model consists of six 1.5 MW wind turbines that cumulatively gives 9 MW power output connected with 25 kV system. Wind turbines have DFIG as core working component composed of a wound rotor induction generator and an AC/DC/AC insulated gate bipolar transistor (IGBT) based pulse width modulation (PWM) converter. The grid operating at 60 Hz is linked directly to the stator winding while the rotor is supplied at fluctuating frequency through the AC/DC/AC IGBT-based PWM converter. The DFIG technology enables us to pluck out maximal energy from the wind despite of low wind speeds by regulating the turbine speed, while diminishing mechanical burdens on the turbine during puff of wind.

The test system considered is shown in Figure 18.6. In this model, the parameters considered are given in Table 18.1. The duration for which samples are taken is in between t=10 ms to t=30 ms (i.e., 4,000 samples). Various operating condition like islanding and faults are introduced after 20 ms.

TABLE 18.1   Parameters of the Test System Considered

Parameters	Value
Wind speed	15 m/s
Torque	1.2 p.u.
Reactive power output	0 MVAR
Sampling time	5 μs
Simulation time	40 ms

18.4.4   WALSH HADAMARD TRANSFORM

The WHT is a form of Fourier transform where a linear, symmetric and orthogonal operation on $2i$ real numbers, as in Eqn. (1), is performed to get H_i, which is a $2i \times 2i$ matrix [36–38]. Here $1/\sqrt{2}$ is normalization. WHT can be used both recursively or by binary representation. In recursive method, WHT matrix H_1 of order 1×1 is generated by taking $H_0 = 1$ and then H_i is calculated using Eqn. (2), where $\emptyset$ denotes the Kronecker product. Similarly, the Hadamard matrix for its $(k, n)^{th}$ entry is written by using Eqn. (3). In Eqn. (4) and (5), we are denoting k_j and n_j as the binary digits, i.e., 0 or 1, of k and n, respectively [39].

$$H_i = \frac{1}{\sqrt{2}}\begin{pmatrix} H_{i-1} & H_{i-1} \\ H_{i-1} & -H_{i-1} \end{pmatrix} \tag{1}$$

$$H_i = H_1 \emptyset H_{i-1}, \; i > 1 \tag{2}$$

$$(H_i)_{k,n} = \frac{1}{2^{\frac{i}{2}}}(-1)^{\sum_j k_j n_j} \tag{3}$$

$$k = \sum_{m=0}^{i-1} k_m 2^m = k_{i-1}2^{i-1} + k_{i-2}2^{i-2} + \ldots + k_1 2 + k_0 \tag{4}$$

$$n = \sum_{m=0}^{i-1} n_m 2^m = n_{i-1}2^{i-1} + \ldots + n_1 2 + n_0 \tag{5}$$

18.4.5   NET ENERGY METER

NEM has the ability to record flow of energy in two directions [3, 40]. As can be seen from the simulated model, NEM is placed at PCC for measuring bi-directional power flow [21]. To feed the household loads or various types of other connected loads, consumer utilizes power from two sources, i.e.,

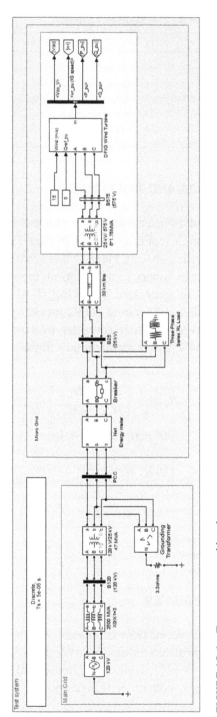

FIGURE 18.6   Test system considered.

from the grid and different DGs. If there is a surplus power generation by DGs than the requirement, the excess power which is not required for the connected loads on DG side will be fed back to the grid through a NEM [41]. It can obtain power measurement in both directions and that provides additional profit to the consumer end. The working formula of NEM is given by Eqn. (6). Here, power fed to or taken from the grid is denoted by P_{grid}, P_{grenDG} denotes power produced by DG and $P_{consumed}$ is the power absorbed by the load. To measure the equivalent power flow, voltage and current are calculated by the NEM [42, 43]. The proposed algorithm uses this measured voltage signals for processing through signal processors and detecting island scenarios. This eliminates the additional requirement for a measuring device separately for island detection.

$$P_{grid} = P_{consumed} - P_{grenDG} \tag{6}$$

18.4.6 PROPOSED METHODOLOGY

The methodology required for islanding detection needs to be very reliable and fast as islanding can cause serious damages to the connected loads and could be proved life threatening for the utility workers. Thus, looking on negative side of islanding and the damages it can cause to the working unit, this chapter proposed a method of islanding detection employing WHT. To demonstrate the applicability of WHT for island detection, normal, islanding and faults scenarios are simulated. Samples of voltage signals are collected from NEM at PCC for processing through signal processor, i.e., WHT and WHT coefficients obtained for the scenarios. The WHT coefficients graphs obtained for the scenarios are plotted in Figures 18.7–18.11. As seen in Figure 18.8, during islanding, the graph shows high density of samples having lower magnitude with respect to normal scenario in which density of lower magnitude samples are quite low as shown in Figure 18.7. The graphs as obtained in Figures 18.9–18.11 for various faults show that this density is different from the normal and islanding scenarios. Therefore, it can be observed that WHT provides different results for above listed scenarios. Thus, the flow chart of the proposed methodology is shown in Figure 18.12. The ANN is properly trained by simulating various islanding and non-islanding scenarios given in Table 18.2. The power mismatch is ensured by load switching or load disconnection with respect to the DG generation just before opening of the circuit breaker for demonstrating islanding scenarios (Figure 18.12).

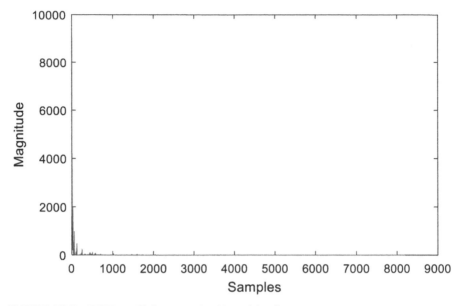

FIGURE 18.7   WHT coefficients graph with no islanding.

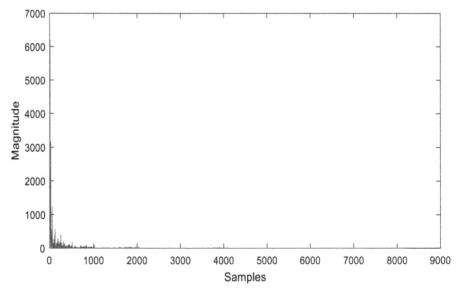

FIGURE 18.8   WHT coefficients graph with islanding.

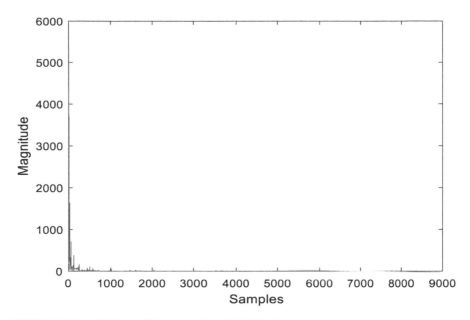

FIGURE 18.9 WHT coefficients graph with LG fault.

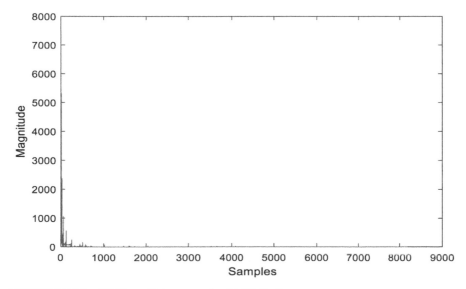

FIGURE 18.10 WHT coefficients graph with LL fault.

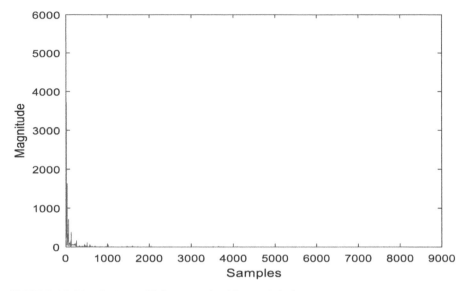

FIGURE 18.11   WHT coefficients graph with LLLG fault.

TABLE 18.2   Scenarios for ANN Training

Scenarios	Number of Scenarios
Island with active power mismatch ranging from +10% to –10% (keeping zero reactive power mismatch).	1,000
Island with reactive power mismatch ranging from +10% to –10% (keeping zero active power mismatch).	1,000
Island with both active as well as reactive power mismatch ranging from +10% to –10%.	1,000
Various type of faults with fault resistance ranging from 50 Ω to 200 Ω at various grid positions.	1,000
Switching on of capacitor ranging from 50 kVAR to 100 kVAR at different locations.	500
Switching on of loads ranging from 50 kW to 100 kW at different locations.	500

18.4.7   RESULTS AND DISCUSSION

To test the performance of the proposed method, the algorithm is tested against various power system scenarios which are not considered during the ANN training process. Table 18.3 shows the effectiveness of the proposed algorithm in segregating island scenarios from various other conditions.

Thus, it is seen that the method is able to detect island cases in a small time span of 20 ms.

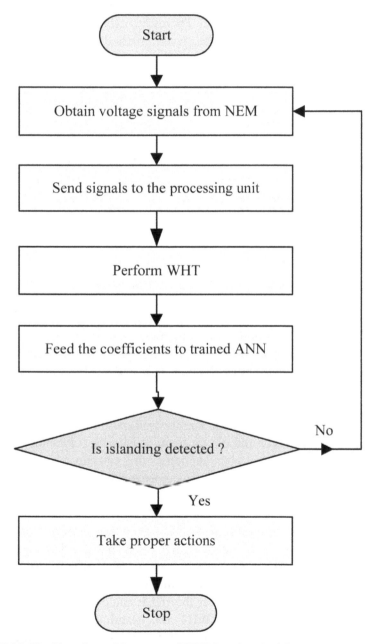

FIGURE 18.12 Flowchart of the proposed WHT-based methodology.

The IEEE Standard 1547-2018 for the connection of DG with the main grid states that each DG must detach itself from the islanded grid within 2s of unintentional islanding. This time includes islanding detection algorithm time and execution time of trip signal. Therefore, a technique having minimal detection time is desirable. Thus, the present method meets this criteria.

TABLE 18.3 Results for Various Scenarios under WHT-based Method

Scenarios	Results
Island with 50% power mismatch.	Island
Island with 150% power mismatch.	Island
Capacitor switching near PCC.	Not island
LG fault near PCC with fault impedance angle = 45°.	Not island
LL fault near PCC with fault resistance of 50 Ω.	Not island

18.5 CONCLUSION

Modern grids are the irrefutable response for managing the complex power grid, catering to the required dynamics to the power system to enhance its efficacy. The usage of DGs have fortified the grid utility's sustainability. Hence, limitations like islanding instants must be dealt with rationally. The penetration of hybrid grid system is on rise, owing to its advantages and flexibilities that provides assurance for continuity of supply. Moreover, hybrid grid composes of inverters and batteries that are crucial for backup power management. On the other hand, besides the advantages of interconnecting DG with the main supply grid, certain drawbacks of these interconnections are there and one of them is islanding. Islanding, in most interconnection is a decisive phenomenon as this can cause severe damages to the grid management system by fluctuating the voltage and frequency or would affect the working of connected loads and safety equipments or can put life of workers to a great risk. Therefore, experts and scholars have constantly dedicated there time in devising islanding detection approaches which are exceedingly effective and quick such that the damaging effects of islanding can be dodged to a huge degree. In continuation with the same trajectory, the chapter is composed of a method which can detect islanding in very short interval of time as per the IEEE standards. The method uses signal processing tools such as WHT for islanding detection, which is comparatively accurate and efficient with the previous available methods as it can detect islanding in 20 ms. Furthermore, the use of NEM reduces the components requirement. This is because the hardware components of the NEM is used for measuring different signals and the software part of NEM is used for the processing of these signals with WHT. Thus, unlike other island detection methods, no extra components are needed to be installed for detection of islanding instants. This in turn reduces the overall cost for island detection.

KEYWORDS

- **distributed generation**
- **island detection**
- **islanding**
- **micro-grid**
- **net energy meter**
- **smart grid**
- **Walsh Hadamard transform**

REFERENCES

1. Khalid, H., & Abdulfetah, S., (2021). Existing developments in adaptive smart grid protection: A review. *Electric Power Systems Research, 191*, 106901.
2. Butt, O. M., Muhammad, Z., & Tallal, M. B., (2021). Recent advancement in smart grid technology: Future prospects in the electrical power network. *Ain Shams Engineering Journal 12*(1), 687–695.
3. Dutta, S., Debomita, G., & Dusmanta, K. M., (2016). Location biased nature of net energy metering. In: *2016 International Conference on Computation of Power, Energy Information and Communication (ICCPEIC)* (pp. 350–355). IEEE.
4. Kappagantu, R., & Arul, D. S., (2018). Challenges and issues of smart grid implementation: A case of Indian scenario. *Journal of Electrical Systems and Information Technology, 5*(3), 453–467.
5. Dutta, S., Pradip, K. S., Maddikara, J. B. R., & Dusmanta, K. M., (2020). Role of microphasor measurement unit for decision making based on enhanced situational awareness of a modern distribution system. In: *Decision Making Applications in Modern Power Systems* (pp. 181–199). Academic Press.
6. Kakran, S., & Saurabh, C., (2018). Smart operations of smart grids integrated with distributed generation: A review. *Renewable and Sustainable Energy Reviews, 81*, 524–535.
7. Singh, B., & Janmejay, S., (2017). A review on distributed generation planning. *Renewable and Sustainable Energy Reviews, 76*, 529–544.

8. Zsiborács, H., Nóra, H. B., András, V., László, Z., Zoltán, B., Kinga, M., & Gábor, P., (2019). Intermittent renewable energy sources: The role of energy storage in the european power system of 2040. *Electronics, 8*(7), 729.

9. HA, M. P., Phung, D. H., & Vigna, K. R., (2017). A review of the optimal allocation of distributed generation: Objectives, constraints, methods, and algorithms. *Renewable and Sustainable Energy Reviews, 75*, 293–312.

10. Bansal, R., (2017). *Handbook of Distributed Generation*. Switzerland Valsan SP, Springer International Publishing, Switzerland.

11. Marwali, M. N., Jin-Woo, J., & Ali, K., (2004). Control of distributed generation systems-Part II: Load sharing control. *IEEE Transactions on Power Electronics, 19*(6), 1551–1561.

12. Marwali, M. N., & Ali, K., (2004). Control of distributed generation systems-Part I: Voltages and currents control. IEEE *Transactions on Power Electronics, 19*(6), 1541–1550.

13. Hussain, A., Van-Hai, B., & Hak-Man, K., (2019). Microgrids as a resilience resource and strategies used by microgrids for enhancing resilience. *Applied Energy 240*, 56–72.

14. Hirsch, A., Yael, P., & Josep, G., (2018). Microgrids: A review of technologies, key drivers, and outstanding issues. *Renewable and Sustainable Energy Reviews, 90*, 402–411.

15. Hossain Md, A., Hemanshu, R. P., Md Jahangir, H., & Frede, B., (2019). Evolution of microgrids with converter-interfaced generations: Challenges and opportunities. *International Journal of Electrical Power & Energy Systems, 109*, 160–186.

16. Wu, H., Yifei, Z., Rui, C., Kai, D., Deqiang, G., Shanglin, D., Xi, W., Yang, Y., & Hao, Y., (2019). A microgrid system with multiple island detection strategies. In: *2019 IEEE PES Asia-Pacific Power and Energy Engineering Conference (APPEEC)* (pp. 1–4). IEEE.

17. Yin, G., Sun, M., & Xiao, L. (2007). Review of island detection methods of distributed generation [J]. *Electronic Measurement Technology, 1.*

18. Dutta, S., Pradip, K. S., Jaya, B. R. M., & Dusmanta, K. M., (2018). Shifting of research trends in islanding detection method-a comprehensive survey. *Protection and Control of Modern Power Systems, 3*(1), 1–20.

19. Dutta, S., Shailesh, V., Pradip, K. S., Maddikara, J. B. R., & Dusmanta, K. M., (2019). Islanding detection in a distribution system: A pattern assessment based approach using Concordia analysis. In: *2019 20th International Conference on Intelligent System Application to Power Systems (ISAP)* (pp. 1–5). IEEE.

20. Kulkarni, N. K., & Khedkar, M. K., (2021). Methods to detect the occurrence of an unintentional island with passive approach: A review. *Journal of The Institution of Engineers (India): Series B*, 1–21.

21. Verma, S., Soham, D., Pradip, K. S., Jaya, B. R. M., & Dusmanta, K. M., (2019). Islanding detection using bi-directional energy meter in a DFIG based active distribution network. In: *2019 International Conference on Computer, Electrical & Communication Engineering (ICCECE)* (pp. 1–4). IEEE.

22. Atram, S., Hitesh, M., Yogesh, L., & Bhombe, R. M. (2013). *Review on Active Islanding Detection Technique in Distributed Generation for Current Controlled Inverter. 28*, 483–493.

23. Singh, S. K., Mayank, R., Mahiraj, S. R., & Tripurari, N. G., (2021). Hybrid islanding detection technique for inverter based microgrid. In: *2021 2nd International Conference for Emerging Technology (INCET)* (pp. 1–5). IEEE.

24. Murugesan, S., & Venkatakirthiga, M., (2018). Hybrid analyzing technique based active islanding detection for multiple DGs. *IEEE Transactions on Industrial Informatics, 15*(3), 1311–1320.

25. Min-Sung, K., Raza, H., Gyu-Jung, C., Chul-Hwan, K., Chung-Yuen, W., & Jong-Seo, C., (2019). Comprehensive review of islanding detection methods for distributed generation systems. *Energies, 12*(5), 837.

26. Chawda, G. S., Abdul, G. S., Mahmood, S., Sanjeevikumar, P., Bo Holm-Nielsen, J., Om Prakash, M., & Palanisamy, K., (2020). Comprehensive review on detection and classification of power quality disturbances in utility grid with renewable energy penetration. *IEEE Access, 8*, 146807–146830.

27. Mahat, P., Zhe, C., & Bak-Jensen, B., (2008). Review of islanding detection methods for distributed generation. In: *2008 Third International Conference on Electric Utility Deregulation and Restructuring and Power Technologies* (pp. 2743–2748). IEEE.

28. Dehghani, M., Mohammad, G., Taher, N., Kavousi-Fard, A., & Sanjeevikumar, P., (2020). False data injection attack detection based on Hilbert-Huang transform in AC smart islands. *IEEE Access, 8*, 179002–179017.

29. Kolli, A. T., & Navid, G., (2020). A novel phaselet-based approach for islanding detection in inverter-based distributed generation systems. *Electric Power Systems Research, 182*, 106226.

30. Ahmadipour, M., Hashim, H., Mohammad, L. O., & Mohd, A. R., (2019). Islanding detection method using ridgelet probabilistic neural network in distributed generation. *Neurocomputing 329*, 188–209.

31. Khamis, A., Yan, X., & Azah, M., (2017). Comparative study in determining features extraction for islanding detection using data mining technique: Correlation and coefficient analysis. *International Journal of Electrical and Computer Engineering, 7*(3), 1112.

32. Kumar, D., & Partha, S. B., (2018). Artificial neural network and phasor data-based islanding detection in smart grid. *IET Generation, Transmission & Distribution, 12*(21), 5843–5850.

33. Kong, X., Xiaoyuan, X., Zheng, Y., Sijie, C., Huoming, Y., & Dong, H., (2018). Deep learning hybrid method for islanding detection in distributed generation. *Applied Energy, 210*, 776–785.

34. Yu, S., & Lulin, Y., (2018). Islanding detection method based on S transform and ANFIS. *Journal of Renewable and Sustainable Energy, 10*(5), 055503.

35. Admasie, S., Syed, B. A. B., Teke, G., Raza, H., & Chul, H. K., (2020). Intelligent islanding detection of multi-distributed generation using artificial neural network based on intrinsic mode function feature. *Journal of Modern Power Systems and Clean Energy, 8*(3), 511–520.

36. Sneha, P. S., Syam, S., & Ashok, S. K., (2020). A chaotic color image encryption scheme combining Walsh–Hadamard transform and Arnold–Tent maps. *Journal of Ambient Intelligence and Humanized Computing, 11*(3), 1289–1308.

37. Subathra, M. S. P., Mazin, A. M., Mashael, S. M., Garcia-Zapirain, B., Sairamya, N. J., & Thomas, G. S., (2020). Detection of focal and non-focal electroencephalogram signals using fast Walsh-Hadamard transform and artificial neural network. *Sensors, 20*(17), 4952.

38. Prabha, K., & Shatheesh, S. I., (2020). A novel blind color image watermarking based on Walsh Hadamard transform. *Multimedia Tools and Applications, 79*(9), 6845–6869.

39. Andrushia, A. D., & Thangarjan, R., (2018). Saliency-based image compression using Walsh–Hadamard transform (WHT). In: *Biologically Rationalized Computing Techniques for Image Processing Applications* (pp. 21–42). Springer, Cham.

40. Syafii, S., Muhammad, I. R., Lovely, S., & Irvan, Z., (2020). Web-based net energy meter for grid connected PV system. *TEM Journal, 9*(1), 37–41.

41. Nguyen, T. A., & Raymond, H. B., (2017). Maximizing the cost-savings for time-of-use and net-metering customers using behind-the-meter energy storage systems. In: *2017 North American Power Symposium (NAPS)* (pp. 1–6). IEEE.
42. Dutta, S., Debomita, G., & Dusmanta, K. M., (2016). Optimum solar panel rating for net energy metering environment. In: *2016 International Conference on Electrical, Electronics, and Optimization Techniques (ICEEOT)* (pp. 2900–2904. IEEE.
43. Anderson, T., Daniel, C., Chloe, F., Harrison, H., Nina, M., Bailey, T., Andres, C., & Arthur, S., (2021). Behind the meter: Implementing distributed energy technologies to balance energy load in Virginia. In: *2021 Systems and Information Engineering Design Symposium (SIEDS)* (pp. 1–6). IEEE.

Index

A

Algorithms, 4, 16, 18, 39, 45, 67–69, 74, 75, 78, 79, 84–87, 100, 234, 269, 284, 285, 314, 316–318, 320, 321, 325, 333
Alternating current (AC), 5, 8, 9, 104, 105, 155, 182, 229, 234, 272, 372
Aluminum tube, 24
Ambient temperature, 30, 31, 40, 196, 200, 260
Ameliorated Harris Hawks optimization (AHHO), 284
Anaerobic bio-gasification, 59
Ant
    colony optimization (ACO), 286, 318, 327
    lion optimizer (ALO), 316, 333
Artificial
    bee colony (ABC), 209, 287, 314, 318, 319
    intelligence (AI), 42, 56, 60, 70, 71, 74, 86, 269, 270, 289, 314, 316
    neural network (ANN), 61, 71, 74, 360, 371, 375, 378
Authentic load forecasting, 84
Automated
    capacitor bank, 279
    opportunities, 51
    selfwatering system, 295
    solar power plant watering system, 295–297
Automotive industry, 251, 256, 263
Average energy not supplied (AENS), 344–346, 348, 352–355

B

Bat
    algorithm (BA), 209, 314, 318
    inspired, 318
Battery, 39, 91–94, 96, 97, 99, 100, 126, 155–157, 172, 174, 196, 197, 253, 255–263, 296, 297, 299–305
    couple water pumping system, 304

electric vehicles (BEVs), 254, 255, 257
    management system (BMS), 91, 92, 100, 255, 256, 258–260
    optimization, 259
    pack model, 99
    system, 301
Bayesian regularization, 73
Binary particle swarm optimization (BPSO), 269
Biogas, 39, 55, 56, 59–62
Biogeography-based optimization (BBO), 288, 318
Bioinformatics
    algorithms, 318
    CI (BioCI), 318
Biology meta-heuristic algorithm, 318
Biomass energy, 37, 39, 55, 196
Block
    chain technology, 61
    diagram (simulation model), 371
Bluetooth, 39, 48, 53
    low energy (BLE), 48
Boost converter, 131, 139, 149
Bosch Global Software Technologies (BGSW), 123, 125, 152
Branch current to bus voltage (BCBV), 214, 340
Buck
    boost converters, 125, 134, 151, 152
    converter, 105, 127, 136, 146
Building
    energy management system (BEMS), 37, 39, 41, 42, 45, 51, 53, 55, 61, 62
    integrated photovoltaic (BIPV), 58
    management information system (BMIS), 43
Bus
    injection to branch current (BIBC), 213, 214, 340, 341
    system, 208, 216, 218, 220–223, 284, 286, 328, 350

C

Cable geometry, 315
Cadmium sulfide (CdS), 298
Capacitance, 142, 143, 152, 155, 158, 160,
    161, 171, 172, 174
    deliverance, 171, 172, 174
Capacitor, 1, 3, 6, 9, 18, 91, 92, 94, 103,
    104, 106, 113, 122, 123, 126, 136, 157,
    163, 166, 173, 174, 242, 268–270,
    274–281, 283–290, 320, 378
    allocation, 288
Carbon
    dioxide (CO2), 45, 51, 59, 60, 361
    monoxide, 45
Cascaded H-bridge
    inverter, 3
    multilevel inverter (CHB-MLI), 1, 3, 4,
        9, 18
Central Power Research Institute (CPRI),
    155, 174
Charge
    discharge protocol, 164
    efficiency, 155, 171, 172, 174
Chemical oxygen demand (COD), 60
Chemistry CI (ChmCI), 318
Circuit averaging (CA), 126, 127, 142, 151,
    152
Cleaner power production, 208
Climate, 69, 70, 251
Cloud software applications, 50
Compare swap algorithm (CAS), 316
Computational
    efficiency, 327
    intelligence (CI), 318
Connected vehicle technology, 263, 264
Constant
    current (CC), 155, 156, 158–160,
        162–167, 169, 171, 174
        constant voltage (CCCV), 159, 163,
            164, 171
    load electrical facilities, 279
    voltage (CV), 105, 141, 155, 156,
        158–160, 164, 166, 169, 173, 174
Consumer
    average interruption duration index
        (CAIDI), 344, 345, 348, 352–355
    electronics, 50, 94

Continuous conduction mode (CCM),
    103–105, 113, 117, 123–127, 131, 142,
    151, 152
Control algorithm, 6, 18, 186, 228, 248
Conventional
    automotive industry, 256
    electricity, 256, 257
    energy, 260
        resources, 2
    machine learning, 76
    two-wheelers, 262
    vehicles, 251, 252, 263, 264
Converter active space vectors, 239
Cost
    analysis, 209, 223
    benefit analysis, 41
    effective energy, 315
    function
        minimization, 180
        reduction, 244
    optimization, 308, 309
Cuckoo search (CS), 286, 314, 318
    algorithm (CSA), 286
Cuk converter, 127, 142
Cyclic voltammetry studies, 157

D

Data
    communication, 48, 53, 61
    load requirements, 48
Dataset, 71
Deep learning, 44, 68, 69, 72, 74, 75, 78,
    86, 87
Delamination, 56
Depth of discharge (DoD), 158
Design
    dc link
        capacitors, 6
        voltage reference, 5
    interfacing inductor, 5
    maintenance, 270
Dielectric medium, 157
Diesel
    fuel-operated diesel generators, 196
    generators, 196
Differential evolution, 319
Digital signal processors (DSPs), 180
Dimethylformamide (DMF), 163

Diode clamped inverter (DCMLI), 3
Direct
  axis current reference, 187
  couple solar pumping systems, 305
  current (DC), 1, 4–13, 15, 16, 18,
    103–105, 114, 120, 125–127, 141, 151,
    182, 197, 201, 229, 234, 258, 259,
    300–302, 305, 338, 362, 372
  link voltage reference, 18
  loss calculation generation, 8
  load flow (DLF), 207, 209, 213, 214, 222,
    337, 339–341, 348, 353, 355
  normal irradiance (DNI), 22
  torque control (DTC), 180, 238–240,
    245–248
Discharge protocol, 155, 156, 164, 169,
  171–175
Discontinuous conduction mode (DCM),
  104, 126, 142
Distributed
  energy resources (DERs), 283, 337–339,
    344, 355
  generation (DG), 207–209, 211, 212,
    215–220, 222, 223, 260, 261, 269, 270,
    283, 285–287, 326–328, 330, 337–339,
    348, 349, 359–367, 370, 371, 375, 380,
    381
  load flow, 337
  network, 2, 268, 271, 287, 288, 326, 327,
    339
  static compensator (DSTATCOM), 1–4,
    6, 9, 10, 15, 18
  system, 207, 208, 213, 267–271, 280,
    282, 283, 285, 287–290, 329, 330, 339,
    340, 349, 362, 364
  deliver energy, 271
Dolphin echolocation (DE), 314, 319
Domestic applications, 21, 34
Doubly-fed induction generator (DFIG),
  360, 372
Dragonfly algorithm (DA), 316, 333
Duty cycle, 104, 105, 107, 116, 122,
  124–127, 143, 151
Dynamic
  load variations, 18
  models (synchronous reference frame), 231
  parameter approximation, 315
  performance, 9, 246, 247, 249

response, 180, 245, 246, 249
waveforms, 15

E

Electric
  double-layer-capacitor (EDLC), 157
  engineering issues, 314
  generation, 39, 54, 55, 57, 58, 196, 197, 260
  load forecasting, 79–83, 86, 87
  market pricing techniques, 283
  mobility, 252, 253, 256, 257, 260–264
  motor, 228, 253, 255, 258, 259
  power
    buses, 263
    cars, 261, 263
    systems, 272
    vehicles, 254, 257, 261, 263
  production, 362
  traction motor, 258, 259
  utilization, 67
  vehicle (EV), 54, 91, 92, 94, 100, 103,
    125, 179–181, 184, 185, 189, 190,
    251–264
  technologies, 262
Electrochemical characteristics, 158
Electromagnetic
  load torque, 190
  torque, 181, 184, 189, 231, 238, 239
Electronic equipment, 307
Encircling prey, 322
End-user electric load demand applications,
  205
Energy
  conservation, 69, 267
  consumption
    efficiency, 38
    optimization, 39
  efficiency, 39, 41, 42, 52, 54, 61, 155,
    158, 160, 169, 171, 174
  insufficiency, 2
  management, 38–41, 62
  network enhancement, 70
  resource consumption, 41
  sources (ES), 2, 21, 39, 55, 56, 58, 156,
    196, 197, 282, 318, 361, 364, 366
  storage system, 155, 156, 158, 161, 172,
    196, 255, 285, 317
  transmission, 361

Engine automobiles, 254
Enhanced multilayer second-order general-
    ized integrator (EMSOGI), 4
Equivalent
    circuit model (ECM), 91, 92, 96, 97, 100
    series resistance (ESR), 103, 104, 107,
        123, 126, 139, 143, 149, 152, 162
E-rickshaw, 253
E-scooters, 263, 264
Evolution programming (EP), 319

F

Faster adoption manufacturing of electric
    (FAME), 261
Feeder reactance, 275
Feed-forward neural network, 75
Field-oriented
    control (FOC), 180, 189, 232, 233, 245–249
    controllers, 227
Firefly optimization, 269
Fixed capacitor group, 279
Flexibility, 365
Flower pollination algorithm (FPA), 288
Flux error, 238
Fly
    capacitor inverter (FCMLI), 3
    optimization algorithm (FOA), 314
Fossil fuel consumption, 260
Framework generation costs, 288
Fruit fly optimization algorithm, 314
Fuel cell (FC), 195–204
Fuzzy logic techniques, 42

G

Gain margin (GM), 120, 143, 149
Gas microturbine, 362
Gasoline-powered vehicle, 254
Gauge workspace utilization, 51
Gaussian mutation perturbation, 316
Generation
    reference current, 187
    unit vector template, 7
Genetic algorithm (GA), 209, 285, 286, 318,
    326–329
Geothermal energy, 55
Global, 317
    climate changes, 256
    electricity generation, 55

horizontal irradiance (GHI), 22, 23
    optimization, 319, 324, 328
    warming, 38, 156, 258, 361, 366
    wireless communication protocol, 48
Gravitational search
    algorithm (GSA), 288
    method, 319
Gray wolf optimization, 283, 314, 316, 319,
    321
Green
    energy, 37, 39, 41, 55–57, 61, 62
    house
        effects, 256
        gas emissions, 61, 208, 258
        Internet of Things (G-IoT), 62
Grey wolf optimizer (GWO), 283, 284, 287,
    313, 314, 316–320, 329–331, 333
    sine cosine and crow search algorithm
        (GWOSCACSA), 283
Grid
    connected modes, 283
    management, 62, 380

H

Hair tension moisture sensors, 44
Hardware
    in loop (HIL), 338–340, 342, 343, 348,
        351, 353, 355
    specification, 305, 309
Harmonic distortion, 279, 370
Harris Hawks optimization (HHO), 284
Heat
    energy, 22
    transfer, 23–25, 27, 31, 34
Heating ventilation air-conditioning
    (HVAC), 41, 43
Higher
    intensity discharge lights, 273
    level heuristic algorithm, 287, 326
    performance power electronic devices, 180
    power distribution systems, 3
Hirschberg-Sinclair algorithm (HS), 316
Hoffman Electronics, 298
Hybrid
    electric vehicles (HEVs), 258
    energy system, 195, 196, 198, 205
    optimization, 269, 284, 333
        design, 269

of multiple energy resources
(HOMER), 195, 197, 199, 200, 204
techniques, 328
ultracapacitor, 175
systems, 156
Hyderabad Electric Supply Company
(HESCO), 280
Hydrogen
fuel cell, 205, 256
tank, 197, 198, 200, 203, 204
Hysteresis predictive control approach, 240

I

Imperialistic competitive algorithm (ICA),
286
Improved
performance management, 51
second-order generalized integrator, 1
Indoor environment quality (IEQ), 43, 44
Induction motor (IM), 179–182, 184, 185,
188–190, 229, 232, 259, 272, 273
Infrared (IR), 21, 43, 45, 60, 275
Instability, 103, 105, 123–125, 149, 151,
152, 181, 369
Institute of Electrical Electronics Engineers
(IEEE), 2, 10, 18, 48, 50, 207–209,
216, 222, 284, 286, 328, 338, 340, 343,
348–354, 380
Insulated gate bipolar
junction transistors (IGBT), 6
transistor (IGBT), 6, 372
Intelligent power consumption, 54
Interior permanent-magnet synchronous
machines (IPMSM), 227, 229, 232, 233,
245, 246
Internal resistance, 156, 161–163, 169, 173,
174
International Energy Agency, 38
Internet
connection, 48
of energy (IoE), 41, 42, 61
of things (IoT), 37–40, 42, 43, 48–56,
58–62, 299
remote monitoring system, 56
Irrigation system, 299, 302, 308
Island, 283, 339, 359–361, 363, 367–372,
375, 376, 380, 381
detection, 369, 371, 375, 381

K

Kaggle database, 71, 78, 86
Kirchhoff's current law, 213
Krill herd (KH), 318

L

Lead
acid batteries, 156, 157, 196, 255, 263
carbon hybrid ultracapacitor (Pb-C HUC),
155–158, 173, 175
oxide, 155, 156
Least square linear curve fitting technique, 160
Li-ion cell model, 92, 99
Line modeling, 210, 223
Linear-quadratic regulator (LQR), 242
Lithium-ion batteries, 91, 255, 260
Load
demand growth, 87
flow, 208, 269, 270, 339, 340, 342, 355
analysis, 223, 269, 340, 342
modeling, 211
Long
range (LoRa), 39, 53, 56, 61, 255
Wide Area Network (LoRaWAN), 48
short-term memory (LSTM), 68, 71, 72,
75–87
term evolution (LTE), 39, 48
for machines (LTE-M), 39

M

Machine
discrete-time model, 243
learning (ML), 38, 45, 74, 76, 86
Material tracking, 50
Mathematical
modeling, 104, 105, 124, 209, 227, 229,
320
morphology, 370
optimization, 289
MATLAB-Simulink, 2, 4, 18, 103–106, 108,
123–127, 152, 179, 181, 182, 188
Maximum torque per
ampere (MTPA), 232, 246
voltage (MTPV), 232
Mean
absolute percentage (MAPE), 68, 72,
77–82, 84–87
electric power production, 200

Meta-heuristic
  algorithm, 287, 314, 318
  approach, 288
  method, 208, 326
  optimization algorithms, 284
  techniques, 288, 321
Methane, 59
Microbial dynamics, 60
Micro
  controller, 232, 296, 299
  grid (MG), 70, 156, 283, 337–340, 342,
    344–348, 353, 355, 361, 365–367, 372,
    381
Model
  predictive
    control (MPC), 180, 189, 240–245, 248
    current control (MPCC), 179–181, 184,
      185, 187–190, 242, 243, 245–248
    measurement, 244
    torque control, 244
  reference adaptive system (MRAS), 179
    current estimator, 190
Modified cultural algorithm (MCA), 288
Modulation
  space vector theory, 249
  vector, 237
Monetary smart grid activities, 67
Monitoring installation, 68
Monte-Carlo simulation (MCS), 340,
  345–348, 352, 355
Moth-flame optimization (MFO), 323, 324
Motivation, 208
Multi
  dimensional search habitat, 321
  layer perceptron (MLP), 68, 72–74,
    78–87
  level inverter, 1, 18

N

Narrow Band-Internet of Things (NB-IoT),
  39
National Aeronautics Space Administration
  (NASA), 198
Negative temperature coefficient (NTC), 43
Net
  energy meter (NEM), 359, 360, 371–373,
    375, 380, 381
  zero energy buildings (nZEBs), 40

Network
  architecture, 38
  connectivity, 256
  protocols used (smart buildings), 45
  reconfiguration, 337, 339, 340, 354, 355
Neural network development, 71
Non
  conventional resources, 208
  detection zone (NDZ), 370, 371
  faradaic reactions, 158
  ideal, 124
    higher-order converters, 127
  linear
    load, 4, 7, 8, 18
    optimization, 328
  rechargeable primary batteries, 253
  technical losses (NTLS), 268, 271
    customer management systems, 271
  trackbound transport systems, 228
Novel heuristic algorithms, 316

O

Objective function, 180, 209, 212, 215, 285,
  289, 329, 333, 338
Occupancy detection, 43, 44
Off-grid Biogas Power Generation Program,
  59
Off-line parameter estimation, 315
Open
  circuit
    potential (OCP), 164
    voltage, 96, 161
  loop configuration, 125, 151
Oppositional
  gray wolf optimization (OGWO), 283
  Harris Hawk Optimization (OHHO), 284
Optical sensors, 45
Optimal
  performance, 163, 165, 175
  positioning, 223
Optimization, 61, 68, 96, 97, 155, 158, 173,
  185, 207, 209, 214, 222, 240–242, 244,
  259, 268, 269, 283–286, 288–290, 313,
  314, 316–319, 321, 324–331, 333
  algorithm, 96, 97, 209, 284
  dependability, 327
  methodologies, 284
  techniques, 268, 289, 290, 317, 330, 333

Optocoupler, 307, 309
Organic fluids, 24
Orthogonal frequency-division multiplexing (OFDM), 48
Overhead transmission line system (OHTL), 329, 330

P

Paper target, 209
Parabolic trough collector (PTC), 23–29, 32–34
Parameter estimation, 92, 96, 98, 100, 314, 315, 329, 333
Particle swarm
    optimization (PSO), 207–209, 214, 217, 222, 223, 269, 284, 285, 318, 326
    optimizer, 285, 326, 327
Peak inverse voltage (PIV), 3
Performance characterization, 175
Permanent magnet, 229–231
    synchronous automobiles, 228
Phase
    lock loops (PLL), 3, 4
    margin (PM), 114, 120, 127, 143, 149, 229, 238
Phasor measuring unit (PMU), 315, 339
Photovoltaic (PV), 1–4, 6, 8–12, 15, 18, 45, 55–58, 60, 105, 156, 195–205, 208, 256, 269, 289, 298, 299, 362
    array, 1–4, 6, 10–12, 15, 18, 195, 200, 201, 204
    tracking system, 205
Physical security, 52
PIC micro-controller, 296, 309
Plugin hybrid electric vehicles (PHEVs), 258
Point of,
    common coupling (PCC), 5, 370, 372, 373, 375, 380
    load (POL), 105
Polyvinylidene fluoride (PVDF), 163
Power
    applications, 180, 229, 314, 329
    electronics devices, 259, 362
    factor, 1, 4, 8, 15, 18, 126, 267–270, 272–275, 277–279, 283, 290
        correction capacitor (PFCC), 278
    flow equations, 284

    generation, 60, 67, 68, 156, 208, 212, 360, 363, 366, 375
    load fluctuations, 70
    loss reduction, 207, 269, 284, 290
    modulation, 259
    production consumption, 87
    quality (PQ), 1–3, 18, 208, 211, 268, 328, 338, 365, 367, 370, 371
    system
        economics, 268
        Simulator Siemens Calculation (PSSSINCAL), 280
Predictive
    flux-weakening algorithm, 246
    maintenance, 51
Pressure regulating valve (PRV), 141
Probabilistic incremental learning, 318
Proportional-integral (PI), 8, 9, 12, 61, 151, 180, 187–189, 234, 243–245, 248
Pulse width
    modulation, 2, 234, 372
        space vector theory, 234
    modulator (PWM), 6, 9, 143, 179, 180, 232, 234, 235, 237, 296, 299, 372

Q

Quadrature-axis current reference, 187
Quantifiable building insights, 51

R

Radial distribution
    network, 223, 285
    system, 207, 269, 341
Radiofrequency (RF), 50
Railway traction drive system, 245
Ramp speed, 249
Rated voltage, 161, 162
Reactive power compensation, 268, 283
Real-time
    applications, 317
    data, 10, 39, 40, 42, 87, 353
    electricity, 52
    irradiation, 10
    monitoring, 42, 56
    power system, 327
Recurrent neural network (RNN), 68, 71, 72, 74–76, 78–87

Reduced
  energy consumption, 51
  operational costs, 52
Reflector, 21–23, 25, 28, 30, 33, 34
Renewable
  energy
    charging infrastructure, 257
    charging systems, 262
    resources (RESs), 22, 53, 196, 366
  fraction, 195, 197, 201, 205
  sources, 59, 61, 70, 366
Resistance temperature detectors (RTD), 43
Right half plane (RHP), 122, 123, 127, 149, 152
Root mean squared (RMSE), 68, 72, 77–82, 84–87
Rotor flux
  estimation, 184
  orientation, 184, 189

S

Safety assurance, 263, 264
Sample hold circuit (S-H), 6, 7
Second order generalized integrator, 3
Self-loop memory blocks, 76
Sensible climatic consignments, 69
Sensitivity analysis approach, 209
Sensorless, 179–181, 184, 185, 189, 190
Sensors network protocols, 62
Shark smell optimization (SSO), 288
Short
  range communication, 48
  term load forecasting (STLF), 69–72, 86, 87
  wave radiation, 22
Shunt
  capacitor, 269, 275, 288
    location, 278
  compensation, 278
Sigfox, 39, 48, 53
Signal injection speed estimation techniques, 181
Simulated
  annealing (SA), 287, 328, 329, 333
  study, 13, 188, 190
Simulink
  design optimization, 197
  parameter estimation, 96

Sine-cosine algorithm (SCA), 288
Small signal modeling, 124, 152
Smart
  building, 37–40, 42–45, 53–57, 59, 61, 62
    grid management, 37
  grid, 38, 54–57, 67, 68, 70, 71, 87, 260, 261, 359–361, 381
    technology, 260, 361
Social
  configuration, 318
  hierarchy, 320
Sodium nickel chloride battery, 255
Software implementation, 295
Soil moisture sensor, 295
Solar
  cell, 205, 300
  energy, 22, 34, 55, 56, 58, 59, 62, 297, 298, 302, 305
  irradiation, 9, 10, 18, 197, 299
  microgrid, 155, 156, 158, 172, 175
  panel pumping system, 300
  photovoltaic (SPV), 156, 196, 299, 339, 350
  resource, 299
  power
    applications, 155, 171, 173, 174
    automatic irrigation system, 302
    automatic power plant watering system, 295
    irrigation system, 299
    meter, 56
    technology, 299
  PV, 3, 309
  radiation, 13, 21, 23, 195, 196, 199, 200, 205
  thermal systems (STSs), 58
  water pumping system, 319
Space
  allocation management, 53
  vector modulation (SVM), 71, 234
Stable closed loop operations, 6
Stainless steel reflectors, 23, 33, 34
Stakeholders, 309
Standard
  optimal performance of HUCS, 172
  test protocol, 155, 156, 171–175
State
  of charge (SoC), 91–93, 97–100, 158, 197

space
averaging (SSA), 126, 127
modeling, 127, 152
Stator phase current, 190
Steady
state
modeling, 340
performance, 18
response, 246, 249
waveforms, 15
voltage discharge curve, 165
Steam engines, 254
Stockwell transform, 370
Structural
engineering design optimization, 286
optimization problems, 286
Supercapacitors, 255
Supervisory control data acquisition
(SCADA), 315
Surface
mount motor rotor designs, 229
permanent magnet synchronous machine
(SPMSM), 229, 230
Swarm
algorithms, 318
intelligence, 314
method, 316
systems, 321
meta-heuristic algorithms, 318
Switch mode power supplies (SMPS), 125,
126, 134
Synchronous
AC machines, 229
converters, 104, 105, 124
machines, 229
reference frame theory (SRFT), 3, 4, 16–18
reluctance, 229
System
average interruption
duration index (SAIDI), 345, 355
frequency index (SAIFI), 344–346,
348, 352–355
integration, 279

T

Tabu search (TS), 287, 314, 316, 326
Teaching-learning-based optimization
(TLBO), 284, 285
Technological breakthroughs, 263

Tech-savvy smart society, 51
Temperature
coefficient, 198
correction resistance, 284
data, 199, 205
sensor, 24
Thermal
efficiency, 22, 23, 25, 26, 28, 29, 32–34
performance, 24, 27, 34
runaway, 260
Time-delay compensation, 243
Top-of-line voltage switching, 239
Torque
error, 189, 190, 238
producing current, 243
Total harmonic distortion (THD), 3, 10, 18,
234, 326, 370
Tracking, 320
Transistor, 308, 309
Transmission line, 53, 54, 273, 283, 284,
314, 315, 317, 319, 320, 329–331, 333,
366, 367, 372
parameter, 314, 317, 320, 329
chain, 320
properties, 284

U

Ultra-wideband (UWB), 39, 48, 53
Uninterruptible power supplies (UPS), 2
Unit vector template, 7, 18
Universal serial bus (USB), 39, 50

V

Vector control (VC), 95, 180, 184
Vehicular pollution, 263, 264
Ventilation system, 52
Voltage
discharge characteristics, 164–167
fluctuations, 259, 271
profile, 207–209, 220, 221, 269, 280,
286, 289, 290, 326, 338, 349, 350
improvement, 289, 290, 338
stability, 287
regulation, 1, 271, 272, 274
regulators, 125, 274
source
converter (VSC), 1
inverter, 180, 190, 238

stability, 209, 268, 287, 338, 365
   index (VSI), 180–183, 185, 186, 238, 288
   variation, 268, 272, 280, 287, 289, 319

W

Walsh Hadamard transform (WHT), 360, 371, 373, 375–381
Waste disposal system, 39
Water conservation, 309
Wavelet transform, 370
Weighting factor tuning, 245

Whale
   optimization algorithm (WOA), 288, 314, 317, 321, 322, 329–331
   time-dependent position, 322
Wireless fidelity (Wi-Fi), 39, 50, 53
Workplace opportunities, 52

Z

Zero current detector circuit (ZCD), 6, 7
Zigbee, 50
   protocol, 50
   technology, 50

9 781774 914182